A Review of the Alumina/Ag-Cu-Ti Active Metal Brazing Process

A Review of the Alumina/Ag-Cu-Ti Active Metal Brazing Process

Tahsin Ali Kassam

CRC Press is an imprint of the
Taylor & Francis Group, an **informa** business

CRC Press
Taylor & Francis Group
6000 Broken Sound Parkway NW, Suite 300
Boca Raton, FL 33487-2742

© 2019 by Taylor & Francis Group, LLC
CRC Press is an imprint of Taylor & Francis Group, an Informa business

No claim to original U.S. Government works

Printed on acid-free paper

International Standard Book Number-13: 978-1-138-60291-5 (Hardback)

This book contains information obtained from authentic and highly regarded sources. Reasonable efforts have been made to publish reliable data and information, but the author and publisher cannot assume responsibility for the validity of all materials or the consequences of their use. The authors and publishers have attempted to trace the copyright holders of all material reproduced in this publication and apologize to copyright holders if permission to publish in this form has not been obtained. If any copyright material has not been acknowledged, please write and let us know so we may rectify in any future reprint.

Except as permitted under U.S. Copyright Law, no part of this book may be reprinted, reproduced, transmitted, or utilized in any form by any electronic, mechanical, or other means, now known or hereafter invented, including photocopying, microfilming and recording, or in any information storage or retrieval system, without written permission from the publishers.

For permission to photocopy or use material electronically from this work, please access www.copyright.com (http://www.copyright.com/) or contact the Copyright Clearance Center, Inc. (CCC), 222 Rosewood Drive, Danvers, MA 01923, 978-750-8400. CCC is a not-for-profit organization that provides licenses and registration for a variety of users. For organizations that have been granted a photocopy license by the CCC, a separate system of payment has been arranged.

Trademark Notice: Product or corporate names may be trademarks or registered trademarks, and are used only for identification and explanation without intent to infringe.

Library of Congress Cataloging-in-Publication Data

Names: Kassam, Tahsin Ali, author.
Title: A review of the Alumina/Ag-Cu-Ti active metal brazing process / Tahsin Ali Kassam.
Description: Boca Raton, FL : CRC Press/Taylor & Francis Group, 2018. |
Revision of author's thesis—Ph.D., Brunel University, London, 2016. |
Includes bibliographical references and index.
Identifiers: LCCN 2018021431| ISBN 9781138602915 (hardback : acid-free paper) |
ISBN 9780429469343 (ebook)
Subjects: LCSH: Aluminum—Brazing—Materials. | Aluminum-copper alloys. |
Titaniam-aluminum alloys. | Aluminum oxide. | Ceramic metals—Welding.
Classification: LCC TS555 .K39 2018 | DDC 661/.0673—dc23
LC record available at https://lccn.loc.gov/2018021431

Visit the Taylor & Francis Web site at
http://www.taylorandfrancis.com

and the CRC Press Web site at
http://www.crcpress.com

Contents

Preface ix
Acknowledgements xi
Author xiii
Abbreviations xv

Chapter 1 Introduction 1

    1.1 Joining of Alumina 1
    1.2 Brazing of Ceramics 1
    1.3 Active Metal Brazing of Ceramics 1
    1.4 Variables in Active Metal Brazing 2
    1.5 The Alumina/Ag-Cu-Ti System 3
    1.6 Ceramic-to-Ceramic Joining 3
    1.7 Industrial Applications and Market Size 4

Chapter 2 Literature Review 7

    2.1 Introduction 7
        2.1.1 Reactive Wetting of Alumina Ceramics 7
        2.1.2 Ag-Cu-Ti Active Braze Alloys 8
            2.1.2.1 Commercially Available Ag-Cu-Ti Braze Alloys 8
        2.1.3 Reaction Layer Formation 9
        2.1.4 Typical Microstructure 10
        2.1.5 Coefficient of Thermal Expansion 11
        2.1.6 Role of the Braze Interlayer 12
        2.1.7 Joining Mechanism 14
    2.2 Variables in the Design of an Ag-Cu-Ti Active Braze Alloy 16
        2.2.1 Ag-Cu Concentrations 16
        2.2.2 Ti Concentration 19
            2.2.2.1 Ag-Cu-Ti Braze Foil Thickness 19
            2.2.2.2 Ti Concentration and Wetting 19
            2.2.2.3 Ti Concentration and the Reaction Layer 22
            2.2.2.4 Ti Concentration and Joint Strength 31
    2.3 Process Parameters in Active Metal Brazing 32
        2.3.1 Brazing Time 32
            2.3.1.1 Brazing Time and Wetting 33
            2.3.1.2 Brazing Time and the Reaction Layer 34
            2.3.1.3 Brazing Time and Joint Strength 36
        2.3.2 Brazing Temperature 38
            2.3.2.1 Brazing Temperature and Wetting 39
            2.3.2.2 Brazing Temperature and the Reaction Layer 40
            2.3.2.3 Brazing Temperature and Joint Strength 44
        2.3.3 Brazing Atmosphere 46
    2.4 Variables Influencing Ceramic Properties 47
        2.4.1 Alumina Purity 47
            2.4.1.1 Alumina Purity and Wetting 48
            2.4.1.2 Alumina Purity and the Reaction Layer 49

		2.4.1.3 Alumina Purity and Joint Strength.................................52
	2.4.2	Alumina Surface Condition 52
		2.4.2.1 Surface Roughness 55
		2.4.2.2 Grinding and Polishing...58
		2.4.2.3 Post-Grinding Heat Treatment............................59
2.5	Mechanical Testing... 61	
	2.5.1	Typical Testing Methods 62
2.6	Gaps Identified in the Literature 64	
	2.6.1	Ag-Cu-Ti Braze Preform Thickness..................................64
	2.6.2	Secondary Phase Interaction 65
	2.6.3	Post-Grinding Heat Treatment 65
2.7	Summary of Objectives 66	

Chapter 3 Experimental Methods...67

3.1	Alumina Materials Selection and Design................................67
3.2	Surface Roughness Measurements 68
3.3	Post-Grinding Heat Treatment 69
3.4	Ag-Cu-Ti Braze Alloy Selection and Design 71
3.5	Brazing Procedure 71
3.6	Mechanical Testing...72
3.7	Macro Images 75
3.8	Mounting and Polishing 75
3.9	Etching Techniques 77
3.10	Optical and Scanning Electron Microscopy..........................78
3.11	Electron Probe Microanalysis 78
3.12	Focussed Ion Beam Milling 78
3.13	Transmission Electron Microscopy 81
3.14	Nanoindentation 81
3.15	Design of Experiments 82

Chapter 4 Alumina Ceramics 85

4.1	Chemical Composition 85	
4.2	Surface Roughness 85	
4.3	Microstructure 89	
	4.3.1	D-96 Alumina 89
	4.3.2	D-100 Alumina..92
4.4	Flexural Strength 94	
	4.4.1	Flexural Strength and Surface Roughness 96
4.5	Post-Grinding Heat Treatment 97	
4.6	Conclusions.. 107	

Chapter 5 Microstructural Evolution 109

5.1	As-Received TICUSIL® Braze Foils 109	
	5.1.1	Cu$_4$Ti$_3$ in TICUSIL® Braze Foil 110
5.2	Microstructures of Brazed Joints in As-Ground Condition 116	
	5.2.1	50-μm-Thick TICUSIL® Braze Preforms 116
	5.2.2	100-μm-Thick TICUSIL® Braze Preforms 120
		5.2.2.1 Cu-Ti Phase Formation 125

Contents vii

			5.2.2.2 Ag-Rich Braze Outflow 128
		5.2.3	150-μm-Thick TICUSIL® Braze Preforms................................. 129
			5.2.3.1 Multi-Layered Cu-Ti Structure 132
		5.2.4	250-μm-Thick TICUSIL® Braze Preforms 134
	5.3	Transmission Electron Microscopy 137
		5.3.1	50-μm-Thick TICUSIL® Braze Preforms 137
		5.3.2	100-μm-Thick TICUSIL® Braze Preforms 145
		5.3.3	Secondary Phase Interaction 153
	5.4	Microstructures of Brazed Joints in Ground-and-Heat-Treated Condition....... 162
		5.4.1	D-96 GHT Brazed Joints... 162
		5.4.2	D-100 GHT Brazed Joints... 167
		5.4.3	Braze Infiltration 169
	5.5	Summary 172
	5.6	Conclusions 174

Chapter 6 Joint Performance.. 175

	6.1	Strengths of Brazed Joints in As-Ground Condition...................................... 175
		6.1.1 50-μm-Thick TICUSIL® Braze Preforms 175
		6.1.2 100-μm-Thick TICUSIL® Braze Preforms 177
		6.1.3 150-μm-Thick TICUSIL® Braze Preforms.................................. 181
		6.1.4 Secondary Phase Interaction 184
		6.1.5 250-μm-Thick TICUSIL® Braze Preforms 186
	6.2	Nanoindentation 188
		6.2.1 Calibration 192
		6.2.2 Targeted Indents in Alumina... 195
		6.2.3 Targeted Indents in the Reaction Layer...................................... 195
		6.2.4 Targeted Indents in the Braze Interlayer 197
		6.2.5 Nanohardness Distribution Plots 201
	6.3	Strengths of Brazed Joints in Ground-and-Heat-Treated Condition 206
		6.3.1 50-μm-Thick TICUSIL® Braze Preforms 206
		6.3.2 100-μm-Thick TICUSIL® Braze Preforms 207
		6.3.3 150-μm-Thick TICUSIL® Braze Preforms.................................. 211
	6.4	Summary 211
	6.5	Conclusions.. 216

Appendix 1: Advanced Ceramics Definition 219

Appendix 2: Macro Images of Brazed Joints 221

Appendix 3: Four-Point Bend Testing 229

Appendix 4: Surface Roughness Measurements... 231

Appendix 5: Brazing Fixture 233

Nomenclature 235

References 239

Index 245

Preface

This book originated as a thesis submitted for the degree of Doctor of Philosophy and was awarded by Brunel University London. The research was carried out between the period of May 2013 and October 2016 under the supervision of Prof. Hari Babu Nadendla at the Brunel Centre for Advanced Solidification Technology (BCAST), Brunel University London and Dr. Nicholas Ludford in the Specialist Materials and Joining Group, TWI Ltd, Cambridge. The author wishes to acknowledge the financial support of the Engineering and Physical Sciences Research Council (EPSRC), Brunel University London and TWI Ltd (EP/K504270/1).

Journal publications tied to this work:

Kassam, T.A., Nadendla, H.B. and Ludford, N. (2015) The effect of alumina purity and Ag-Cu-Ti braze preform thickness on the microstructure and mechanical properties of alumina-to-alumina brazed joints, *Journal of Materials Engineering and Performance*, 25(8), 3218–3230.

Kassam, T.A., Nadendla, H.B. and Ludford, N. (2016) The effect of Ag-Cu-Ti braze perform thickness on the nanomechanical properties of alumina-to-alumina brazed joints, *Brazing, High Temperature Brazing and Diffusion Bonding*, DVS Proceedings, Aachen, June 2016, pp. 32–43.

Kassam, T.A., Nadendla, H.B., Ludford, N., Yan, S. and Howkins, A. (2018) Secondary phase interaction at the interfaces of high strength brazed joints made using liquid phase sintered alumina ceramics and Ag-Cu-Ti braze alloys, *Nature Scientific Reports*, 8, Article number 3352.

Acknowledgements

First of all, I thank God for giving me life and blessing me with the opportunity to study His creations. With His mercy, He has gifted me the comfort of loving parents, Azadali and Zehra (Razi), who have sacrificed and dedicated their lives for me. They are my guardian angels sent from heaven, and to them I am forever indebted and grateful. I thank God for providing me with the fragrant flower that is my wife, Suhaylah, who has supported me on all our journeys together. Her selfless and caring commitment towards me has made this experience fruitful and enjoyable, and her family have also shown much encouragement and care towards me. I thank God for bringing me into this world to grow alongside my sister, Naseera, who has always been an inspiration to me from the moment I opened my eyes. It is to my wonderful family that this work is dedicated.

I thank the Engineering and Physical Sciences Research Council (EPSRC), TWI Ltd and Brunel University London for the financial support of this work. At TWI Ltd, I thank Dr. Nick Ludford, my industrial supervisor, and my colleagues, including but not limited to, Maxime Mrovcak, Dr. Raja Khan, Dr. James Kern, Steve Willis, Ashley Spencer, Thomas Adams, Simon Ingram, Ramin Taheri, Paul Evans, Lee Mills, Jerry Godden, Mark Tinkler, Simon Turner, Steve Mycock, Alex Russell, Cliff Hart, Prof. Alan Taylor, Dr. Alec Gunner, Gillian Dixon-Payne, Dr. Damaso De Bono, Paul Jones, Linda Dumper, Dr. Muhammad Shaheer, Dr. Renaud Bourga, Dr. Francisco Arteche, Dr. Antonio Camacho, Dr. Jianlin Tang, Dr. Wee Liam Khor, Dr. David Williams, Dr. Muntasir Hashim, Dr. Farshad Salamat-Zadeh, Dr. Amir Khamsehnezhad, Dr. Graham Wylde, Dr. Roger Wise, Dr. Usani Ofem, Prof. Tat-Hean Gan, Prof. Aamir Khalid, Dr. Marcus Warwick, Dr. Clément Bühr, Dr. Simon Smith, Sheila Stevens, Dr. Michael Dodge, Dr. Geoff Lunn, Dr. David Griffiths, Dr. Michal Lewandowski, Simon Condie, Laura Murfin and Prof. Wamadeva Balachandran for their support.

At Brunel University London, I thank Prof. Hari Babu Nadendla, my academic supervisor who has been like a father figure to me. Many of the opportunities I was fortunate to experience during my PhD would have been impossible without him. I also thank my second academic supervisor Dr. Lorna Anguilano for her support and Dr. Ashley Howkins for his assistance with TEM analysis. My thanks also to other staff members at BCAST and ETC, including Dr. Bryan McKay, Prof. Dmitry Eskin, Dr. Jesus Ojeda, Stephen Cook, Dr. Nico Nelson, Paul Yates, Matthew Ralph, Philip Saunders, Lauren Wigmore, Prof. Zhongyun Fan, and my colleagues Dr. Alireza Valizadeh, Dr. Edward Djan, Dr. Kawther Al-Helal, Dr. Jayesh Patel, Dr. Utsavi Joshi and Dr. Ashutosh Bhagurkar. My sincere gratitude to the Dean of the College of Engineering, Design and Physical Sciences at Brunel – Prof. Stefaan Simons, who provided extra support to the first NSIRC students.

I thank Jim Boff, Martin Hammler and Dr. Nashim Imam, Phillips & Leigh LLP, UK, for supporting me during the publication of this book. My thanks also to Allison Shatkin and the editorial and production team at CRC Press and Taylor & Francis Group, US, for their very professional and helpful approach to the whole publication process, in particular, Jeanine Furino at codeMantra, Iris Fahrer, Cynthia Klivecka and Camilla Michael.

I thank Dr. Roger Morrell, Dr. Bryan Roebuck and Andrew Duncan, National Physical Laboratory, UK, and Prof. Rajiv Asthana, University of Wisconsin-Stout, USA, for their support; Dr. Ude. Hangen and Raymond Peng, Hysitron, Germany, for assistance in nanoindentation experiments; Dr. Scott Doak and Dr. Sabrina Yan, Loughborough University, UK, for TEM sample preparation using FIB milling; and Dr. Iris Buisman, University of Cambridge, UK, for assistance with EPMA measurements. My thanks also to Dr. Charles Marsden, Coorstek Ltd, UK for providing the ceramic materials, and Anja Kaunert and Dr. Toshi Oyama, Morgan Advanced Materials plc, for providing the braze alloy materials, used in this work. Special thanks to Francesca Minett and the Braze Team, Morgan Advanced Materials plc, for providing the image used in the front cover of this book. A word of thanks also goes to all my teachers, particularly to the late Mr. Nicholson,

Watford Grammar School for Boys, UK, who was instrumental in my academic development and Dr. Adam Wojcik, University College London, UK, whose lectures inspired me to pursue a career in this field. I thank The KSIMC of London for the years of multifaceted support provided to both local and global communities. I also thank the authors who publish in this field from whose work I have drawn and without which this book would not be possible. Lastly, I thank Dr. Amir Shirzadi, The Open University/University of Cambridge, UK, and Prof. Hamid Assadi, Brunel University London for examining my PhD thesis, on which this book is based.

Tahsin Ali Kassam MEng MRes PhD CEng MWeldI MIMMM
London, UK

Author

Tahsin Ali Kassam MEng MRes PhD CEng MWeldI MIMMM:

Following graduation from University College London with an Integrated Masters (MEng) degree in Engineering with Business Finance in 2009, Tahsin Ali completed a Masters (MRes) degree in the Science and Engineering of Materials at the University of Birmingham in 2010. Thereafter, Tahsin Ali worked in International Business Development for a leading materials testing manufacturer before commencing a role as a Materials Development Engineer in the hardfacing repair of forging dies. Tahsin Ali qualified as a Chartered Engineer (CEng) in January 2016, and in October 2017, he completed a Doctorate (PhD) in the 'Active Metal Brazing of Ceramics' at Brunel University London, based full-time at TWI Ltd in Cambridge. This book is based on Tahsin's PhD thesis entitled "The Effects of Alumina Purity, Ticusil® Braze Preform Thickness, and Post-Grinding Heat Treatment, on the Microstructure, Mechanical and Nanomechanical Properties of Alumina-to-Alumina Brazed Joints" which was awarded by Imperial College London's CASC Steering Group as the 2017 recipient of the Professor Sir Richard Brook Prize (sponsored by Morgan Advanced Materials plc) for Best Ceramics PhD Thesis in the UK and also received the Dean's Prize for Innovation and Impact and the Vice-Chancellor's Prize for Doctoral Research at Brunel University London in 2017.

Tahsin Ali is currently a Patent Scientist & Trainee Patent Attorney at Phillips & Leigh LLP, London, UK, where he prepares, drafts and prosecutes patent applications at the European Patent Office for inventions in the field of materials science and engineering.

Abbreviations

~	Approximately
<	Less than
>	Greater than
±	Plus or minus
3PB	Three-point bend testing
4PB	Four-point bend testing
ABA	Active braze alloy
AES	Auger electron spectroscopy
AG	As-ground
AMB	Active metal brazing
ASTM	American Society for Testing and Materials
ASTM C1161-13	ASTM C1161-13 flexural testing standard
ASTM F-19	ASTM F-19 tensile test standard
AWS	American Welding Society
AWS C3.2M/C3.2:2008	Series of AWS testing standards including four-point bend testing
BSE	Backscattered electron
CCD	Charge-coupled device
CTE	Coefficient of thermal expansion
Cusil ABA®	Commercially available silver–copper–titanium braze alloy
DBS	Double-bonded shear testing
DCB	Double-cantilever beam testing
D-96	Dynallox 96 grade alumina
D-100	Dynallox 100 grade alumina
D-96 AG	Dynallox 96 grade alumina in as-ground condition
D-96 GHT	Dynallox 96 grade alumina in ground-and-heat-treated condition
D-100 AG	Dynallox 100 grade alumina in as-ground condition
D-100 GHT	Dynallox 100 grade alumina in ground-and-heat-treated condition
EDX	Energy-dispersive X-ray spectroscopy
EPMA	Electron probe microanalysis
FEG-SEM	Field Emission Gun Scanning Electron Microscopy
FG	Fine grinding
FIB	Focussed ion beam
GHT	Ground-and-heat-treated
HAADF	High-angle annular dark field
ISO	International Organisation for Standardisation
ISO 4288	ISO standard for surface roughness measurements using profilometry
LOCTITE SF 7900	Boron-nitride-based stop-off spray
MD-	Product code for Struers grinding and polishing discs
OPU	Oxide (final) polishing using a standard colloidal silica suspension
P	Polishing
PG	Plane grinding
SEM	Scanning electron microscopy
SENB	Single-edge-notched beam testing
SH	Shear testing

STEM	Scanning transmission electron microscopy
TICUSIL®	Commercially available silver–copper–titanium braze alloy
TEM	Transmission electron microscopy
WAM	Wetting angle measurements
XRD	X-ray diffraction

1 Introduction

1.1 JOINING OF ALUMINA

Polycrystalline alumina (Al_2O_3, aluminium oxide in its crystalline form corundum) has been the most commonly used material in ceramic-to-ceramic and ceramic-to-metal joining studies (Locatelli et al., 1997). Alumina as an engineering ceramic was first developed in the 1940s as an insulator for spark plugs in aircraft piston engines (Foley and Anders, 1994). In the 1950s, alumina was used in ceramic-to-metal joining for the development of vacuum envelopes. These new alumina tubes that replaced glass provided higher bake-out temperature capabilities, better mechanical strength as well as the ability to handle greater electrical power (Mizuhara and Huebel, 1986).

In the 1980s, major efforts were undertaken to use ceramics in structural applications, which had significant tensile and multiaxial stresses. Automotive manufacturers considered ceramics such as alumina as potential candidates to replace power-consuming cooling systems in engines with higher efficiency. Due to their high hardness and wear resistance properties, alumina ceramics were also candidates for use in the manufacture of other automotive parts such as cylinders, pistons, caps and wear pads (Moorhead and Keating, 1986) (see Appendix 1).

The inherent brittleness of ceramics such as alumina, however, made it challenging and uncertain as to whether these materials could actually withstand the high stresses exhibited in demanding industrial applications. Ceramic-to-metal joining, therefore, was seen as a potential solution to exploit the desirable properties of ceramics, in an industrial context, using a wide range of joining technologies which have been summarised in the literature (Bahrani, 1992; Gauthier, 1995; Fernie et al., 2009). Brazing was seen as one of the industrially preferred joining methods which could enable the formation of strong, hermetic and reliable ceramic-to-metal joints.

1.2 BRAZING OF CERAMICS

One of the preferred industrial methods for joining ceramics is brazing, whereby a braze alloy is melted on a ceramic surface at high temperatures (>450°C). The two most critical problems in this process include the poor wetting of chemically inert ceramic surfaces by metallic braze alloys and thermally induced residual stresses generated by a coefficient of thermal expansion (CTE) mismatch at the joint interfaces. Active metal brazing (AMB) can be used to overcome these challenges.

1.3 ACTIVE METAL BRAZING OF CERAMICS

AMB is a single-step liquid-state joining process (usually conducted in vacuum), whereby a braze alloy that contains a highly reactive element such as titanium (Ti), hafnium (Hf) or zirconium (Zr), can wet an otherwise chemically inert ceramic surface (Eustathopoulos, 1998) (see Section 2.1.1). The reactive element can reduce and chemically bond with the ceramic surface, which is usually an oxide, carbide, nitride or silicide, resulting in the formation of a reaction layer at the joint interface. This adds a higher degree of metallic character to the ceramic surface enabling the braze alloy to wet and spread effectively (see Section 2.1.3). The reaction layer can also provide a gradual transition in the CTE mismatch across the joint interface thereby reducing thermally induced residual stresses (Moorhead et al., 1987) (see Section 2.1.5). An optimum reaction layer thickness exists, however, since it is usually a brittle phase; hence, an excessively thick reaction layer may weaken the joint interface and degrade the strength of the joint.

The simple addition of a highly reactive element in the formulation of an active braze alloy (ABA), therefore, makes the AMB of ceramics a relatively simple and cost-effective joining technique (Asthana and Singh, 2008; Ghosh et al., 2012). Therefore, the AMB process is referred to as a suitable joining technique for the mass production of brazed ceramic parts (Paulasto and Kivilahti, 1998).

1.4 VARIABLES IN ACTIVE METAL BRAZING

This work has identified that variables in the AMB process, which can affect the strength of the ceramic brazed interface may be divided into three main categories: (i) Variables which can influence the properties of the ceramic material, (ii) Variables concerning the design of a suitable ABA and (iii) Process parameters in AMB experiments (Figure 1.1).

Interdependencies between some of the variables shown in Figure 1.1 can directly influence each other; for example, a given mechanical testing method may determine the geometries of parts to be joined as well as the braze preform dimensions, and thereby the design of a brazing fixture. Similarly, the composition of a braze alloy, which depends on the type of ceramic to be brazed, may determine the peak brazing temperature selected in the brazing cycle. These interdependencies between variables in the AMB process can make the formation of reliably strong joints a relatively complex task.

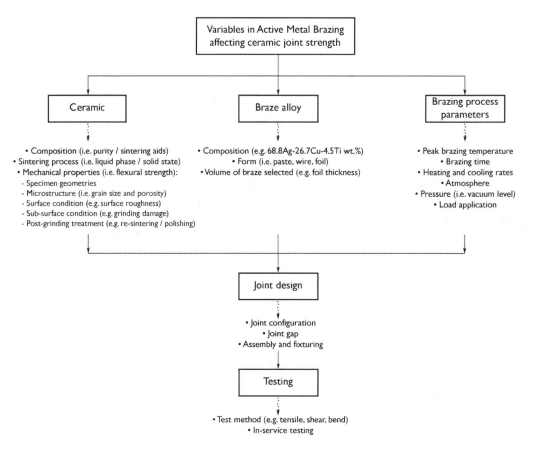

FIGURE 1.1 Variables in the active metal brazing process which can affect the strength of the ceramic/braze joint interface.

Introduction

Discrepancies in the results of similar ceramic brazing studies in the literature may be due to variations in the experimental methodologies adopted with incoherent considerations of interdependencies between the variables. Reliability in the formation of strong joints may be improved, therefore, by identifying the predominant effects of each variable in the AMB process (shown in Figure 1.1), including those variables less commonly studied, on the reactive wetting process, reaction layer formation and ultimately joint strength. This is discussed further in Section 2.

1.5 THE ALUMINA/Ag-Cu-Ti SYSTEM

The alumina/Ag-Cu-Ti system has been established in the literature as a model system against which the effects of variables in the AMB process and on the development and testing of other ceramic/braze systems have been compared (Lin et al., 2001; Mohammed Jasim et al., 2010; Peytour et al., 1990; Shiue et al., 2000; Valette et al., 2005; Vianco et al., 2003).

This work has focussed on the alumina/Ag-Cu-Ti system in an attempt to clarify the direct effects of variables in the AMB process, based on the literature. This has helped to identify variables in the AMB process, which have not previously been widely investigated or reported. Some of these, namely, (i) alumina purity, (ii) braze preform thickness and (iii) post-grinding heat treatment, have subsequently been explored in this work.

Findings based on the literature and from investigative work may not only help to improve understanding of the alumina/Ag-Cu-Ti system, but may also be helpful in the development and/or refinement of other ceramic/braze systems.

The alumina/Ag-Cu-Ti system is the most established system in the AMB of ceramics owing to the fact that alumina is the most commonly used advanced ceramic material and Ag-Cu-Ti braze alloys are the most commonly used ABAs (Lee and Rainforth, 1994; Phillips, 1991; Verband der Keramischen Industrie, 2004).

1.6 CERAMIC-TO-CERAMIC JOINING

This study has focussed on ceramic-to-ceramic joining, as opposed to ceramic-to-metal joining, in order to review the effects of variables in the AMB on the strength of the ceramic/braze interface.

The current state-of-the-art in ceramic-to-metal joining still requires the undertaking of systematic experiments in order to aid understanding of the joining mechanism and the optimisation of joint strength and reliability. In the earlier stages of developing or refining a ceramic-to-metal system, the effects of a given variable in the AMB process may be more easily understood by performing brazing trials and thereafter analysing the ceramic/braze interface in ceramic-to-ceramic joints (Conquest, 2003; Mizuhara and Huebel, 1986).

The effects of variables in the AMB process may be more clearly observed in ceramic-to-ceramic joints due to the elimination of additional chemical reactions associated with the braze/parent metal interface as in ceramic-to-metal joints (Kozlova et al., 2010; Valette et al., 2005; Vianco et al., 2003). In ceramic-to-ceramic joining, chemical reactions are limited to those which are fundamental to the formation of chemical bonds at the relatively weaker ceramic/braze interface.

The interfaces of ceramic-to-ceramic brazed joints can exhibit significantly lower residual stresses than the interfaces of ceramic-to-metal brazed joints. Differences in the CTE and elastic modulus between ceramics and metals coupled with the limited fracture toughness of ceramics can cause rupture of a brazed joint upon cooling from the brazing temperature (Figure 1.2). The inherent symmetry of a ceramic-to-ceramic brazed joint may provide a more uniform stress distribution thereby enabling greater probability in clearly evaluating joint strength and drawing relationships to variables in the AMB process.

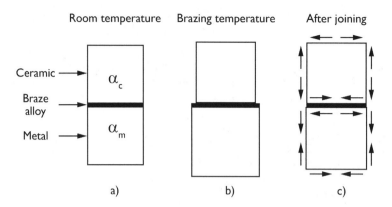

FIGURE 1.2 The development of residual stresses (indicated by arrows) in a ceramic-to-metal brazed joint at (a) room temperature, (b) brazing temperature and (c) after joining (Adapted from do Nascimento et al. 2003).

The benefits of ceramic-to-ceramic joining are not only limited to the early stage development of ceramic-to-metal systems. Applications for ceramic-to-ceramic brazed joints, commonly reported in the literature, include those whereby reduction in the expense of machining complex ceramic parts is enabled by the joining of simpler-shaped ceramic parts together (Ghosh et al., 2012; Hahn et al., 1998; Paiva and Barbosa, 2000; Rohde et al., 2009; Yang et al., 2012).

1.7 INDUSTRIAL APPLICATIONS AND MARKET SIZE

Alumina dominates the global advanced ceramics market. It is relatively inexpensive to manufacture owing to the abundantly available mineral bauxite, from which it is produced via the Bayer method (Boch and Niepce, 2007; Heiman, 2010). Three of the largest bauxite reserves are in Guinea, Australia and Brazil and are each estimated to contain 7.4, 6.5 and 2.6 billion tonnes, respectively (Statista, 2014).

Refractoriness, electrical insulation, wear- and corrosion-resistance are desirable properties of alumina which have made it suitable for use in a wide range of applications including abrasives, protective linings, endoprostheses and electronic circuit packaging.

In 1986, the global engineering ceramics market was estimated to be worth £3 billion (Bahrani, 1992). In 2015, this had grown to an estimated worth of £38.4 billion with a forecasted compound annual growth rate of ~6% to 2020 (Profound, 2014). In 2015, alumina was reported to represent nearly 46% of the global advanced ceramics market (Markets and Markets, 2015) with a total estimated global production of 10 million tonnes for the purpose of manufacturing ceramic parts (Heiman, 2010).

Since the late 1990s, the reliability of ceramic processing techniques has been improved considerably and this has coincided with a surge of interest in the use of advanced ceramics for a wide range of applications (Fernie et al., 2009).

Brazed joints made using alumina ceramics are commonly used in ultra-high vacuum coaxial feedthroughs (Figure 1.3). These vacuum feedthroughs, used in signal transmission, particle physics, thin film deposition and ion beam applications, make use of the dielectric properties of alumina which provides high-voltage insulation with little signal attenuation. These brazed components can be inexpensive to manufacture via AMB and perform well to meet the demands of these applications.

Recently, alumina has been considered for use in the ultra-high vacuum chambers of rapid cycle proton synchrotron machines since it can minimise eddy current losses in rapidly varying magnetic

Introduction

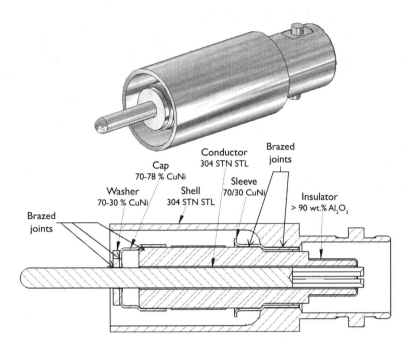

FIGURE 1.3 An ultra-high vacuum BNC design coaxial feedthrough with a grounded shield produced via the active metal brazing of alumina (Courtesy of CeramTec).

fields while also maintaining good flexural strength (Yadav et al., 2014). Joining of multiple alumina tube segments through the formation of alumina-to-alumina brazed joints is reported to be highly efficient since manufacturing the entire chamber length using conventional ceramic processing techniques is neither practical nor commercially viable.

2 Literature Review

2.1 INTRODUCTION

This review compares findings from literature studies concerning the AMB of alumina using Ag-Cu-Ti braze alloys. The primary objective is to derive how variables in the AMB process can affect wetting, microstructure (of the reaction layer and braze interlayer) and joint strength in the alumina/Ag-Cu-Ti system. Most studies cited have focussed on how the braze alloy composition, brazing temperature and brazing time can affect interfacial chemistry and joint strength. Studies relating to other less obvious variables; for example, the volume of Ag-Cu-Ti braze selected and its effects on the microstructure and mechanical properties of alumina-to-alumina and alumina-to-metal brazed joints, have been less widely reported. Furthermore, few studies have investigated how the purity of alumina or its surface condition may also affect interfacial chemistry and joint strength.

In this work, the alumina/Ag-Cu-Ti/alumina system has been used as a model system to provide a more complete understanding of the roles of variables in the AMB process. This review compares methodologies and findings from over 40 studies which have investigated the formation of alumina-to-alumina brazed joints made using Ag-Cu-Ti ABAs. This review may help to reduce future effort in the establishment of the alumina/Ag-Cu-Ti system and may aid the development of other ceramic/braze systems. Important gaps in the current state-of-the-art have been identified, and some of these have been subsequently investigated in later sections of this book (see Section 2.6).

2.1.1 Reactive Wetting of Alumina Ceramics

A successful brazing operation, by definition, depends on the ability of a braze alloy to wet the surfaces of parts being joined (Foley and Anders, 1994; Pak et al., 1990). Wetting occurs when a liquid droplet in contact with a solid surface, in a given atmosphere, produces an equilibrium contact angle that is less than 90°. This predominantly requires the interfacial energy, in the formation of a solid–liquid interface (γ_{SL}), to be less than the interfacial energy of the solid–vapour interface (γ_{SV}), according to the following Young–Dupré equation:

$$\cos\theta = \frac{\gamma_{SV} - \gamma_{SL}}{\gamma_{LV}} \tag{2.1}$$

Differences in the chemical bonding between the free electron state of metals and the stable ionic and/or covalent bonding of ceramics makes the wetting of ceramics challenging. Chemical bonding in alumina is both ionic and covalent; however, since two thirds of its octahedral interstices are occupied by Al^{3+} cations, it is said to be predominantly ionic (Heiman, 2010). It is not thermodynamically favourable, therefore, for conventional braze alloys which usually consist of pure metals, such as Sn, Au, Ni, Cu and Ag, to wet ceramics effectively (Fernie et al., 2009). In AMB, the reactive wetting of ceramics is thermodynamically favourable due to the addition of a reactive element in the braze alloy and due to the chemically inert atmosphere in which the procedure is performed.

Ceramics typically comprise metal oxides, metal carbides, metal nitrides and metal silicides. Reactive element additions, commonly Ti, Hf and Zr, and less commonly Nb, Cr, Y and V, can reduce the ceramic surface during the brazing procedure to chemically bond with the oxygen, carbon, nitrogen or silicon in the ceramic surface to form oxide, carbide, nitride or silicide reaction products, respectively, at the joint interface. Hence, the reactive element in an ABA can induce an irreversible change in the physicochemical nature of the joint interface, improving wettability (Asthana and Singh, 2008). Furthermore, reaction product formation can significantly decrease the free energy and interfacial tension that is otherwise prevalent at the ceramic/braze interface (Kristalis et al., 1991).

2.1.2 Ag-Cu-Ti Active Braze Alloys

The most commonly used ABAs in the AMB of alumina have been based on the 72Ag–28Cu wt.% eutectic composition, with small additions of Ti as the active element. This Ag-Cu-Ti ternary alloy can provide several advantages, summarised as follows:

1. In a chemically inert atmosphere, diffusion of Ti to the joint interface leads to the reduction of the alumina surface and the formation of chemical bonds.
2. The eutectic temperature in the Ag-Cu binary system of 795°C indicates relatively low brazing temperatures, which reduces the extent of thermally induced residual stresses at the joint interface.
3. Ti has a solubility of less than 2 at.% in the Ag-Cu eutectic alloy. The Ag-Cu lattice can host Ti atoms without inhibiting on its chemical activity. Thus, the diffusivity of Ti to the joint interface can be enhanced by these elements.
4. With complete diffusion of Ti to the joint interface, the resulting ductile Ag-Cu braze interlayer can plastically deform to accommodate thermally induced residual stresses and applied stresses.
5. Ag provides both good flowability and corrosion resistance which are desirable properties of a braze alloy, whereas Cu is relatively inexpensive and enables cost-effective manufacturing of Ag-Cu braze alloys.

It is likely, that due to the above-mentioned reasons, commercial braze alloy development has focussed on the ternary Ag-Cu-Ti system (Nicholas, 1998). One of the major drawbacks in the use of Ag-Cu-Ti braze alloys, however, is the relatively poor corrosion resistance it exhibits at elevated temperatures. Hence, brazed components made using Ag-Cu-Ti braze alloys are usually limited, in an industrial context, to temperatures of less than ~500°C above which degradation in joint strength can occur (Locatelli et al., 1997).

Brazed joints made using alumina ceramics and Ag-Cu-Ti braze alloys are typically used in applications with in-service temperatures lower than 500°C. In these applications, the wear resistance and high-voltage insulation properties of alumina can be exploited if strong and hermetic joints are reliably produced. Alternative braze alloy compositions, for example, Nickel-based braze alloys, can be used to exploit the high-temperature properties of ceramics; however, these systems are relatively less established than those which incorporate the use of Ag-Cu-Ti braze alloys (Conquest, 2003).

2.1.2.1 Commercially Available Ag-Cu-Ti Braze Alloys

The two most commonly used commercially available Ag-Cu-Ti braze alloys are Cusil ABA® composed of 1.75 wt.% Ti and TICUSIL® composed of 4.5 wt.% Ti, both manufactured by Wesgo Metals, Hayward, CA, USA. While both of these ABAs are available in powder, paste and foil forms, braze foils are the most common form in which these braze alloys are typically used in the AMB of alumina.

In Cusil ABA® braze foil, the 1.75 wt.% Ti concentration is distributed as Cu_4Ti compounds in an Ag-Cu eutectic microstructure (Figure 2.1a). The higher 4.5 wt.% Ti concentration in TICUSIL® braze foil, however, is not completely soluble in the Ag-Cu microstructure. Therefore, TICUSIL® is manufactured by a cladding process whereby a Ti ribbon is sandwiched in an Ag-Cu eutectic that is subsequently roll-formed and rapidly solidified. During the rapid solidification process, Cu_4Ti_3 phases can form adjacent to the central Ti ribbon (Figure 2.1b). Nevertheless, this process allows the higher Ti concentration to be physically contained in the Ag-Cu-Ti braze foil without any further segregation of phases which ensures good wettability of the alumina surface (Jacobson and Humpston, 2005).

Literature Review

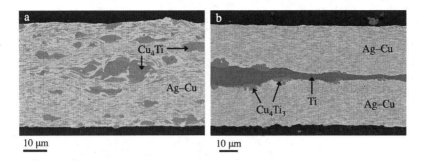

FIGURE 2.1 Backscattered electron images showing cross-sections of the commercially available Ag-Cu-Ti braze alloys: (a) Cusil ABA® and (b) TICUSIL® in the form of 50 μm-thick braze foils (Ali et al., 2015).

2.1.3 Reaction Layer Formation

According to the reaction product control model in Eustathopoulos (2005), reaction layer formation during reactive wetting can lead to a decrease in dynamic contact angle (θ_d). An initial contact angle (θ_i) is formed when an Ag-Cu-Ti braze droplet is first melted onto an alumina substrate (Figure 2.2a). Upon heating to the brazing temperature, Ti diffuses to the joint interface where it reduces the alumina surface to form Ti-O reaction products on the alumina side of the joint interface (layer I) and Ti-Cu-O reaction products on the braze side of the joint interface (layer II). With time at the brazing temperature, these reaction products precipitate to form continuous layers; Eustathopoulos (2005) observed these to be an nm-thick Ti_2O layer (layer I) and a μm-thick Ti_3Cu_3O layer (layer II).

Reaction layer formation can depend on several factors which govern the chemical activity of Ti and its diffusivity towards the joint interface. These factors include (i) the relative amounts of Ag and Cu in the braze alloy which chemically interact with Ti, (ii) the Ti concentration in the braze alloy and (iii) the brazing temperature, which can enhance the diffusivity of the available Ti towards the joint interface – higher brazing temperatures typically lead to the formation of reaction products that are increasingly richer in Ti. The effects of these and other variables in the AMB process on wetting, reaction layer formation and joint strength in alumina/Ag-Cu-Ti system are discussed in Sections 2.2 to 2.4.

Layer II is commonly reported to consist of a M_6O-type Ti-Cu-O layer which imparts metallic character to an otherwise non-wettable alumina surface. This enables the braze alloy to wet and spread across the alumina surface. Kelkar and Carim (1993) measured the electrical resistivity of Ti_3Cu_3O to be 5×10^{-6} Ω·m, which is similar to that of Ti alloys (~1.5×10^{-6} Ω·m) and ~20 orders of magnitude lower than that of alumina. Hence, the formation of a continuous Ti-Cu-O layer leads to a notable decrease in the dynamic contact angle such that a limited quasi-steady contact angle (θ_q) is achieved (Figure 2.2b). In this quasi-steady state, further spreading of the Ag-Cu-Ti braze

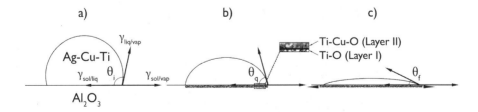

FIGURE 2.2 Reaction product control model: (a) an initial contact angle (θ_i) is formed when the Ag-Cu-Ti braze droplet is melted on an alumina substrate. (b) With time at the brazing temperature, reaction layer formation leads to a quasi-steady contact angle (θ_q) in which the triple line is pinned by the non-wettable alumina surface. (c) A steady-state final equilibrium contact angle (θ_f) is achieved following the lateral growth of the M_6O-type reaction layer (adapted from Eustathopoulos 2005).

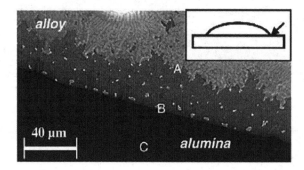

FIGURE 2.3 Backscattered electron image showing the triple line of an Ag-Cu-2.9 at.% Ti braze droplet (A) spreading over a Ti$_3$Cu$_3$O reaction layer (B) on a monocrystalline alumina substrate (C). The wetting experiment was performed at a brazing temperature of 900°C for 3 min (Voytovych et al., 2006).

droplet is inhibited by the presence of non-wettable alumina surface at the triple line. The reaction product control model suggests, therefore, that the final equilibrium contact angle (θ_f) is achieved with further spreading of the braze alloy which relies on continued lateral growth of the Ti-Cu-O layer (Figure 2.2c).

Voytovych et al. (2006) characterised this triple line following a wetting experiment on monocrystalline alumina using an Ag-Cu-2.9 at.% Ti braze alloy. Lateral growth of the Ti$_3$Cu$_3$O layer as per the reaction product control model in Eustathopoulos (2005) was clearly observed at the joint edges (Figure 2.3). These results support the reaction product control model therefore, and illustrate how reaction layer formation can provide a 'stable chemical thermodynamic equilibrium' whereby the wettability of alumina can be promoted by Ag-Cu-Ti braze alloys (Bang and Liu, 1994).

2.1.4 Typical Microstructure

The terms used to refer to the features of a typical brazed joint in the alumina/Ag-Cu-Ti system are hereby defined. The constituents of a typical alumina/Ag-Cu-Ti/alumina brazed joint comprise the following: two alumina ceramics, two reaction layers and a braze interlayer. Each reaction layer typically comprises a Ti-O layer (layer I) and a Ti-Cu-O layer (layer II), while the braze interlayer typically comprises an Ag-rich phase and a Cu-rich phase (Figure 2.4).

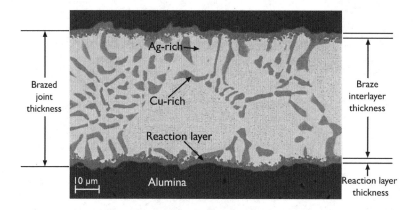

FIGURE 2.4 Backscattered electron image showing the typical constituents of an alumina-to-alumina brazed joint made using a TICUSIL® braze alloy, including the reaction layer thickness and the braze interlayer thickness. Complete diffusion of Ti to the joint interfaces in the formation of the reaction layers is notable.

2.1.5 Coefficient of Thermal Expansion

The coefficient of thermal expansion (CTE) mismatch in a ceramic-to-metal brazed joint is manifested during cooling as thermally induced residual stresses at the joint interface. If these stresses cannot be alleviated, cracking in the brittle ceramic and premature joint failure can occur. While joint failure due to thermally induced residual stress may not always occur, the residual stress state of the joint may be adversely affected. In this case, the net level of applied stress which a brazed joint in a weakened stress state may withstand in service, therefore, may be somewhat reduced.

The CTE ($\propto_l$) is a material property that indicates the extent to which a material expands when heated. It can be expressed in terms of the change in length (Δl) of a solid material with temperature (ΔT) (Cverna, 2002):

$$\frac{\Delta l}{l_0} = \propto_l \Delta T \tag{2.2}$$

Generally, ceramic materials have lower CTE values than metals. An approximate relationship between the CTE and the melting temperature (T_m) of a solid material is given according to Equation 2.3 (Ho and Taylor, 1998). This relationship shows that typically, high melting temperature materials such as ceramics tend to exhibit lower CTE values than relatively low melting temperature materials such as metals, which tend to exhibit higher CTE values.

$$\propto = \frac{0.020}{T_m} \tag{2.3}$$

The CTE of alumina ranges from 8.1 to 8.5 × 10^{-6}.°C^{-1} (Ghosh et al., 2012; Kar et al., 2006; Mandal et al., 2004b). The CTE of a typical Ag-Cu-Ti braze alloy such as TICUSIL® ranges from 18.5 × 10^{-6}.°C^{-1} for temperatures up to ~500°C (Morgan Advanced Materials plc) and 26.7 × 10^{-6}.°C^{-1} for temperatures up to ~600°C (Lin et al., 2001). This large difference in the CTE between alumina and TICUSIL® can lead to premature joint failure following a brazing experiment, due to thermal stresses generated during cooling at the joint interface. Figure 2.5 shows a typical 'dome crack' which occurred as a result of thermally induced residual stress following a wetting experiment on a monocrystalline alumina ceramic using an Ag-Cu-2.9 at.% Ti braze alloy (Voytovych et al., 2006) (Table 2.1).

Reaction layer formation can provide thermo-elastic compatibility between alumina and the Ag-Cu braze interlayer by introducing a gradual transition in the CTE across the joint interface. This is achieved through the formation of the Ti-rich reaction layers at the joint interface. The Ag-rich and Cu-rich phases in the braze interlayer have relatively high CTE values of 19.2 × 10^{-6}.°C^{-1} and 22.0 × 10^{-6}.°C^{-1}, respectively (Table 2.2) (Kar et al., 2006; Mandal et al., 2004b). Gradation in CTE is provided by the TiO and Ti$_3$Cu$_3$O layers at the joint interface which have intermediate CTE values

FIGURE 2.5 Backscattered electron image showing a typical 'dome crack' as a result of a CTE mismatch, following a wetting experiment at 900°C using an Ag-Cu-2.9 at.% Ti braze alloy on monocrystalline alumina (Voytovych et al., 2006).

TABLE 2.1
Properties of the Commercially Available Cusil ABA® and TICUSIL® Braze Alloys

	Composition (wt.%)			Liquidus (°C)	Solidus (°C)	CTE (25°C) ($\times 10^{-6} \cdot °C^{-1}$)	CTE (600°C) ($\times 10^{-6} \cdot °C^{-1}$)
	Ag	Cu	Ti				
Cusil ABA®	63.00	35.25	1.75	815	780	–	–
TICUSIL®	68.80	26.70	4.50	900	780	18.7	26.7

Source: Morgan Advanced Materials plc and Lin et al. (2001).

TABLE 2.2
CTE Values of Phases in a Typical Alumina/Ag-Cu-Ti/Alumina Brazed Joint

Coefficient of Thermal Expansion (CTE) ($\times 10^{-6} \cdot °C^{-1}$)				
Alumina	TiO	Ti_3Cu_3O	Ag-Rich	Cu-Rich
8.1	9.1	15.1	19.2	22.0
Ghosh et al. (2012)	Moorhead et al. (1987)(1)	Kar et al. (2006)	Kar et al. (2006)	Kar et al. (2006)
		Mandal et al. (2004b)	Mandal et al. (2004b)	Mandal et al. (2004b)
8.5	9.2	15.2		
Kar et al. (2006)	Kar et al. (2006)	Lin et al. (2014)		
Mandal et al. (2004b)	Mandal et al. (2004b)			
	Lin et al. (2014)			

of $9.2 \times 10^{-6} \cdot °C^{-1}$ and $15.2 \times 10^{-6} \cdot °C^{-1}$, respectively (Kar et al., 2006; Lin et al., 2014; Mandal et al., 2004b; Moorhead et al., 1987). Therefore, the ordering of compound layers at the interface occurs in a manner that not only creates a gradation in metallic character (i.e. interfacial energies) but also minimises the strain energy of the system by grading the CTE (Asthana, 2017).

2.1.6 Role of the Braze Interlayer

The CTE mismatch between alumina and the Ag-Cu braze interlayer is typically much higher than that between alumina and the reaction layer. Therefore, the presence of the braze interlayer generates greater thermally induced residual stresses, which can be detrimental to joint strength. However, the ductility of the braze interlayer is far greater than that of the reaction layer, which is a brittle phase.

The braze interlayer is solely responsible for accommodating the sum of all stresses which are generated due to CTE mismatch in ceramic brazed joints. While the braze interlayer can plastically deform to accommodate stress which is generated due to its presence, the reaction layer in turn cannot provide the same. Instead, strong chemical bonding at the joint interface which leads to reaction layer formation can be useful in transmitting both residual stresses and applied stresses into the braze interlayer – the only ductile constituent of the brazed joint (Foley and Anders, 1994). Weak chemical bonding at the joint interface, however, can lead to joint failure initiating at the brittle reaction layer phase. This shows that both the reaction layer and the braze interlayer are critical components in the brazed joint microstructure.

While the joining mechanism does not strictly require a braze interlayer, several studies have shown that the absence of a braze interlayer can lead to joints that are often too weak to withstand even thermally induced residual stresses. Therefore, the role of the braze interlayer is to impart mechanical strength to a brazed joint through its ability to plastically deform.

Literature Review

Carim and Mohr (1997) formed alumina-to-alumina brazed joints using 300- to 500-μm-thick Ti$_3$Cu$_3$O and Ti$_4$Cu$_2$O foils. Brazing was conducted at 1,180°C for 90 min. Joint strength was evaluated using four-point bend testing. These joints were compared to another set of joints made using Cusil ABA® braze foils. Despite the reduced CTE mismatch in joints made using the M$_6$O-type foils, most of the joints were too weak to even survive handling. As for the joints which had survived handling, their average strength was found to be an order of magnitude less than that of joints made using the Cusil ABA® braze foil. It can be concluded that the braze interlayer is an essential constituent of a brazed joint due to the ductility which it can provide. The braze interlayer has been evaluated using micro-indentation (Asthana and Singh, 2008) and nanoindentation techniques (Ghosh et al., 2012) to be highly ductile relative to the other constituents in the alumina/Ag-Cu-Ti system.

Ghosh et al. (2012) formed alumina-to-alumina brazed joints using 150-μm-thick TICUSIL® braze foils. Brazing was conducted at 910°C for 5 min. The nanohardness of alumina, the reaction layer and the braze interlayer were all evaluated using nanoindentation. A Berkovich indenter of 150-nm tip radius was used to apply a 100-mN load with a loading/unloading time of 30 s. The nanohardness of alumina was measured to be 15.3 ± 2.7 GPa whereas the nanohardness of the reaction layer and braze interlayer were found to be 5.7 ± 1.1 GPa and 1.76 ± 0.58 GPa, respectively (Figure 2.6). It was concluded that the reaction layer, with an intermediate nanohardness as compared with alumina and the braze interlayer, provided a gradual transition in properties across the interface which enabled reliable joint performance. Therefore, the somewhat lacking ductility of the reaction layer phase may be inherent to its role as a load transfer medium. These results also confirmed that the braze interlayer was the most ductile constituent in the alumina/Ag-Cu-Ti system.

In ceramic-to-metal brazed joints, the braze interlayer thickness has been shown to significantly affect joint strength. Zhu and Chung (1997) formed alumina-to-stainless-steel brazed joints using a 63.0Ag-34.25Cu-1.0Sn-1.75Ti wt.% braze alloy. The brazed thickness was controlled by the amount of braze paste that was applied. Brazing was conducted between 800°C and 850°C for 10 min and joint strength was evaluated using shear testing. As the brazed joint thickness was increased from ~22.5 to ~37.0 μm, the joint strength was also observed to increase from ~77 to ~90 MPa. This increase in strength was correlated to the increase in the thickness of the braze interlayer, which was reported to have absorbed thermal stresses to a greater extent. With a further increase in the

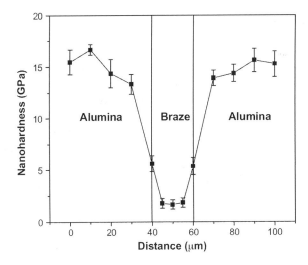

FIGURE 2.6 The nanohardness distribution in an alumina-to-alumina brazed joint made using a 150 μm-thick TICUSIL® braze foil (Ghosh et al., 2012).

brazed thickness to 110 μm, however, the joint strength decreased to ~80 MPa. This degradation in strength was suggested to be due to an excessive amount of Ti, and hence excessive reactions at the joint interface which increased with the amount of braze applied. These results showed that maximum joint strength was controllable by formulating joints with optimum braze interlayer and reaction layer thicknesses.

2.1.7 Joining Mechanism

Several studies have hypothesised a series of chemical reactions which thermodynamically explain reaction product formation in the alumina/Ag-Cu-Ti system. These equations describe the 'joining mechanism' and have been proposed on the basis of empirical observations.

It is well established that Ti from the Ag-Cu-Ti braze alloy diffuses to the alumina surface and thereby promotes the dissociation of Al and oxygen during brazing (do Nascimento et al., 2003). This reduction reaction, limited to the alumina surface, leads to the formation of a TiO phase (α-TiO, β-TiO, γ-TiO or β-Ti$_{1-x}$O) as Ti bonds with oxygen:

$$\frac{1}{3}Al_2O_3 + Ti \rightarrow TiO + \frac{2}{3}Al \qquad (2.4)$$

This reduction reaction rapidly coats the alumina surface with a TiO reaction product which typically forms as a continuous layer (layer I). At lower Ti activity levels, progressively less Ti-rich oxides can form (Ghosh et al., 2012). This can be influenced by the Ti concentration in the braze alloy (see Section 2.2.2.3) as well as the brazing temperature (see Section 2.3.2.2). These Ti-O compounds can be arranged in the order of increasing stability as follows: Ti_2O_3, Ti_3O_5, Ti_4O_7 and TiO_2 (Akselsen, 1992). At higher Ti activity levels, therefore, Ti oxides richer in Ti such as Ti_2O rather than TiO can form (Wiese, 2001).

The Gibbs free energy required for Equation 2.4 is reported to be positive which makes its occurrence thermodynamically unfavourable. Ali et al. (2016) reported the Gibbs free energy of Equation 2.4 to be 5.50 kJ/mole of Ti at 830°C; Stephens et al. (2003) reported it to be 4.66 kJ/mole of Ti at 845°C, and Loehman and Tomsia (1992) reported it to be 3.59 kJ/mole of Ti at 1,027°C. While in all cases, the positive Gibbs free energy means the surface reduction reaction of Equation 2.4 is thermodynamically unfavourable; the TiO reaction layer has been accurately characterized using TEM techniques in literature studies (See Section 2.2.2.3). Thus, the occurrence of the reaction in Equation 2.4 has been explained by the following mechanisms: first, a reduction of the alumina surface requires significantly less energy than that of bulk alumina. Despite alumina being more stable than TiO, therefore, Equation 2.4 becomes thermodynamically favourable if the formation of TiO at the joint interface produces a more stable phase than the immediate alumina surface. In other words, reduction of the alumina surface requires less energy than reduction of the bulk ceramic material. Considering this, reactions between Ti and O alone are sufficient to cause the dissociation of Al and oxygen from alumina at 1,000°C (Moorhead et al., 1987). Secondly, Stephens et al. (2003), in reference to the work by Kelkar et al. (1994), reported that Ti_3Cu_3O can form according to Equation 2.5 and, thus, this reaction can be coupled to that of Equation 2.4. The sum of the Gibbs free energies when these reactions are combined, i.e. the reactant in one equation is the product of the other equation, is negative which thereby makes the formation of these reaction products thermodynamically favourable:

$$Cu + Ti + \frac{1}{6}O_2 \rightarrow \frac{1}{3}Ti_3Cu_3O \qquad (2.5)$$

Stephens et al. (2003) reported the Gibbs free energy required for Equation 2.5 to be −170 kJ/mole of Ti, indicating that Ti_3Cu_3O is a stable phase and its formation is thermodynamically favourable.

Literature Review

Combining Equations 2.4 and 2.5 provides an overall negative Gibbs free energy at temperatures above ~227°C, which therefore supports the formation of TiO according to Equation 2.4.

The unspecified destination of the Al in Equation 2.4 and the source of the oxygen in Equation 2.5, however, suggested that this proposed mechanism required further clarification. Stephens et al. (2003) attempted to clarify the joining mechanism and proposed Equation 2.6 as an alternative to Equation 2.5 incorporating TiO as a direct reactant in the formation of Ti_3Cu_3O. This coupled the two reactions (Equations 2.4 and 2.6), which together express the formation of TiO and Ti_3Cu_3O reaction products:

$$Ti + \frac{3}{2}Cu + \frac{1}{2}TiO \rightarrow \frac{1}{2}Ti_3Cu_3O \tag{2.6}$$

In this mechanism, the reaction shown in Equation 2.6 comprises the sole supply of oxygen from TiO as a reactant, while TiO is also a product of Equation 2.4. This suggests that the formation of Ti_3Cu_3O relies on oxygen from TiO and no other oxygen diffuses through the reaction layer. Oxygen must, therefore, be present in stoichiometric amounts to fit both equations. The consumption of TiO in the formation of Ti_3Cu_3O may explain the thinness of the TiO layer as observed in the literature, as well as the lack of any further reaction products during the reactive wetting process.

The liberation of Al in Equation 2.4 was considered to replace Cu in the Ti_3Cu_3O compound, expressed by Equation 2.7. This is consistent with other studies that have found the solubility of Al in Ti_3Cu_3O to be as high at 15 at.% Al (Kelkar and Carim, 1995). Therefore, Al could also be considered as a reactant in the formation of Ti_3Cu_3O, expressed by Equation 2.8. Thus, Al was considered to be interchangeable with Cu in the $Ti_3(Cu + Al)_3O$ compound. The presence of Al in the M_6O-type reaction layer enables chemical mass balance to be maintained:

$$Ti + Cu + \frac{1}{2}TiO + \frac{1}{2}Al \rightarrow \frac{1}{2}Ti_3Cu_2AlO \tag{2.7}$$

$$\frac{1}{2}Ti_3Cu_2AlO + \frac{1}{2}Cu \rightarrow \frac{1}{2}Ti_3Cu_3O + \frac{1}{2}Al \tag{2.8}$$

At low concentrations of Ti in the Ag-Cu-Ti braze alloy, a single continuous Ti-O layer is often characterised at the joint interface (see Section 2.2.2). Combining Equations 2.4 and 2.5 to provide an overall negative Gibbs whereby TiO requires Ti_3Cu_3O in order to be thermodynamically favourable is not, therefore, completely agreed with in the literature.

Furthermore, several studies have observed the formation of a Ti_2O layer on the alumina side of the joint interface (layer I). Suenaga et al. (1997) proposed that following Equation 2.4, Ti_2O and Ti_3Cu_3O could form according to Equation 2.9. This required a Cu-Ti compound, from the braze alloy to react with TiO, which was reported to form during cooling:

$$3Cu_4Ti + 2TiO \rightarrow Ti_3Cu_3O + Ti_2O + 9Cu \tag{2.9}$$

Similarly, Lin et al. (2014) observed a Ti_2O layer on the alumina side of the joint interface (layer I) and a Ti_3Cu_3O layer on the braze side of the joint interface (layer II). The formation of Ti_2O and Ti_3Cu_3O according to Equation 2.9 was proposed as the likely mechanism. The unspecified destination of Al however, which is a product of Equation 2.4 suggested that further clarification was required. Lin et al. (2014) attempted to clarify this and proposed Equation 2.10 as an alternative to Equation 2.4. In this mechanism, Ti_2O could be a product of the initial reduction reaction of alumina whereby Al could be liberated to the braze alloy. Cu from the braze alloy could then react with Ti_2O to form Ti_3Cu_3O during cooling, according to Equation 2.11:

$$Al_2O_3 + 6Ti \rightarrow 3Ti_2O + 2Al \tag{2.10}$$

$$6Cu + 3Ti_2O \rightarrow 2Ti_3Cu_3O + O \qquad (2.11)$$

The formation of these reaction products during cooling is not entirely agreed with in the literature. Voytovych et al. (2004) reported the thickening of the M_6O-type layer at the brazing temperature as the brazing time was increased (see Section 2.3.1.2). This showed that Ti_3Cu_3O does not form during cooling and instead forms at the brazing temperature. Considering this, Ali et al. (2015) proposed Equation 2.12 as an alternative to Equations 2.4 and 2.10 whereby Ti and Cu could both diffuse to the joint interface and participate in the reduction of the alumina surface. The liberation of Al from the reduction reaction could replace Cu in the otherwise Ti_3Cu_3O compound to form $Ti_9Cu_7Al_2O_3$:

$$Al_2O_3 + 9Ti + 7Cu \rightarrow Ti_9Cu_7Al_2O_3 \qquad (2.12)$$

Ali et al. (2015) observed ~6.5 at.% Al in the M_6O-type layer using TEM. This was significantly less than the maximum solubility of 15 at.% Al in Ti_3Cu_3O as found by Kelkar and Carim (1995). Ali et al. (2015) suggested that Al could also form solid solutions with Ag near the joint interface.

The presence of Al in the reaction layer has been commonly reported in the literature (Byun and Kim, 1994; Kelkar and Carim, 1995; Kozlova et al., 2011; Lee and Kwon, 1995; Lin et al., 2014; Santella et al., 1990; Stephens et al., 2003; Valette et al., 2005; Voytovych et al., 2004). It should be noted that the presence of Al in the reaction products confirms the reduction of alumina, since the ceramic is the only likely source of Al in the alumina/Ag-Cu-Ti system. The presence of Al in the reaction layers provides mass balance in the reduction reaction expressed by Equation 2.4.

In a later study, Ali et al. (2016) performed systematic brazing experiments whereby alumina-to-alumina brazed joints were formed using the commercially available Ag-Cu-Ti braze alloy Cusil ABA®, with a fixed brazing time of just 1 min (see Section 2.3.1). Based on their observations, Ali et al. (2016) proposed that following the reaction in Equation 2.10 and the dissolution of Al in the braze alloy particularly in the Ag-rich phase, Ti_2O could react with Ti and Cu to form Ti_3Cu_3O. The lateral growth of Ti_2O was suggested to proceed until a continuous layer on the alumina side of the joint interface had formed. The formation of Ti_3Cu_3O could thereby occur according to Equation 2.13:

$$Ti_2O + Ti + 3Cu \rightarrow Ti_3Cu_3O \qquad (2.13)$$

Ali et al. (2016) found the Ti_2O phase to be transient and unstable. With an increase in the brazing temperature, the Ti_2O phase was found to be replaced by γ-TiO, a high-temperature TiO phase. The transformation of Ti_2O to γ-TiO, however, could not be supported experimentally.

This summarises a thorough understanding of the joining mechanism in the alumina/Ag-Cu-Ti system. Further investigation is needed to build on this and it is thought that a clearer understanding of the effects of variables in the AMB process on reaction layer formation may provide some support.

2.2 VARIABLES IN THE DESIGN OF AN Ag-Cu-Ti ACTIVE BRAZE ALLOY

Inconsistent observations and proposed joining mechanisms may have stemmed from differences in experimental methodologies adopted in studies concerning the AMB of alumina. Variables associated to the design of an ABA include the braze alloy composition, which affects interfacial chemistry (Figure 1.1).

2.2.1 Ag-Cu Concentrations

The Ag and Cu concentrations in an Ag-Cu-Ti braze alloy form the host lattice in which Ti atoms must diffuse, at the brazing temperature, in order to reach the alumina surface where reactions subsequently occur. The Ag and Cu concentrations in the braze alloy can influence the Ti activity level, i.e. the energy with which Ti diffuses to the alumina surface which thereby governs reaction kinetics at the joint interface.

The Ti activity level hereby refers to the chemical activity of Ti atoms in the host lattice, such as an Ag-Cu lattice. It is a dimensionless quantity given as a fraction of the chemical activity of pure Ti, which is equal to 1. The Ti activity level is always less than 1 therefore, due to interactions between Ti atoms and the Ag-Cu lattice (Kozlova et al., 2010; Valette et al., 2005).

It is commonly reported in the literature that the Ti activity level can greatly affect the stoichiometry of reaction products which form at the joint interface (do Nascimento et al., 2003; Ghosh et al., 2012; Kelkar et al., 1994).

The effect of Ag and Cu concentrations in the Ag-Cu-Ti braze alloy on the Ti activity level must be considered in the design of a brazing experiment. For example, a reduced Ti activity level may lead to incomplete diffusion of Ti to the joint interface. This may affect reaction product formation and subsequently, the joint strength. Furthermore, the ductility of the braze interlayer typically relies on the complete diffusion of Ti to the joint interface (see Section 5.3.2).

As the Ag concentration in an Ag-Cu-Ti braze alloy increases, the Ti activity level also increases. On the contrary, as the Cu concentration in an Ag-Cu-Ti braze alloy increases, the Ti activity level decreases. The solubility of Ti in Ag is significantly lower than that of Ti in Cu. For example, at 1,150°C the solubility of Ti in Ag is ~3.7 wt.%, whereas the solubility of Ti in Cu is ~67 wt.% (Hansen and Anderko, 1958). These interactions are evidenced by comparing the enthalpies of mixing of Cu in Ti or Ag with that of Ag in Cu or Ti (Table 2.3). It can be seen that while the formation of Cu-Ti compounds is favourable, the formation of Ag-Ti compounds is not. This may be used to explain the formation of Cu-Ti phases observed in the as-received Cusil ABA® and TICUSIL® braze foils (Figure 2.1).

Pak et al. (1990) studied the effect of increasing Ag concentration on the Ti activity level at 1,000°C in an Ag-Cu-1 wt.% Ti braze alloy. As the Ag concentration was increased from 72 to 99 wt.% Ag, the Ti activity coefficient (γ_{Ti}) increased by a factor of ~20 (Figure 2.7). This coincided with a decrease in the final molar fraction of Ti in the melt (χ_{Ti}) due to the consumption of Ti in the formation of a reaction layer. It was found that as the Ag concentration in the Ag-Cu-Ti braze alloy increased, the Ti activity level also increased.

These results are consistent with suggestions in the literature that the higher Ag concentration of TICUSIL®, as compared with Cusil ABA®, enhances the activity of Ti in the braze alloy and, therefore, TICUSIL® exhibits relatively better wetting and spreading behaviour on alumina (Asthana and Singh, 2008).

Several studies have shown that an increase in the Cu concentration in an Ag-Cu-Ti braze alloy, and hence a corresponding decrease in the Ag concentration, can significantly reduce the

TABLE 2.3
Partial Enthalpies of Mixing (ΔH_i^∞) of Solute *i* Infinitely Diluted in Me for Binary Liquid Alloys of the Ag-Cu-Ti System Showing That the Ti Affinity to Cu Is Much Higher Than the Ti Affinity to Ag (Kozlova et al., 2010)

Me-*i* Alloy	ΔH_i^∞ (kJ/mole)
Cu-Ti	−10
Cu-Ag	16
Ag-Cu	23
Ag-Ti	25

Source: Kozlova et al. (2010).

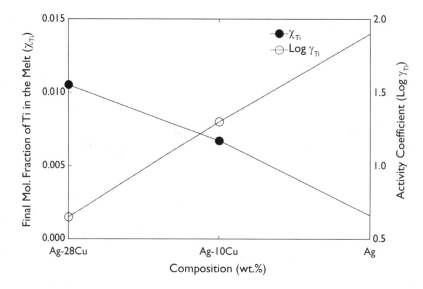

FIGURE 2.7 The effect of Ag concentration on the Ti activity coefficient (γ_{Ti}) at 1,000°C in an Ag-Cu-1 wt.% Ti braze alloy. An increase in the Ag concentration from 72 to 99 wt.% Ag led to an increase in the Ti activity coefficient with a corresponding decrease in the final molar fraction of Ti in the melt (χ_{Ti}) (Pak et al., 1990).

Ti activity level in an alumina-to-alumina brazed joint. Mandal et al. (2004a) studied the effect of increasing Cu concentration on the microstructure of alumina-to-alumina brazed joints. The braze compositions selected in order of increasing Cu concentrations were 69.84Ag-27.16Cu-3Ti wt.%, 58.2Ag-38.8Cu-3Ti wt.% and 48.5Ag-48.5Cu-3Ti wt.%. Brazing was conducted at 1,000°C for 15 min. As the Cu concentration was increased from 27.6 to 38.8 and then to 48.5 wt.% Cu, the diffusivity of Ti to the joint interface was consistently inhibited. This observation suggested that Ti could be retained in the braze interlayer in amounts corresponding to the Cu concentration in the Ag-Cu-Ti braze alloy (Figure 2.8).

These results showed that despite a fixed Ti concentration, therefore, Cu concentrations in the Ag-Cu-Ti braze alloy above the eutectic composition of 72Ag-28Cu wt.% could adversely affect the diffusivity of Ti towards the joint interface. In contrast, an increase in the Ag concentration, which enhances the Ti activity level, led to improved reaction kinetics and thereby reaction layer formation at the joint interface.

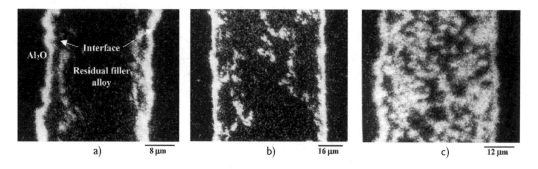

FIGURE 2.8 SEM-EDX elemental maps showing the distribution of Ti in the microstructure of alumina-to-alumina brazed joints made using (a) 97(Ag-28Cu)$_3$Ti, (b) 97(Ag-40Cu)$_3$Ti and (c) 97(Ag-50Cu)$_3$Ti wt.% braze alloys. As the Cu content in the Ag-Cu-Ti braze alloy increased, the amount of Ti which was retained in the braze interlayer was also found to increase (Mandal et al., 2004a).

2.2.2 Ti Concentration

The concentration of Ti (number of Ti atoms per unit volume) in an Ag-Cu-Ti braze alloy can counter any adverse influence of reduced Ag or increased Cu concentrations on the Ti activity level. According to Fick's first law, the diffusivity of Ti atoms in an Ag-Cu lattice, towards the joint interface, depends on its concentration gradient:

$$J = -D\frac{\partial C}{\partial x} \tag{2.14}$$

Fick's first law states that the net flux (J) of atoms in a given cross-sectional area in a given unit of time is proportional to the concentration gradient $\left(\frac{\partial C}{\partial x}\right)$ across the distance x, where D is the diffusion coefficient of Ti, a measure of the mobility of the diffusing Ti atoms in the Ag-Cu lattice. This assumes that the flux depends on a linear concentration gradient i.e. there is a constant variation in the Ti concentration; however, this is not the case in AMB due to the build-up of a Ti-rich reaction product at the joint interface. Since the concentration gradient changes with time $\left(\frac{\partial C}{\partial t}\right)$, Fick's second law becomes more applicable:

$$\frac{\partial C}{\partial t} = D\frac{\partial^2 C}{\partial x^2} \tag{2.15}$$

During the brazing process, a large concentration gradient initially exists between the central Ti-rich region of the TICUSIL® braze foil and the alumina surface (Figure 2.1b). This gradient decreases with the formation of Ti-rich reaction products at the joint interface. Conversely, the gradient increases as the starting Ti concentration in the braze foil is increased. An increase in the Ti concentration, therefore, creates a greater driving force for Ti diffusion towards the joint interface. Reaction layer formation may slow the potential of this driving force as the gradient equilibrates; the reaction layer also forms a diffusion barrier between Ti (in the melt) and the alumina surface.

In order to fairly determine the effect of Ti concentration on wetting, reaction layer formation and joint strength, it is important to consider the effect of the volume of braze selected on the Ti activity level. The starting braze foil thickness can govern the amount of Ti available to diffuse to a given surface area of alumina.

2.2.2.1 Ag-Cu-Ti Braze Foil Thickness

The volume of braze selected, or braze preform dimensions, can influence the amount of Ti which is available to diffuse to the joint interface. However, these dimensions are seldom reported in the literature, with a few exceptions (Carim, 1991; Carter, 2004; Hao et al., 1994; Kar et al., 2007; Lee and Kwon, 1995; Mandal et al., 2004b). Typically, the starting braze foil thicknesses are reported. Figure 2.9 shows a comparison of all the starting Ag-Cu-Ti braze foil thicknesses reported in the formation of alumina-to-alumina brazed joints. While 50- to 200-μm-thick Ag-Cu-Ti braze foils have been used, the effect of braze foil thickness on the strength of alumina-to-alumina brazed joints could not be found in the literature.

2.2.2.2 Ti Concentration and Wetting

Ag-Cu braze alloys do not wet alumina ceramics well, producing equilibrium contact angles as high as 150° in wetting experiments. When Ti is added to an Ag-Cu braze alloy, however, wetting can improve significantly. The extent of this improvement depends on the amount of Ti added. Studies which have performed wetting experiments on alumina can be divided into three

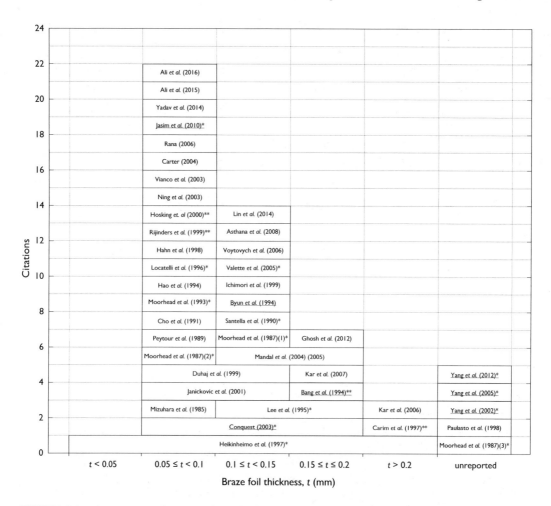

FIGURE 2.9 A summary of the starting Ag-Cu-Ti braze preform thicknesses used in the formation of alumina-to-alumina brazed joints. Underline denotes Ag-Cu-Ti braze alloy was applied in the form of a paste. * denotes the Ag-Cu-Ti braze composition contained other alloying additions and/or alternative braze alloy compositions, in addition to Ag-Cu-Ti braze compositions were reported in the study. ** denotes the braze composition reported was not exactly Ag-Cu-Ti based but an alternative composition, e.g. Cu-Ti or Au–based.

main groups, based on the relative Ti concentrations in the Ag-Cu-Ti braze alloys selected. These groups are hereby defined as Ag-Cu-Ti braze alloys with (i) low Ti concentrations (<1.75 wt.% Ti), (ii) intermediate Ti concentrations (1.75 < wt.% Ti < 4.5) and (iii) high Ti concentrations (> 4.5 wt.% Ti).

At low Ti concentrations (< 1.75 wt.% Ti), relatively high equilibrium contact angles that are greater than ~60° typically form (Eustathopoulos, 2015). At intermediate Ti concentrations (1.75 < wt.% Ti < 4.5), this decreases significantly to ~10° (Asthana and Singh, 2008; Lin et al., 2001; Mandal et al., 2004b; Nicholas and Peteves, 1994; Santella et al., 1990; Shiue et al., 2000; Voytovych et al., 2004).

Voytovych et al. (2006) performed wetting experiments on monocrystalline alumina using three Ag-Cu-Ti braze alloys with increasing Ti concentrations. The braze alloy compositions were Ag-Cu-0.7Ti at.%, Ag-Cu-2.9Ti at.% and Ag-Cu-8.0Ti at.%. Brazing was conducted at 900°C for 30 min. As the Ti concentration increased from 0.7 to 2.9 and then to 8.0 at.% Ti, the equilibrium contact angles decreased from 63°, 10° and then to 7°, respectively. It was concluded that a Ti concentration of at least 2.9 at.% was required to enable the adequate wetting of alumina.

Literature Review

The addition of ~2 wt.% Ti to an Ag-Cu eutectic alloy is significant in improving the wettability of alumina because of a miscibility gap in the Ag-Cu-Ti system, which leads to the formation of an Ag-rich liquid (Ag-27Cu-2Ti wt.%) and a Ti-rich liquid (Ag-66Cu-22Ti wt.%). It is this latter Ti-rich liquid with 22 wt.% Ti that enables an improvement in the wetting of alumina (Jacobson and Humpston, 2005)

At high Ti concentrations (> 4.5 wt.% Ti), the equilibrium contact angle can either remain at ~10°, or improve slightly, or it can become adversely affected increasing to above 30°. The latter phenomenon occurs due to properties of the braze interlayer as opposed to the metallic character of the reaction layer, which otherwise typically governs the reactive wetting process.

Conquest (2003) performed wetting experiments on 99.7 wt.% Al_2O_3 alumina using TICUSIL® paste. Additions of up to 9.5 wt.% Ti were made to the TICUSIL® paste, increasing the total Ti concentration to 14.0 wt.% Ti. Brazing was conducted at 950°C for 10 min. As the total Ti concentration increased from 4.5 to 14.0 wt.% Ti, the equilibrium contact angle increased from 11° to 31°. A Ti-rich region was observed at the centre of the brazed microstructure which indicated incomplete diffusion of Ti to the joint interfaces (Figure 2.10). It was suggested that an increase in the Ti concentration had increased the liquidus temperature of the braze alloy and at a fixed brazing temperature of 950°C insufficient melting and sluggish spreading kinetics had adversely affected the wetting behaviour. Differences in the rheology between the phases in the Ti-rich region and the Ti-depleted region, however, were not analysed.

Moorhead et al. (1987) used a 40.54Cu-19.69Ti wt.% braze alloy when performing wetting experiments on 99.5 and 99.8 wt.% Al_2O_3 alumina ceramics. Despite the extremely high Ti concentration, an equilibrium contact angle of just ~15° was reported. Brazing was conducted at between 980°C and 1,090°C. This somewhat contradicts the observations made by Conquest (2003); however, it is likely that Moorhead et al. (1987) did not observe retained Ti in the brazed microstructure owing to either longer brazing times (see Section 2.3.1) or higher brazing temperatures (see Section 2.3.2). These results show that incomplete diffusion of Ti to the joint interface can adversely affect wetting behaviour of Ag-Cu-Ti braze alloys on alumina.

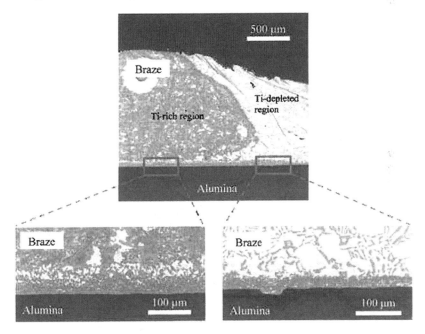

FIGURE 2.10 Backscattered electron image showing a Ti-rich region in the braze interlayer, following a wetting experiment at 950°C for 10 min on 99.7 wt.% Al_2O_3 alumina using TICUSIL® paste. Spreading of the braze alloy was led by an Ag-rich phase (Ti-depleted region), observed at the triple line (Conquest, 2003).

2.2.2.3 Ti Concentration and the Reaction Layer

According to reactive wetting theory, the improved wettability of alumina by Ag-Cu-Ti braze alloys, in comparison to Ag-Cu braze alloys, is achieved through the formation of a reaction layer, at the joint interface, which exhibits metallic character. Thus, increasing the Ti concentration of an Ag-Cu-Ti braze alloy often leads to an improvement in the wettability of alumina. Therefore, the concentration of Ti in an Ag-Cu-Ti braze alloy, which affects its diffusivity, may directly influence the reaction layer composition including its metallic character.

The improved wettability of alumina with increasing Ti concentration has been reported to depend on the formation of M_6O-type compounds; hence, changes to both the reaction layer composition and thickness have been observed in the literature where Ag-Cu-Ti braze alloys with increased Ti concentration have been selected. It is important to establish that if the Ti concentration provides the potential for the Ag-Cu-Ti braze alloy to reactively wet the alumina surface, then the brazing temperature and time selections govern this potential by activating the diffusion of Ti to the alumina surface for a given period.

Voytovych et al. (2006) observed a decrease in the equilibrium contact angles on monocrystalline alumina from 63° to 10° and then to 7° as the Ti concentration in an Ag-Cu-Ti braze alloy increased from 0.7 to 2.9 and then to 8.0 at.% Ti. EPMA and XRD techniques were used to characterise a Ti_2O layer on the alumina side of the joint interface (layer I) when 0.7 at.% Ti was used. At a Ti concentration of 2.9 at.% Ti, two continuous layers were observed. These were a $Ti_{1.75}O$ layer on the alumina side of the joint interface (layer I) and a Ti_3Cu_3O layer on the braze side of the joint interface (layer II). At a Ti concentration of 8.0 at.% Ti, a 3- to 5-µm-thick reaction layer composed of both Ti_3Cu_3O and Ti_4Cu_2O phases was observed (Figure 2.11).

Table 2.4a outlines several studies in the literature whereby the effect of increasing Ti concentration in an Ag-Cu-Ti braze alloy on the composition of reaction products at the alumina/Ag-Cu-Ti joint interface have been reported. Table 2.4b outlines several other studies which have also characterised the reaction layer in the alumina/Ag-Cu-Ti system, but where a single Ag-Cu-Ti braze alloy was selected. These studies, when compared, provide a better understanding of the effect of Ti concentration on the properties of the reaction layer.

The studies listed in Table 2.4 can be divided into three main groups, based on the Ti concentration in the Ag-Cu-Ti braze alloys selected. These groups, as defined earlier in Section 2.2.2.2, are Ag-Cu-Ti braze alloys with (i) low Ti concentrations (< 1.75 wt.% Ti), (ii) intermediate Ti concentrations (1.75 < wt.% Ti < 4.5) and (iii) high Ti concentrations (>4.5 wt.% Ti).

Ti concentration (at.%)	0.7	2.9	8.0
Equilibrium contact angle θ_f,	63°	10°	7°
Layer I	Ti_2O (1)	$Ti_{1.75}O$ (2)	$Ti_3(Cu, Al)_3O$ (4)
Layer II		$Ti_3(Cu, Al)_3O$ (3)	$Ti_4(Cu, Al)_2O$ (5)

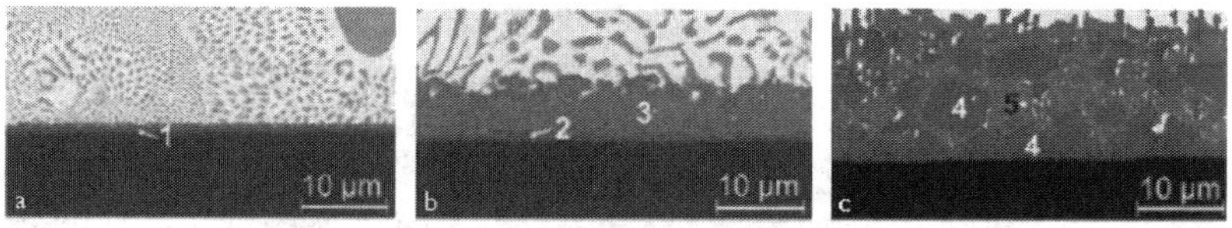

FIGURE 2.11 The effect of Ti concentration on the wettability of alumina and the reaction layer composition, using (a) Ag-Cu-0.7Ti at.%, (b) Ag-Cu-2.9Ti at.% and (c) Ag-Cu-8.0Ti at.% braze alloys. Brazing was conducted at 900°C for 30 min (Voytovych et al., 2006).

TABLE 2.4
Details of Experimental Methods and Characterisation Techniques of Studies Discussed in Section 2.2

Publication	Experiment	Al$_2$O$_3$ Purity (wt.%)	Ag-Cu-Ti Composition (wt.%)	Braze Quantity/ Thickness (μm)	Vacuum (mbar)	Brazing Temp, T (°C)	Brazing Time (min)	Techniques
(a) Studies which have investigated the effect of Ti concentration on the properties of the reaction layer at the Alumina/Ag-Cu-Ti joint interface									
Voytovych et al. (2006)	Wetting trials	Mono-crystalline	Ag-Cu-0.7Ti (at.%) Ag-Cu-2.9Ti (at.%) Ag-Cu-8.0Ti (at.%)	0.1–0.3 g	1.0×10^{-7}	900	30	SEM EPMA XRD
Loehman and Tomsia (1992)	Wetting trials	95.0 99.8	71.3Ag-27.7Cu-1.0Ti 68.6Ag-26.6Cu-4.8Ti	~0.5 g	$6. \times 10^{-13}$ Ar atm	1,000	60	EPMA
Conquest (2003)	Wetting trials	99.7	68.8Ag-26.7Cu-4.5Ti TICUSIL® + 1.5Ti TICUSIL® + 3.5Ti TICUSIL® + 9.5Ti	—	4.0×10^{-4}	950	10	SEM
Suenaga et al. (1997)	Wetting trials	Mono-crystalline	Ag-Cu/2.5Ti Ag-Cu/6.3Ti Ag-Cu/10Ti	0.8	6.0×10^{-5}	850	10	XRD AES TEM
Lin et al. (2014)	Brazed joints	High purity (0.05 wt.% MgO)	63.0Ag-35.25Cu-1.75Ti 68.8Ag-26.7Cu-4.5Ti	100–120	1.3×10^{-6}	830 915	20 20	XRD TEM
(b) Selected studies in which the reaction layer at the Alumina/Ag-Cu-Ti joint interface has been characterised, ranked in ascending order of Ti concentration									
Cho et al. (1992)	Brazed joints	High purity (0.15 wt.% MgO)	71.1Ag-28.4Cu-0.5Ti	50	1.3×10^{-5}	850–950	10	SEM XRD
Stephens et al. (2003)	Wetting trials	Mono-crystalline	63.0Ag-35.25Cu-1.75Ti	76	H atm	845	6	TEM
Valette et al. (2005)	Brazed joints	99.5	63.0Ag-35.25Cu-1.75Ti	100	1.0×10^{-6}	900	15	EDX EPMA

(Continued)

TABLE 2.4 (Continued)
Details of Experimental Methods and Characterisation Techniques of Studies Discussed in Section 2.2

Publication	Experiment	Al$_2$O$_3$ Purity (wt.%)	Ag-Cu-Ti Composition (wt.%)	Braze Quantity/ Thickness (μm)	Vacuum (mbar)	Brazing Temp, T (°C)	Brazing Time (min)	Techniques
Kozlova et al. (2011)	Wetting trials	Mono-crystalline	63.0Ag-35.25Cu-1.75Ti	—	6.0×10^{-7}	850–900	0–20	SEM EDX
Mizuhara and Mally (1985)	Brazed joints	97.6 99.5	(72Ag+28Cu)+1.5Ti (72Ag+28Cu)+2.5Ti (72Ag+28Cu)+3.2Ti	50	1.3×10^{-5}	—	—	—
Santella et al. (1990)	Brazed joints	99.5	56.0Ag-36.0Cu-6.0Sn-2.0Ti	100	1.0×10^{-5}	900	20	TEM
Hahn et al. (1998)	Brazed joints	96.0	70.1Ag-27.3Cu-2.6Ti	~50	1.3×10^{-5}	830	10	SEM-EDX TEM
Lee and Kwon (1995)	Brazed joints	High purity (0.1–0.15 wt.% MgO)	48.1Ag-48.1Cu-3.8Ti	120–150	5.0×10^{-5}	920	20	EDX XRD
Ichimori et al. (1999)	Brazed joints	Mono-crystalline	66.7Ag-28.4Cu-4.9Ti	100	5.0×10^{-7}	900	5	TEM
Jasim et al. (2010)	Brazed joints	High purity Al$_2$O$_3$	68.0Ag-27.0Cu-5.0Ti	70	6.7×10^{-5}	980	—	SEM XRD
Peytour et al. (1990)	Brazed joints	99.5	40.0Ag-55.0Cu-5.0Ti	80	6.7×10^{-5}	850–890	2–10	EDX EPMA
Moorhead et al. (1987)	Brazed joints	99.5 99.8	39.8Ag-40.5Cu-19.7Ti	—	6.7×10^{-5}	980–1090	—	—

With low Ti concentrations (<1.75 wt.% Ti) in an Ag-Cu-Ti braze alloy, the reaction layer typically consists of a single Ti-O layer on the alumina side of the joint interface (layer I). The continuity of this layer can vary, with lower Ti concentrations (<1.0 wt.% Ti) being deficient and leading to the Ti-O layers which exhibit a degree of discontinuity (Table 2.5). As the Ti concentration increases, therefore, the continuity of the Ti-O layer improves.

Cho and Yu (1992) formed alumina-to-alumina brazed joints using 50-μm-thick 71.1Ag-28.4Cu-0.5Ti wt.% braze foils. SEM and XRD techniques were used to characterise a single discontinuous reaction layer composed of γ-TiO, with isolated δ-TiO particles, on the alumina side of the joint interface (layer I). This discontinuity observed was suggested to be due to the low Ti concentration in the braze alloy.

Voytovych et al. (2006) performed wetting experiments on monocrystalline alumina using 0.1 to 0.3 g of an Ag-Cu-0.7Ti at.% braze alloy. Using EPMA and XRD techniques, a single continuous Ti_2O layer was observed on the alumina side of the joint interface (layer I).

Loehman and Tomsia (1992) performed wetting experiments on 95.0 and 99.8 wt.% Al_2O_3 alumina using 0.5 g of a 71.3Ag-27.7Cu-1.0Ti wt.% braze alloy. EPMA was used to characterise a single continuous TiO layer on the alumina side of the joint interface (layer I); the oxygen to Ti ratio was reported to be between 0.4 and 0.6.

Suenaga et al. (1997) performed wetting experiments on monocrystalline alumina using a 0.8-μm-thick Ag-Cu/2.5Ti wt.% bilayer film. XRD was used to characterise a single Ti_3O_2 layer on the alumina surface; however, joining was reported to be unsuccessful.

Overall, these results show that in the low Ti concentration range, the continuity of the single Ti-O layer improves as the Ti concentration is increased. Furthermore, this Ti-O layer, which forms on the alumina side of the joint interface, comprises of Ti-O reaction products that become increasingly richer in Ti as the Ti concentration in the Ag-Cu-Ti braze alloy is increased (Table 2.5). Furthermore, there appears to be a lack of any M_6O-type layer on the braze side of the joint interface.

With intermediate Ti concentrations (1.75 < wt.% Ti < 4.5) in an Ag-Cu-Ti braze alloy, the reaction layer typically consists of a Ti-O layer on the alumina side of the joint interface (layer I) and an M_6O-type Ti-Cu-O layer on the braze side of the joint interface (layer II). While the stoichiometry of the Ti-O layer can vary, according to comparisons made between literature studies, the M_6O-type layer has been commonly reported as Ti_3Cu_3O.

Voytovych et al. (2006) performed wetting experiments on monocrystalline alumina using 0.1 to 0.3 g of an Ag-Cu-2.9Ti at.% braze alloy. A $Ti_{1.75}O$ layer on the alumina side of the joint interface (layer I) and a Ti_3Cu_3O layer on the braze side of the joint interface (layer II) were characterised using SEM, EPMA and XRD techniques.

Stephens et al. (2003) performed wetting experiments on monocrystalline alumina using 76-μm-thick Cusil ABA® braze foils. TEM was used to characterise a γ-TiO layer, with isolated Ti_2O particles, on the alumina side of the joint interface (layer I) and a Ti_3Cu_3O layer on the braze side of the joint interface (layer II). A TEM image of the joint interface is shown in Figure 2.12.

Valette et al. (2005) formed alumina-to-alumina brazed joints using 99.5 wt.% Al_2O_3 alumina and 100-μm-thick Cusil ABA® braze foils. SEM and EDX techniques were used to characterise a Ti_2O layer on the alumina side of the joint interface (layer I) and a Ti_3Cu_3O layer on the braze side of the joint interface (layer II).

Lin et al. (2014) formed alumina-to-alumina brazed joints using ~110-μm-thick Cusil ABA® and TICUSIL® braze foils. In both sets of joints, TEM and XRD techniques were used to characterise a Ti_2O on the alumina side of the joint interface (layer I) and a Ti_3Cu_3O layer on the braze side of the joint interface (layer II). The average reaction layer thickness in joints made using Cusil ABA® was reported to be ~2.6 μm, significantly thinner than the ~6.75-μm-thick average reaction layer thickness that was reported for joints made using TICUSIL®.

Santella et al. (1990) formed alumina-to-alumina brazed joints using 99.5 wt.% Al_2O_3 alumina and 100-μm-thick 56.0Ag-36.0Cu-6.0Sn-2.0Ti wt.% braze foils. TEM was used to characterise

TABLE 2.5
Reaction Layer Compositions Reported in the Literature for Brazing Experiments Conducted on Alumina Using Low Ti Concentrations (<1.75 wt.% Ti) in Ag-Cu-Ti Braze Alloys

Publication	Experiment	Ag-Cu-Ti Composition (wt.%)	Braze Quantity/ Thickness (μm)	Brazing Temp, T (°C)	Brazing Time (min)	Techniques	Layer I	Layer II
Cho et al. (1992)	Brazed joints	71.1Ag-28.4Cu-0.5Ti	50 μm	850–950	10	SEM XRD	γ-TiO + isolate δ-TiO	–
Voytovych et al. (2006)	Wetting trials	Ag-Cu-0.7Ti (at.%)	0.1–0.3 g	900	30	EPMA XRD	Ti_2O	–
Loehman and Tomsia (1992)	Wetting trials	71.3Ag-27.7Cu-1.0Ti	~0.5 g	1,000	60	EPMA	$TiO_{0.5}$	–
Suenaga et al. (1997)	Wetting trials	Ag-Cu/2.5Ti	0.8 μm	850	10	XRD AES TEM	Ti_3O_2	–

Literature Review 27

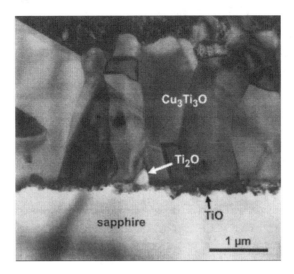

FIGURE 2.12 TEM image showing a continuous γ-TiO layer, with isolated Ti$_2$O particles, on the alumina side of the joint interface (layer I) and a Ti$_3$Cu$_3$O layer on the braze side of the joint interface (layer II). Brazing was conducted at 900°C for 30 min using Cusil ABA® on monocrystalline alumina (Stephens et al., 2003).

a 0.1- to 0.2-μm-thick TiO layer on the alumina side of the joint interface (layer I) and a 2- to 3-μm-thick Ti$_3$Cu$_3$O layer on the braze side of the joint interface (layer II).

Hahn et al. (1998) formed alumina-to-alumina brazed joints using 96.0 wt.% Al$_2$O$_3$ alumina and ~50-μm-thick 70.1Ag-27.3Cu-2.6Ti braze foils. SEM, TEM and EDX techniques were used to characterise a TiO$_{1.04}$ layer on the alumina side of the joint interface (layer I) and a Ti$_4$Cu$_2$O on the braze side of the joint interface (layer II).

Lee and Kwon (1995) formed alumina-to-alumina brazed joints using ~135-μm-thick 48.1Ag-48.1Cu-3.8Ti wt.% braze foils. SEM and XRD techniques were used to characterise a 1.2-μm-thick TiO layer on the alumina side of the joint interface (layer I) and a 2.4-μm-thick Ti$_4$Cu$_2$O layer on the braze side of the joint interface (layer II).

Ichimori et al. (1999) formed alumina-to-alumina brazed joints using monocrystalline alumina and 100-μm-thick 66.7Ag-28.4Cu-4.9Ti braze foils. TEM and EDX techniques were used to characterise a 10 to 50 μm thick TiO layer, with isolated Ti$_2$O particles, on the alumina side of the joint interface (layer I) and a 1- to 2-μm-thick Ti$_3$Cu$_3$O layer on the braze side of the joint interface (layer II). A Cu-Al-O compound, suggested to have formed due to a reaction between alumina and Cu, was also observed between layer I and the alumina surface.

These results suggest good correlation between the Ti concentration range 1.75 < wt.% Ti < 4.5 and the reaction layer composition, which generally comprises of Ti-O and Ti-Cu-O layers. In this intermediate Ti concentration range therefore, an increase in the Ti concentration can lead to reaction products in both layers that are richer in Ti i.e. Ti$_2$O rather than TiO and Ti$_4$Cu$_2$O rather than Ti$_3$Cu$_3$O, owing to higher Ti activity levels (Kelkar et al., 1994). Generally, the Ti-O layer appears to be more sensitive to an increase in the Ti concentration than the M$_6$O-type layer, typically Ti$_3$Cu$_3$O and less commonly Ti$_4$Cu$_2$O – both reported to consist of small concentrations of Al. The intermediate Ti concentration range produces a reaction layer that is overall thicker than that which forms in the low Ti concentration range. This is primarily due to layer II - the M$_6$O-type layer, which forms on the braze side of the joint interface. Another consideration for the thicker reaction layer is the increased Ti concentration in the braze alloy, which physically provides more Ti to diffuse to the joint interface.

With high Ti concentrations (> 4.5 wt.% Ti) in an Ag-Cu-Ti braze alloy, the reaction layer can comprise Ti$_4$Cu$_2$O rather than being completely composed of Ti$_3$Cu$_3$O on the braze side of the joint interface (layer II). According to some studies, layer I is no longer observed; however, this may be due to its thinness coupled with the limitations of the analytical techniques used in the literature (Table 2.7), or its consumption in the formation of layer II as per proposed joining mechanisms (see Section 2.1.7).

Voytovych et al. (2006) performed wetting experiments on monocrystalline alumina using 0.1 to 0.3 g of an Ag-Cu-8.0Ti at.% braze alloy. SEM, EPMA and XRD techniques were used to characterise a 3- to 5-μm-thick layer, composed of both Ti_3Cu_3O and Ti_4Cu_2O on the braze side of the joint interface (layer II) (Figure 2.11c).

Jasim et al. (2010) formed alumina-to-alumina brazed joints using 70-μm-thick 68.0Ag-27.0Cu-5.0Ti braze foils. XRD was used to characterise a Ti_4Cu_2O layer on the braze side of the joint interface (layer II).

Peytour et al. (1990) formed alumina-to-alumina brazed joints using 99.5 wt.% Al_2O_3 alumina and 80-μm-thick 40.0Ag-55.0Cu-5.0Ti braze foils. SEM, EDX and EPMA techniques were used to characterise a 2-μm-thick Ti_4Cu_2O layer on the braze side of the joint interface (layer II).

These results show that in the high Ti concentration range, an increase in the Ti concentration can lead a single Ti-Cu-O reaction layer that is richer in Ti, i.e. Ti_4Cu_2O, rather than Ti_3Cu_3O. Some studies which have reported the formation of Ti_4Cu_2O, however, have relied on XRD techniques (Lee and Kwon, 1995). Coinciding major peaks in XRD measurements have been suggested to cause confusion when attempting to distinguish between Ti_3Cu_3O and Ti_4Cu_2O (Kar et al., 2006; Lin et al., 2014; Mandal et al., 2004b). The lattice parameter for Ti-Cu-O phases can vary according to their relative Ti/Cu ratios (Karlsson, 1951). The lattice parameters for Ti_3Cu_3O and Ti_4Cu_2O are similar; 1.124 and 1.149 nm, respectively, and both phases share the same diamond cubic crystal structure (Karlsson, 1951; Suenaga et al., 1993, 1997).

The maximum solubility of Al in Ti_4Cu_2O is 1.5 at.% Al, whereas the maximum solubility of Al in Ti_3Cu_3O can be up to 15 at.% Al (Kelkar and Carim, 1995). The concentration of Al in the M_6O-type layer has commonly been reported to be well above 1.5 at.% Al and this has been argued to support the formation of Ti_3Cu_3O rather than Ti_4Cu_2O.

Although the limitations of using XRD in distinguishing between Ti_3Cu_3O and Ti_4Cu_2O have been expressed in the literature, atomic contrast in BSE images of the reaction layer clearly show two phases, reported to be Ti_3Cu_3O and Ti_4Cu_2O in a study by Voytovych et al. (2006) (Figure 2.11c). Voytovych et al. (2006) performed wetting experiments on alumina and found, using SEM, XRD and EPMA techniques that an increase in the Ti concentration from 2.9 to 8.0 at.% Ti led to the formation of both Ti_3Cu_3O and Ti_4Cu_2O phases in the reaction layer. The thickness of the reaction layer was also observed to increase with the increase in Ti concentration. These results show that under conditions where the diffusivity of Ti towards the joint interface is enhanced, Ti_4Cu_2O can form in layer II, on the braze side of the joint interface.

Using a similar set of brazing conditions, Lin et al. (2014) formed alumina-to-alumina brazed joints and found using TEM techniques that an increase in the Ti concentration from 1.75 to 4.5 wt.% Ti did not lead to any such microstructural change in the M_6O-type layer, which was composed completely of Ti_3Cu_3O. The thickness of the reaction layer did increase, however, from 2.6 to 6.8 μm. In reference to the work by Voytovych et al. (2006), Lin et al. (2014) suggested that the absence of Ti_4Cu_2O may have been due to differences in the brazing times and/or brazing atmospheres selected in the two studies (Tables 2.6 and 2.7). There appears to be insufficient evidence in the literature to suggest that the brazing time or the brazing atmosphere can affect the reaction layer composition (see Sections 2.3.1 and 2.3.3). Instead, Lin et al. (2014) formed brazed joints which comprises two faying alumina surfaces whereas Voytovych et al. (2006) performed wetting experiments comprising a single alumina surface. That said, the average height of a sessile drop in a typical wetting experiment is ~2 orders of magnitude greater than the thickness of a typical braze foil. The relative amount of Ti available to diffuse to the joint interface in the wetting experiment performed by Voytovych et al. (2006) may have been significantly greater than that available to diffuse to the two joint interfaces in the brazing experiments performed by Lin et al. (2014). The formation of Ti_4Cu_2O

TABLE 2.6
Reaction Layer Compositions Reported in the Literature for Brazing Experiments Conducted on Alumina Using Intermediate Ti Concentrations (1.75 < wt.% Ti < 4.5) in Ag-Cu-Ti Braze Alloys

Publication	Experiment	Ag-Cu-Ti Composition (wt.%)	Braze Quantity/ Thickness (μm)	Brazing Temp, T (°C)	Brazing Time (min)	Techniques	Layer I	Layer II
Voytovych et al. (2006)	Wetting trials	Ag-Cu-2.9Ti (at.%)	0.1–0.3 g	900	30	EPMA XRD	$Ti_{1.75}O$	Ti_3Cu_3O
Stephens et al. (2003)	Wetting trials	63.0Ag-35.25Cu-1.75Ti	76	845	6	TEM	γ-TiO + isolated Ti_2O	Ti_3Cu_3O
Valette et al. (2005)	Brazed joints	63.0Ag-35.25Cu-1.75Ti	100	900	15	EDX EPMA	Ti_2O	Ti_3Cu_3O
Kozlova et al. (2011)	Wetting trials	63.0Ag-35.25Cu-1.75Ti	–	850–900	0–20	SEM EDX	Ti_2O	Ti_3Cu_3O
Lin et al. (2014)	Brazed joints	63.0Ag-35.25Cu-1.75Ti 68.8Ag-26.7Cu-4.5Ti	100–120	830 915	20 20	XRD TEM	Ti_2O	Ti_3Cu_3O
Santella et al. (1990)	Brazed joints	56.0Ag-36.0Cu-6.0Sn-2.0Ti	100	900	20	TEM	TiO	Ti_3Cu_3O
Hahn et al. (1998)	Brazed joints	70.1Ag-27.3Cu-2.6Ti	~50	830	10	SEM-EDX TEM	$TiO_{1.04}$	Ti_4Cu_2O
Lee and Kwon (1995)	Brazed joints	48.1Ag-48.1Cu-3.8Ti	120–150	920	20	EDX XRD	TiO	Ti_4Cu_2O
Conquest (2003)	Wetting trials	68.8Ag-26.7Cu-4.5Ti	–	950	10	SEM	–	–
Ichimori et al. (1999)	Brazed joints	66.7Ag-28.4Cu-4.9Ti	100	900	5	TEM	TiO + isolated Ti_2O	Ti_3Cu_3O

TABLE 2.7
Reaction Layer Compositions Reported in the Literature for Brazing Experiments Conducted on Alumina Using High Ti Concentrations (>4.5 wt.% Ti) in Ag-Cu-Ti Braze Alloys

Publication	Experiment	Ag-Ti Composition (wt.%)	Braze Quantity/ Thickness (μm)	Brazing Temp, T (°C)	Brazing Time (min)	Techniques	Layer I	Layer II
Voytovych et al. (2006)	Wetting trials	Ag-Cu-8.0Ti (at.%)	0.1–0.3 g	900	30	SEM EPMA XRD	—	Ti_3Cu_3O Ti_4Cu_2O
Jasim et al. (2010)	Brazed joints	68.0Ag-27.0Cu-5.0Ti	70	980	—	SEM XRD	—	Ti_4Cu_2O
Peytour et al. (1990)	Brazed joints	40.0Ag-55.0Cu-5.0Ti	80	850–890	2–10	SEM EDX EPMA	—	Ti_4Cu_2O

Literature Review

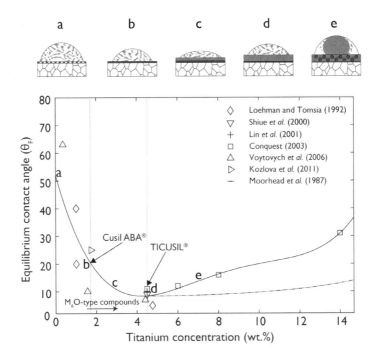

FIGURE 2.13 A comparison of the reported equilibrium contact angles (θ_f) on alumina, with increasing Ti concentration (wt.%) in Ag-Cu-Ti braze alloys. Brazing was conducted at 900°C (Kozlova et al., 2011; Lin et al., 2001; Shiue et al., 2000; Voytovych et al., 2006), 950°C (Conquest, 2003), 1,000°C (Loehman and Tomsia, 1992) and 1,090°C (Moorhead, 1987). The corresponding reaction layer properties are as follows: (a) discontinuous Ti-O (layer I), (b) continuous Ti-O (layer I), (c) Ti-O (layer I) and Ti_3Cu_3O (layer II), (d) Ti-O (layer I) and thicker Ti_3Cu_3O (layer II) and (e) Ti-O (layer I) and Ti_3Cu_3O/Ti_4Cu_2O (layer II) with an increase in θ_f owing to Ti retained in the braze interlayer.

observed by Voytovych et al. (2006) has been shown to occur at higher Ti activity levels (Kelkar et al., 1994).

These differences in the reaction layer composition despite similar brazing conditions may suggest that the volume of braze selected, and hence the amount of Ti available to diffuse to a given surface area of alumina, may also affect the properties of the reaction layer.

The use of high Ti concentration Ag-Cu-Ti braze alloys generally leads to an increase in the reaction layer thickness, as compared with the thicknesses of reaction layers which form when Ag-Cu-Ti braze alloys with intermediate Ti concentrations are used.

The effect of an increase in the Ti concentration in an Ag-Cu-Ti braze alloy on the wettability of alumina, and the corresponding reaction layer formation at the joint interface, is summarised in Figure 2.13. This comparison shows that under typical brazing conditions, the formation of M_6O-type compounds coincides with an improvement in wetting behaviour. This typically occurs when the Ti concentration in an Ag-Cu-Ti braze alloy exceeds 1.75 wt.%, which is the Ti concentration of Cusil ABA®. Furthermore, an optimum equilibrium contact angle appears to be achieved with a Ti concentration of ~4.5 wt.% Ti, which is the Ti concentration in TICUSIL®.

2.2.2.4 Ti Concentration and Joint Strength

The Ti concentration in an Ag-Cu-Ti braze alloy may affect chemical bonding and joint strength since it has consistently been reported to affect the properties of the reaction layer at the alumina/Ag-Cu-Ti joint interface. Few studies have investigated the effect of the Ti concentration on the strength of alumina brazed joints made using Ag-Cu-Ti braze alloys. Instead, most studies have focussed on how the brazing temperature and brazing time can affect the reaction layer properties, which has been strongly correlated to joint strength (see Sections 2.3.1.3 and 2.3.2.3).

The strength of an alumina-to-alumina brazed joint has been reported to increase with the Ti concentration in a Cu-Ti braze alloy. Jasim et al. (2010) found that as the Ti concentration in a Cu-Ti braze alloy was increased from 2 to 10 wt.% Ti, the strength of alumina-to-alumina brazed joints evaluated using shear testing increased, from 15 to 24 MPa. This was still lower than the strength of another set of alumina-to-alumina brazed joints that were made using a 68Ag-27Cu-5Ti wt.% braze alloy, which achieved 42 MPa.

The effect of Ti concentration on the strength of alumina/Ag-Cu-Ti/Kovar® brazed joints has been reported to depend on the surface condition of the alumina ceramics used (Kovar® is a Fe-Ni-Co alloy with a CTE similar to borosilicate glass designed for ceramic-to-metal joining). Mizuhara and Mally (1985) formed alumina-to-Kovar® brazed joints using 97.6 and 99.5 wt.% Al_2O_3 alumina and three different 50-µm-thick Ag-Cu-Ti braze foils with increasing Ti concentration (Table 2.4). Joint strength was evaluated using peel testing. As the Ti concentration was increased from 1.5 to 2.5 and then to 3.2 wt.% Ti, joint strength was consistently found to increase. This increase was observed in joints made using both grades of alumina, in as-sintered, ground/polished and ground/heat-treated conditions (Figure 2.33). When the alumina specimens were brazed in as-ground condition, however, an increase in the Ti concentration led to a degradation in joint strength. The effect of surface condition is discussed in Section 2.4.2.

2.3 PROCESS PARAMETERS IN ACTIVE METAL BRAZING

The variables associated with the brazing process parameters include brazing time, brazing temperature, heating and cooling rates and brazing atmosphere (Figure 1.1). Of these, brazing time and temperature, which can significantly affect the diffusivity of Ti in an Ag-Cu-Ti braze alloy, have been the most commonly studied.

2.3.1 BRAZING TIME

The brazing time refers to the isothermal dwell period at the brazing temperature. The mean distance ($\bar{x}$) which Ti atoms may diffuse in a given time (t) along a random path can be expressed by Equation 2.16, where λ is the jump distance and ν is the successful jump frequency of solute Ti atoms in a host Ag-Cu lattice. Hence, the mean distance ($\bar{x}$) which Ti atoms may diffuse in an Ag-Cu-Ti braze alloy is proportional to the brazing time. The diffusion coefficient D of Ti (a measure of the mobility of Ti atoms) depends on the jump distance (λ) and the successful jump frequency (ν) of solute Ti atoms in the host Ag-Cu lattice:

$$\bar{x} = \lambda\sqrt{\nu t} \qquad (2.16)$$

$$D = \frac{1}{6}\nu\lambda^2 \qquad (2.17)$$

The approximate mean distance ($\bar{x}$) which Ti atoms may diffuse in an Ag-Cu-Ti braze alloy can be expressed as follows:

$$\bar{x} = \sqrt{6Dt} \qquad (2.18)$$

Typical brazing times reported in the literature vary between 10 and 15 min. These times are adequate for the complete diffusion of Ti to the joint interfaces. This has been shown experimentally

Literature Review

in a study by Kozlova et al. (2010) in which alumina-to-alumina brazed joints were formed using monocrystalline alumina and an Ag-Cu eutectic braze alloy. With one of the faying alumina surfaces coated with a Ti paste prior to brazing, the Ag-Cu braze alloy was sandwiched between two alumina parts. Brazing was conducted at 840°C for 15 min. SEM and EDX techniques were used to characterise the reaction layers at both joint interfaces, which were found to be identical in both thickness and composition. It was concluded that a brazing time of 15 min was adequate for the dissolution of Ti in the Ag-Cu braze alloy and its subsequent diffusion from one side of the joint interface to the other.

The brazing time is critical in determining the reaction layer thickness. Of course, the brazing time is a dwell period at a suitably selected brazing temperature, which provides the thermal activation energy for the diffusion process. As the diffusion of Ti atoms increases with brazing time, a parabolic increase in the reaction layer thickness is observed (Shiue et al., 2000):

$$x = \sqrt{k_p t} \tag{2.19}$$

where x is the reaction layer thickness and k_p is the parabolic growth rate constant.

2.3.1.1 Brazing Time and Wetting

Reaction product formation is initiated at the onset of the brazing time period. An immediate improvement in the wetting behaviour and a significant decrease in the dynamic contact angle are observed at the start of the brazing process. The equilibrium contact angle is achieved once the composition of the reaction layer has stabilised ($t_1 = \sim 100$ s). The brazing time technically includes this initial period; however, as a variable in the AMB process, it typically refers more so to the overall isothermal dwell period at the brazing temperature. Within this, the criticality of the initial period is somewhat obscured, e.g. brazing times of 10 or 15 or 30 min are often compared.

Following the initial period, the brazing time leads to further thickening of the reaction layer while there is little further improvement in wetting behaviour. This is due to the composition and hence metallic character of the reaction layer having stabilised at the onset of the brazing time period. This shows that the effect of brazing time on wetting, therefore, is predominantly limited to the initial stages of brazing. So far as there is sufficient time for reaction product formation, the brazing time does not appear to otherwise significantly affect the reaction layer composition or wetting behaviour. Instead, an increase in the reaction layer thickness is the most significant outcome of prolonged brazing times, although some spreading continues.

This effect of the brazing time understood from the literature is shown in Figure 2.14, whereby at $t < t_1$ s, reaction products form and a significant decrease in the dynamic contact angle is observed. At $t = t_1$ s, the reaction layer composition is stable and the equilibrium contact angle is more or less achieved. At $t = t_2$ s, at the end of the brazing time period, a parabolic increase in the reaction layer thickness has occurred without any change in the composition of the reaction layer and only a slight further improvement in wetting behaviour.

Kozlova et al. (2011) performed wetting experiments on monocrystalline alumina using Cusil ABA®. A charge couple device (CCD) camera was used to record the experiment and the dynamic contact angle was measured. After 1×10^{-3} s, the dynamic contact angle decreased from 180° to 135°. This was the first stage of wetting where spreading was non-reactive, limited only by friction at the triple line and no reaction layer had formed. After 1×10^{-2} s, the dynamic contact angle decreased further from 135° to 80°. This was the second stage of wetting where spreading was reactive, limited only by the diffusion of Ti. A single continuous Ti_2O layer was observed on the alumina side of the joint interface (layer I). After 1×10^2 s the dynamic contact angle again decreased further, from 80° to the equilibrium contact angle of 30°. This represented the third stage of the wetting process where spreading was reactive, limited only by local chemical processes at the triple line. The reaction layer consisted of a Ti_2O layer on the alumina side of the joint interface (layer I) and a Ti_3Cu_3O layer on the braze side of the joint interface (layer II) (Table 2.8).

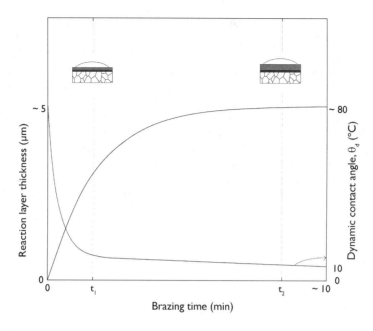

FIGURE 2.14 The effect of brazing time on the wettability of alumina and the reaction layer thickness, using an Ag-Cu-Ti braze alloy.

TABLE 2.8
The Effect of Brazing Time on the Dynamic Contact Angle (θ_d) Formed by an Ag-Cu-Ti Braze Alloy on Monocrystalline Alumina

Publication	Experiment	Al$_2$O$_3$ purity (wt.%)	Ag-Cu-Ti Composition (wt.%)	Brazing Temp, T (°C)	Brazing Time (s)	Dynamic Contact angle (θ_d)	Layer I	Layer II
Kozlova et al. (2011)	Wetting trials	Mono-crystalline	63.0Ag-35.25Cu-1.75Ti	850	0 s	180	–	–
					0.001 s	135	–	–
					0.01 s	80	Ti$_2$O	–
					100 s	30	Ti$_2$O	Ti$_3$Cu$_3$O

These results show that despite typical brazing times of 10 to 15 min, the equilibrium contact angle and hence, a stable reaction layer composition, is actually achieved after just ~100 s, or ~600 s as found elsewhere in the literature (see Figure 2.18). This time period is distinguishable from the remaining duration of the brazing time, during which an increase in the reaction layer thickness is observed. Typically, the brazing time as a variable in the AMB process refers to this latter time period, beyond $t = t_1$ according to Figure 2.14, either side of which different processes actually occur.

2.3.1.2 Brazing Time and the Reaction Layer

There is little evidence in the literature to suggest that the reaction layer composition is affected by the brazing time, however, there is ample evidence to show that an increase in the brazing time leads to a thickening of the reaction layer. This is supported by several studies, in which the brazing times selected have ranged from 1 to 600 min, with brazing temperatures ranging from 815°C to 1,000°C (Table 2.9).

Voytovych et al. (2004) performed wetting experiments on monocrystalline alumina using 0.1 g of Cusil ABA®. The reaction layer, characterised using SEM, EPMA and XRD techniques, consisted of a Ti$_2$O layer on the alumina side of the joint interface (layer I) (Figure 2.15b, 1) and a

TABLE 2.9
Reaction Layer Compositions Reported in the Literature for Brazing Experiments Conducted on Alumina Using Ag-Cu-Ti Braze Alloys, with Various Brazing Times

Publication	Experiment	Ag-Cu-Ti Composition (wt. %)	Braze Quantity/ Thickness (μm)	Brazing Temp, T (°C)	Brazing Time (min)	Techniques	Layer I	Layer II
Voytovych et al. (2004)	Wetting trials	63.0Ag-35.25Cu-1.75Ti	0.1 g	900	3	EPMA	TiO/Ti$_2$O	Ti$_3$Cu$_3$O
					30	XRD	Ti$_2$O	Ti$_3$Cu$_3$O
					600	WAM	Ti$_2$O	Ti$_3$Cu$_3$O
Byun and Kim (1994)	Brazed joints	70.4Ag-28.1Cu-1.5Ti	100	950	1	SEM	α-TiO/δ-TiO	Ti$_3$Cu$_3$O
					5	EDX	α-TiO/δ-TiO	Ti$_3$Cu$_3$O
					10	XRD	α-TiO/δ-TiO	Ti$_3$Cu$_3$O
Ali et al. (2015)	Brazed joints	63.0Ag-35.25Cu-1.75Ti	50	815	2	SEM	Ti$_2$O	Ti$_3$Cu$_3$O
				845	15	TEM-EDS	Ti$_2$O	Ti$_3$Cu$_3$O
				875	30		Ti$_2$O	Ti$_3$Cu$_3$O
					45		Ti$_2$O	Ti$_3$Cu$_3$O
Janickovic et al. (2001)	Brazed joints	53.4Ag-44.8Cu-1.8Ti	50	950–1,000	15	SEM	TiO	Ti$_2$O$_3$/Cu$_2$O
					30	EDX	TiO	Ti$_2$O$_3$/Cu$_2$O
						TEM		

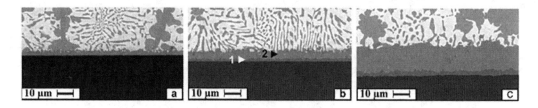

FIGURE 2.15 The effect of brazing time on the thickness of the reaction layer after (a) 3, (b) 30 and (c) 600 min, in which the reaction layer composition (1 – Ti$_2$O and 2 – Ti$_3$(Cu, Al)$_3$O) remained unchanged (Voytovych et al., 2004).

Ti$_3$(Cu, Al)$_3$O layer on the braze side of the joint interface (layer II) (Figure 2.15b, 2). As the brazing time was increased from 3 to 30 and then to 600 min the average reaction layer thickness was also observed to increase from 1.7 to 3.2 and then to 20.0 μm. No change in the reaction layer composition was observed with the increase in brazing time.

Byun and Kim (1994) formed alumina-to-alumina brazed joints using 100-μm-thick 70.4Ag-28.1Cu-1.5Ti wt.% braze foils. Brazing times of 1, 5 and 10 min were used. SEM, EDX and XRD techniques were used to characterise an α-TiO layer, with isolated δ-TiO particles, on the alumina side of the joint interface (layer I) and a Ti$_3$Cu$_3$O layer on the braze side of the joint interface (layer II). The reaction layer composition was unchanged despite the increase in brazing time.

Janickovic et al. (2001) formed alumina-to-alumina brazed joints using 50-μm-thick 53.4Ag-44.8Cu-1.8Ti wt.% braze foils. Brazing times of 15 and 30 min were used. SEM, TEM and EDX techniques were used to characterise a TiO layer on the alumina side of the joint interface (layer I) and a Cu$_2$O/Ti$_2$O$_3$ layer on the braze side of the joint interface (layer II). Despite the increase in the brazing time, no change in the reaction layer composition was observed.

These results agree that reaction product growth occurs at the brazing temperature, rather than during cooling, as suggested in other work (Ichimori et al., 1999; Kristalis et al., 1991). These results have been used to clarify the joining mechanism (see Equation 2.11, in Section 2.1.7).

Recently, an increase in the brazing time from 2 to 45 min was found to lead to a thickening of the TiO layer (layer I) at the expense of a structural degradation in the M$_6$O-type layer (layer II) in alumina-to-alumina brazed joints made using Ag-Cu-Ti braze alloys (Ali et al., 2015). This effect was reported to occur over a range of brazing temperatures, from 815°C to 875°C.

Ali et al. (2015) formed alumina-to-alumina brazed joints using 95.0 and 99.7 wt.% Al$_2$O$_3$ alumina and 50-μm-thick Cusil ABA® braze foils. Brazing times of 2, 15, 30 and 45 min were used. SEM and TEM techniques were used to characterise a γ-TiO layer, with isolated Ti$_3$O$_2$ particles, on the alumina side of the joint interface (layer I) and a Ti$_3$Cu$_3$O layer on the braze side of the joint interface (layer II). With the increase in the brazing time, no change in the reaction layer composition was observed; instead, the thickness of the TiO layer increased while the Ti$_3$Cu$_3$O layer was found to structurally degrade (Table 2.10). Furthermore, after a brazing time of 45 min, the reaction layer was reported to consist of inhomogeneous interfacial structures (Figure 2.16). It was suggested that Ti$_3$Cu$_3$O is a relatively unstable phase as compared with γ-TiO; hence, it can degrade with prolonged brazing times.

2.3.1.3 Brazing Time and Joint Strength

Joint strength has been correlated to the thickness of the reaction layer, a product of chemical bonding at the alumina/Ag-Cu-Ti joint interface. The reaction layer thickness increases proportionally with the brazing time. Therefore, an increase in the brazing time can lead to an increase in joint strength.

The increase in joint strength achieved with brazing time is limited, however, due to the inherent brittleness of the reaction layer phase; hence, an excessively thick reaction layer can lead to a decrease in the joint strength (Figure 2.17). Thus, maximum joint strength is achieved when the

TABLE 2.10
The Effect of Brazing Temperature and Time on the Structures and Thicknesses of TiO and Ti$_3$Cu$_3$O layers Reported at the Alumina/Ag-Cu-Ti Joint Interface using Cusil ABA® (Ali et al., 2015)

	Brazing Temperature (°C)/Reaction Layer Thicknesses (µm)					
	815°C		845°C		875°C	
Brazing Time (min)	TiO	Ti$_3$Cu$_3$O	TiO	Ti$_3$Cu$_3$O	TiO	Ti$_3$Cu$_3$O
2	0.06–0.09	0.9–1.3	0.05–0.10	0.8–1.6	0.06–0.12	1.0–1.9
15	0.07–0.11	0.95–1.3	0.07–0.14	Broken layer	0.07–0.19	Broken layer
30	0.12–0.35	Broken layer	0.12–0.35	Broken layer	0.12–0.35	Broken layer
45			Inhomogeneous interfacial structures			

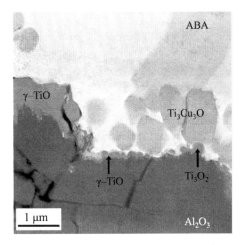

FIGURE 2.16 TEM image showing an inhomogeneous interfacial structure (reaction layer) at the alumina/Ag-Cu-Ti joint interface following brazing for 300 min at 845°C (Ali et al., 2015).

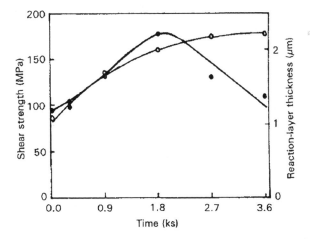

FIGURE 2.17 The effect of brazing time on the reaction layer thickness and shear strength of alumina-to-alumina brazed joints made using a 57.0 Ag-38.0 Cu-5.0 Ti wt.% braze alloy (Hao et al., 1994).

reaction layer is of an optimum thickness, which is controllable through careful selection of the brazing time. Of course, other variables in the AMB process such as the brazing temperature, braze alloy composition, volume of braze selected, ceramic surface area etc. (Figure 1.1) need to be considered alongside this, due to the interdependencies that exist between them, when selecting the appropriate brazing time. Ideal conditions are very much tailored to a given assembly and system.

Hao et al. (1994) formed alumina-to-alumina brazed joints using 99.0 wt.% Al_2O_3 alumina and 50-μm-thick 57.0Ag-38.0Cu-5.0Ti wt.% braze foils. A brazing temperature of 850°C was used, and the brazing time was increased from 1 to 60 min. Joint strength was evaluated using shear testing. An increase in brazing time from 1 to 30 min led to both an increase in the reaction layer thickness from 1.2 to 2.0 μm and an increase in the joint strength from 90 to 175 MPa. While a further increase in the brazing time, from 30 to 60 min, led to a further increase in the reaction layer thickness from 2.0 to 2.2 μm, a significant decrease in joint strength, from 175 to 100 MPa, was observed (Figure 2.17). It was suggested that a maximum joint strength of 175 MPa was achieved for this system, with a reaction layer thickness of 2.0 μm which resulted when a brazing time of 30 min was selected.

2.3.2 Brazing Temperature

The brazing temperature which can govern the diffusivity of Ti to the alumina surface during brazing has been found to affect the composition of reaction products which form at the joint interface (do Nascimento et al., 2003; Ghosh et al., 2012; Kelkar et al., 1994; Nicholas and Mortimer, 1985; Pak et al., 1990).

The brazing temperature refers to the peak temperature in the brazing cycle and is usually at least 15°C above the liquidus temperature of the ABA. It is the temperature at which reactions are expected to occur, and hence the temperature at which an isothermal dwell period (brazing time) is applied. The diffusion of Ti atoms in an Ag-Cu lattice is a thermally activated process. The brazing temperature is the control of thermal energy which is supplied to the Ti atoms and thereby governs their kinetic energies. Therefore, an increase in the brazing temperature can in turn increase the diffusivity of Ti atoms towards the joint interface.

By definition, the diffusion of Ti requires a successful jump frequency of Ti atoms in the host Ag-Cu vacancy sites. According to the Arrhenius equation, this successful jump frequency (v) is dependent on the attempted jump frequency which is proportional to the frequency of atomic vibrations (v_0) and on the probability of successful jumps $\left\{-\dfrac{G^*}{kT}\right\}$, where G^* is the activation free energy of the barrier and kT is the thermal energy of an atom at a given temperature:

$$v = v_0 \exp\left(-\frac{G^*}{kT}\right) \qquad (2.20)$$

The activation free energy term G^* can be expanded to include a temperature-independent activation enthalpy term (S^*) and a temperature-dependent activation entropy term (H^*) (Equation 2.21). Thus, the diffusion coefficient (D), which is proportional to the successful jump frequency (v), can be estimated. The diffusion coefficient follows the Arrhenius equation and it therefore depends on temperature:

$$v = v_0 \exp\left(\frac{S^*}{k}\right) \times \exp\left(-\frac{H^*}{kT}\right) \qquad (2.21)$$

$$D = D_0 \exp\left(-\frac{H^*}{kT}\right) \qquad (2.22)$$

These equations show that the brazing temperature can influence the diffusivity of Ti towards the joint interface. It is expected, therefore, that the brazing temperature affects the reaction kinetics at the joint interface and thereby the composition of the reaction layer. This may depend however on the amount of Ti which is available in the starting braze alloy composition to diffuse to the alumina surface.

2.3.2.1 Brazing Temperature and Wetting

The wetting behaviour of Ag-Cu-Ti braze alloys on alumina surfaces has been shown to improve with an increase in the brazing temperature. Lin et al. (2001) and Shiue et al. (2000) performed wetting trials on 99.5 wt.% Al_2O_3 alumina using TICUSIL®. In both studies, an increase in the brazing temperature, from 880°C to 900°C, led to a decrease in the equilibrium contact angle, from ~20° to 9° (Figure 2.18).

Kozlova et al. (2011) performed wetting trials on monocrystalline alumina using Cusil ABA®. An increase in the brazing temperature from 850°C to 900°C led to a decrease in the equilibrium contact angle, from 38° to 25°. These results were consistent with those reported by Lin et al. (2001) and Shiue et al. (2000); however, the equilibrium contact angles reported were significantly higher (Table 2.11).

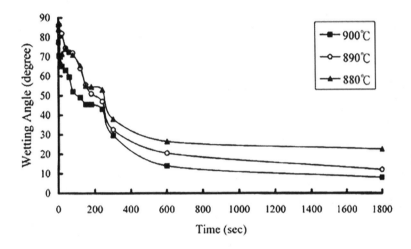

FIGURE 2.18 The effect of brazing temperature on the wettability of alumina using TICUSIL® braze paste (Shiue et al., 2000).

TABLE 2.11
The Effect of Brazing Temperature on the Equilibrium Contact Angles Reported in Wetting Experiments Performed on Alumina Using Ag-Cu-Ti Braze Alloys

Publication	Experiment	Al_2O_3 purity (wt.%)	Ag-Cu-Ti Composition (wt.%)	Braze Quantity/ Thickness (μm)	Brazing Temp, T (°C)	Brazing Time (min)	Equilibrium Contact angle, (θ_t)
Lin et al. (2001)	Wetting trials	99.5	68.8Ag-26.7Cu-4.5Ti	2.5 g	880	0–30	~22
					900		9
Shiue et al. (2000)	Wetting trials	99.5	68.8Ag-26.7Cu-4.5Ti	0.15 g	880	0–30	~22
					890		~13
					900		~9
Kozlova et al. (2011)	Wetting trials	Mono-crystalline	63.0Ag-35.25Cu-1.75Ti	—	850	0–20	38
					900		25

Kozlova et al. (2011) used a dispensed drop technique whereby the ABA was melted in an alumina crucible prior to being deposited onto the alumina surface. This technique differed from the conventional method of applying paste or preplacing a braze preform onto the alumina surface, prior to the wetting experiment as incorporated by Lin et al. (2001) and Shiue et al. (2000).

Reactions between Ti in the braze alloy and the surface of the alumina crucible reduced the instantaneous Ti concentration of the ABA droplet on the alumina substrate from 1.75 wt.% Ti to between 0.76 and 0.97 wt.% Ti. This reduction in the Ti concentration may have produced the unexpectedly high equilibrium contact angle observed. Despite the reduced Ti concentration to between 0.76 and 0.97 wt.% Ti, SEM and EDX techniques were used to characterise a Ti_2O layer on the alumina side of the joint interface (layer I) and a Ti_3Cu_3O layer on the braze side of the joint interface (layer II). At low Ti concentrations (<1.75 wt.% Ti), the formation of an M_6O-type layer is in contradiction with the majority of studies reviewed thus far (see Section 2.2.2.3). Furthermore, the formation of an M_6O-type layer is typically correlated to a significantly lower equilibrium contact angle than that which was observed.

These studies show that the brazing temperature can be used to control the diffusivity of Ti towards the joint interface. While an increase in the brazing temperature can lead to an improvement in the wetting behaviour of alumina, its overall effect is subject to the starting Ti concentration in the Ag-Cu-Ti braze alloy, or more specifically, the amount of Ti available.

Table 2.12 outlines the experimental parameters adopted by studies in the literature that are referenced here, in Section 2.3.2, where the effect of the brazing temperature on wetting, reaction layer formation and joint strength in the alumina/Ag-Cu-Ti system is discussed.

2.3.2.2 Brazing Temperature and the Reaction Layer

The diffusion of Ti in Ag-Cu-Ti braze foils can occur below the solidus temperature of the braze alloy; 780°C for both Cusil ABA® and TICUSIL® (Table 2.1). Ali et al. (2016) observed microstructural changes in both Cusil ABA® and TICUSIL® braze foils heated to 750°C for 1 min (Figure 2.19). In Cusil ABA®, Ti became more uniformly distributed as Cu_4Ti whereas in TICUSIL® a continuous Cu_4Ti_3 layer formed around the Ti ribbon at the centre of the braze foil.

Comparing studies in the literature which have selected brazing temperatures between 830°C and 900°C shows, at first glance, that the reaction layer composition is not affected by an increase in the brazing temperature when similar Ag-Cu-Ti braze alloy compositions are used (Table 2.13). These results may suggest that the reaction layer composition is relatively less sensitive to changes in the brazing temperature as compared with changes in the Ti concentration of an Ag-Cu-Ti braze alloy.

Due to the interdependencies between variables in the AMB process, such as that between the brazing temperature and Ti concentration, conclusions drawn from comparisons can be somewhat misleading. For example, by comparing studies in the literature which have investigated the effect of brazing temperature it can be shown that while the composition of the Ti-O layer (layer I) on the alumina side of the joint interface is affected; the composition of the Ti-Cu-O layer (layer II) remains unchanged (Table 2.14).

In order to better understand the effect of brazing temperature on the composition of the reaction layer, it is necessary to compare a wider range of studies while also considering the effect of Ti concentration in the braze alloy. Furthermore, it is also necessary to consider the effects of these variables on the continuity and structure of the reaction layer. The extent to which an increase in the brazing temperature may affect the properties of the reaction layer at the alumina/Ag-Cu-Ti joint interface depends on the amount of Ti in the Ag-Cu-Ti braze alloy that is available to diffuse to the alumina surface. The thermal activation energy for Ti diffusion is controlled the brazing temperature. Therefore, while the Ti concentration provides the potential for reaction product formation, the brazing temperature is the driving force for the process.

At low Ti concentrations (< 1.75 wt.% Ti) in an Ag-Cu-Ti braze alloy, the continuity of the Ti-O reaction layer formed at the alumina/Ag-Cu-Ti joint interface improves significantly as the result of an increase in the brazing temperature.

TABLE 2.12
Details of Experimental Methods and Characterisation Techniques of Studies Discussed in Section 2.3

Publication	Experiment	Al_2O_3 purity (wt.%)	Ag-Cu-Ti Composition (wt.%)	Braze Quantity/ Thickness (µm)	Vacuum (mbar)	Brazing Temp (°C)	Brazing Time (min)	Techniques
Lin et al. (2001)	Wetting trials	99.5	68.8Ag-26.7Cu-4.5Ti	2.5 g	5.0×10^{-5}	880–900	0–30	WAM
Shiue et al. (2000)	Wetting trials	99.5	68.8Ag-26.7Cu-4.5Ti	0.15 g	–	880–900	0–30	WAM
Kozlova et al. (2011)	Wetting trials	Mono-crystalline	63.0Ag-35.25Cu-1.75Ti	–	6.0×10^{-7}	850–900	0–20	WAM EDX
Stephens et al. (2003)	Wetting trials	Mono-crystalline	63.0Ag-35.25Cu-1.75Ti	76	H atm	845	6	TEM
Valette et al. (2005)	Brazed joints	99.5	63.0Ag-35.25Cu-1.75Ti	100	1.0×10^{-6}	900	15	EPMA EDX
Lin et al. (2014)	Brazed joints	High purity (0.05 wt.% MgO)	63.0Ag-35.25Cu-1.75Ti	100–120	1.3×10^{-6}	830	20	TEM XRD
Ali et al. (2016)	Brazed joints	Mono-crystalline	63.0Ag-35.25Cu-1.75Ti	50	–	800–845	1	TEM EDX
Cho et al. (1992)	Brazed joints	High purity (0.15 wt.% MgO)	71.1Ag-28.4Cu-0.5Ti	50	1.3×10^{-5}	850–950	10	SEM XRD
Byun and Kim (1994)	Brazed joints	99.0	70.4Ag-28.1Cu-1.5Ti	100	3.0×10^{-5}	950–1,100	1–10	SEM-EDS XRD
Hao et al. (1997)	Wetting trials	99.0	57.0Ag-38.0Cu-5.0Ti	700	7.0×10^{-5}	800–900	30	SEM EDX
Hao et al. (1994)	Brazed joints	99.0	57.0Ag-38.0Cu-5.0Ti	50	7.0×10^{-5}	800–1,050	0–60	SEM SH
Ning et al. (2003)	Brazed joints	99.0/Mono-crystalline	70.5Ag-27.5Cu-2.0Ti	500	1.0×10^{-6}	844–860	–	SEM XRD F-19
Moorhead et al. (1987)	Brazed joints	99.5	44.0Ag-48.0Cu-4.0Ti-4.0Sn	100	6.7×10^{-6}	800–900	20	3PB

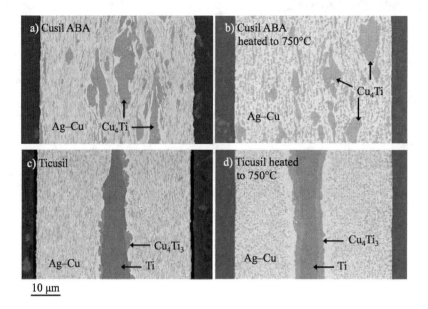

FIGURE 2.19 Microstructural changes in Ag-Cu-Ti braze foils after heating to 750°C for 1 min: (a) as-received Cusil ABA® braze foil, (b) Cusil ABA® braze foil following heat treatment, (c) as-received TICUSIL® braze foil and (d) TICUSIL® braze foil following heat treatment (Ali et al., 2016).

TABLE 2.13
Similar Reaction Layer Compositions Reported in Studies Following the AMB of Alumina using Cusil ABA® with Different Brazing Temperatures

Publication	Experiment	Ag-Cu-Ti Composition (wt.%)	Braze Quantity/ Thickness (μm)	Brazing Temp, T (°C)	Brazing Time (min)	Techniques	Layer I	Layer II
Stephens et al. (2003)	Wetting trials	63.0Ag-35.25Cu-1.75Ti	76	845	6	TEM	γ-TiO + isolated Ti$_2$O	Ti$_3$Cu$_3$O
Valette et al. (2005)	Brazed joints	63.0Ag-35.25Cu-1.75Ti	100	900	15	EPMA EDX	Ti$_2$O	Ti$_3$Cu$_3$O
Kozlova et al. (2011)	Wetting trials	63.0Ag-35.25Cu-1.75Ti	–	850–900	0–20	SEM EDX	Ti$_2$O	Ti$_3$Cu$_3$O
Lin et al. (2014)	Brazed joints	63.0Ag-35.25Cu-1.75Ti	100–120	830	20	TEM XRD	Ti$_2$O	Ti$_3$Cu$_3$O

Cho and Yu (1992) formed alumina-to-alumina brazed joints using 50-μm-thick 71.1Ag-28.4Cu-0.5Ti wt.% braze foils. SEM and XRD techniques were used to characterise a discontinuous γ-TiO layer, with isolated δ-TiO particles, on the alumina side of the joint interface (layer I). As the brazing temperature was increased from 850°C to 950°C, an increase in the volume fraction of γ-TiO as well as an improvement in the overall continuity of the layer was observed.

At intermediate Ti concentrations (1.75 < wt.% Ti < 4.5) in an Ag-Cu-Ti braze alloy, the Ti-O layer (layer I) becomes richer in Ti and the M_6O-type Ti-Cu-O layer (layer II) becomes thicker as a result of an increase in the brazing temperature.

Byun and Kim (1994) formed alumina-to-alumina brazed joints using 99.0 wt.% Al$_2$O$_3$ alumina and 100-μm-thick 70.4Ag-28.1Cu-1.5Ti wt.% braze foils. SEM, EDX and XRD techniques were used to characterise an α-TiO layer, with isolated δ-TiO particles, on the alumina side of the joint interface

TABLE 2.14
Reaction Layer Compositions Reported in Studies Following the AMB of Alumina with Different Brazing Temperatures using Various Ag-Cu-Ti Braze Alloys

Publication	Experiment	Ag-Cu-Ti Composition (wt.%)	Braze Quantity/ Thickness (µm)	Brazing Temp, T (°C)	Brazing Time (min)	Techniques	Layer I	Layer II
Cho et al. (1992)	Brazed joints	71.1Ag-28.4Cu-0.5Ti	50	850	10	SEM XRD	γ-TiO + isolated δ-TiO	–
				950			γ-TiO	–
Ali et al. (2016)	Brazed joints	63.0Ag-35.25Cu-1.75Ti	50	800	1	TEM EDX	Ti_2O_{1-x}	Ti_3Cu_3O
				815			Ti_2O_{1-x}	Ti_3Cu_3O
				845			γ-TiO	Ti_3Cu_3O
Byun and Kim (1994)	Brazed joints	70.4Ag-28.1Cu-1.5Ti	100	900	10	SEM EDX XRD	α-TiO/δ-TiO	Ti_3Cu_3O
				1,000			α-TiO	Ti_3Cu_3O
				1,100			TiO	Ti_3Cu_3O
Hao et al. (1997)	Wetting trials	57.0Ag-38.0Cu-5.0Ti	700	< 850	30	SEM EDX	AlTi	Ti_4Cu_2O
				> 850			TiO/Ti_2O	Ti_2Cu

(layer I) and a Ti_3Cu_3O layer on the braze side of the joint interface (layer II). Brazing temperatures of 950°C, 1,000°C and 1,100°C were selected. At brazing temperatures above 950°C, the isolated δ-TiO particles were no longer observed. It was concluded that the δ-TiO phase was a transient phase that did not form at the relatively higher brazing temperatures. At 1,100°C, a noticeable increase in the overall reaction layer thickness, which was reported to be 5 µm thick, was observed.

Ali et al. (2016) formed alumina-to-alumina brazed joints using monocrystalline alumina and 50-µm-thick Cusil ABA® braze foils. Brazing temperatures of 800°C, 815°C and 845°C were selected with a brazing time of just 1 min. TEM and EDX techniques were used to characterise the reaction layer at the joint interface. Following brazing at 800°C, a 130-nm-thick Ti_2O_{1-x} layer was observed on the alumina side of the joint interface (layer I) and a discontinuous 100-nm-thick Ti_3Cu_3O layer was observed on the braze side of the joint interface (layer II). At 815°C, layer I consisted of Ti_2O_{1-x} particles with a bimodal size distribution. This included 5 to 30 nm diameter Ti_2O_{1-x} particles on the alumina side of the interface and 500 nm diameter Ti_2O_{1-x} particles on the Ti_3Cu_3O side of the interface. The change in the morphology of layer I coincided with an increase in the thickness of the Ti_3Cu_3O layer. Following brazing at 845°C, a 50-nm thick γ-TiO layer was observed on the alumina side of the joint interface while the thickness of the Ti_3Cu_3O layer increased further to 860 nm (Table 2.15).

TABLE 2.15
The Effect of Brazing Temperature on the Composition and Thickness of the Reaction Layer, at the Alumina/Ag-Cu-Ti Joint Interface (Ali et al., 2016)

	Reaction Layer Properties			
	Layer I		Layer II	
Brazing Temperature (°C)	Composition	Thickness (µm)	Composition	Thickness (µm)
800	Ti_2O_{1-x}	0.130	Ti_3Cu_3O	0.100
815	Ti_2O_{1-x}	0.030–0.500	Ti_3Cu_3O	> 0.100
845	γ-TiO	0.500	Ti_3Cu_3O	0.860

Source: Ali et al. (2016).

At high Ti concentrations (> 4.5 wt.% Ti) in an Ag-Cu-Ti braze alloy the thickness of the M_6O-type layer (layer II) significantly increases as a result of an increase in the brazing temperature.

Hao et al. (1994) formed alumina-to-alumina brazed joints using 99.0 wt.% Al_2O_3 alumina and 0.7 mm thick 57.0Ag-38.0Cu-5.0Ti wt.% braze foils. The reaction layer thickness was found to increase from 1.2 to 2.0 μm as the brazing temperature was increased from 800°C to 850°C. With a further increase in the brazing temperature to 1,050°C, the reaction layer thickness increased to 8.6 μm (Figure 2.21a).

Figure 2.20 shows a general trend drawn from the literature for the effect of brazing temperature, limited by the Ti concentration, on the properties of the reaction layer at the alumina/Ag-Cu-Ti joint interface. The values provided are for general guidance purposes and may not necessarily apply as the model is drawn with common findings from different studies in the literature. For example, following brazing at 1,100°C, using an Ag-Cu-Ti braze alloy with 1.5 wt.% Ti, Byun and Kim (1994) observed a 5-μm-thick reaction layer; these observations are anomalous with respect to the general trend shown in Figure 2.20.

2.3.2.3 Brazing Temperature and Joint Strength

According to the literature, the brazing temperature can affect the continuity, composition and thickness of the reaction layer which forms at the at the alumina/Ag-Cu-Ti joint interface; hence, the brazing temperature can also affect joint strength.

Hao et al. (1994) formed alumina-to-alumina brazed joints using 99.0 wt.% Al_2O_3 alumina and 50-μm-thick 57.0Ag-38.0Cu-5.0Ti wt.% braze foils. Joint strength was evaluated using shear testing. Following brazing at 800°C, a discontinuous reaction layer which was measured to be 1.2 μm

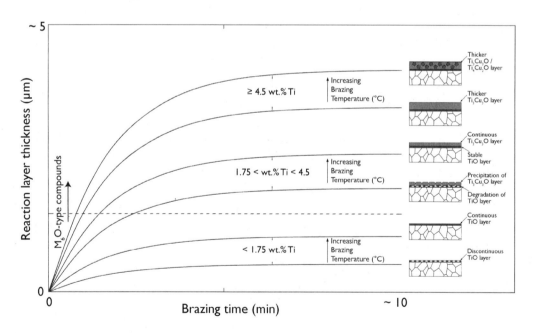

FIGURE 2.20 A comparison of the reported reaction layer compositions, at the alumina/Ag-Cu-Ti interface, with increases in the brazing temperature. At low Ti concentrations (<1.75 wt.% Ti), an increase in the brazing temperature leads to improved continuity of the Ti-O layer (layer I) (Cho and Yu, 1992). At intermediate Ti concentrations (1.75 < wt.% Ti < 4.5), an increase in the brazing temperature can affect the composition of the Ti-O layer (layer I) which becomes richer in Ti, and the thickness of the M_6O-type Ti-Cu-O layer (layer II) increases (Ali et al., 2016; Byun and Kim, 1994). At high Ti concentrations (>4.5 wt.% Ti), an increase in the brazing temperature can lead to an increase in the thickness of the M_6O-type Ti-Cu-O layer (layer II) (Hao et al., 1994), which may consist of Ti_4Cu_2O (Voytovych et al., 2006).

Literature Review

thick was observed at the joint interfaces. These joints were relatively weak and achieved strengths of ~80 MPa with poor interfacial bonding observed at the joint interfaces. Maximum joint strength was achieved following brazing at 850°C; joint strength increased to ~175 MPa with improved interfacial bonding comprising a continuous 2.0-μm-thick reaction layer at the joint interfaces. As the brazing temperature was increased further to 1,050°C, an 8.6-μm-thick reaction layer was observed at the joint interfaces; however, the joint strength decreased to ~70 MPa. An excessively thick reaction layer was suggested to have embrittled the joint interfaces (Figure 2.21a).

Ning et al. (2003) formed alumina-to-alumina brazed joints using 99.0 wt.% Al_2O_3 alumina and 0.5 mm thick 70.5Ag-27.5Cu-2.0Ti wt.% braze foils. The centre of each joint configuration consisted of an additional monocrystalline alumina spacer. Joint strength was evaluated using tensile testing and the hermeticity of the joints was evaluated using helium leak testing. Joints brazed at 845°C achieved an average strength of 30.0 MPa and exhibited poor hermeticity with a leak rate of 2.0×10^{-10} Pa.m³/S. Maximum joint strength was achieved following brazing at 855°C; joint strength increased to 62.5 MPa. These joints exhibited excellent hermeticity with a leak rate of just 5.0×10^{-13} Pa.m³/S. With a further increase in the brazing temperature to 895°C, the joint strength decreased to ~53 MPa. The hermeticity of theses joints were also adversely affected as the leak rate increased to 5.00×10^{-12} Pa.m³/S (Figure 2.21b).

These results show that maximum joint strength is achieved following brazing at an optimum temperature. The brazing temperature can affect the continuity, composition, and thickness of the reaction layer; these properties, therefore, also have optimum values. In addition to its effect on joint strength, the continuity of the reaction layer affects the hermiticity of the brazed joints. The

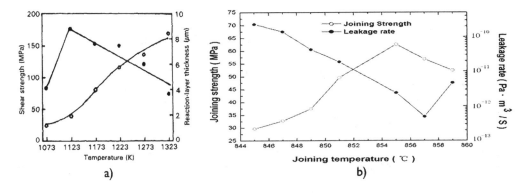

FIGURE 2.21 The effect of brazing temperature on the strength of alumina-to-alumina brazed joints made using Ag-Cu-Ti braze alloys correlated to (a) reaction layer thickness (Hao et al., 1994) and (b) hermeticity (Ning et al., 2003).

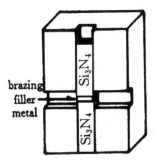

FIGURE 2.22 Brazing fixture used to support a silicon nitride-to-silicon nitride-brazed joint assembly during brazing experiments, prior to bend testing (Xiong et al., 1998; Xian and Si, 1990).

composition of the reaction layer is an indication of interfacial bonding and the reaction layer thickness corresponds to the degree of this bonding. Since the reaction layer comprises a brittle phase, an optimum thickness is required to achieve maximum joint strength.

Joint strength is also influenced by other variables in the AMB process (Figure 1.1). Moorhead et al. (1987) formed alumina-to-alumina brazed joints using 99.5 wt.% Al_2O_3 alumina and 100-μm-thick 44.0Ag-48.0Cu-4.0Ti-4.0Sn wt.% braze foils. Joint strength was evaluated using three-point bend testing. Bend test specimens were machined from two alumina plates that were brazed together. The bend test specimens were lapped, prior to mechanical testing. As the brazing temperature was increased from 800°C to 900°C, joint strengths of 114 and 119 MPa were achieved, respectively. Despite an increase in the reaction layer thickness as a result of an increase in the brazing temperature, joint strength was reported to be unaffected. The post-brazing machining step may have damaged the surfaces of the bend test specimens, prior to mechanical testing. Therefore, the surface condition of the ceramic is an important variable in the AMB process that may limit the strength of a brazed component by overriding any improvement in joint strength as a result of an increase in brazing temperature, or otherwise (see Section 2.4.2.2). The matter of a post-brazing machining step has been somewhat overcome in other studies whereby the formation of brazed bend test specimens has been faciliated through the use of a bespoke brazing fixture (Figure 2.22).

2.3.3 Brazing Atmosphere

In studies relating to AMB, experiments are typically performed in vacuum furnaces at pressures ranging between 4.0×10^{-4} and 1.0×10^{-7} mbar (see Tables 2.4 and 2.11). It is not uncommon for brazing experiments to be performed in other chemically inert environments such as those comprising Ar or He or H gasses (Ali et al., 2015; Loehman and Tomsia, 1992; Stephens et al., 2003; Voytovych et al., 2006). The evacuation of oxygen in the furnace chamber is required to promote the diffusion of Ti to the joint interface, where it reduces and chemically bonds with the ceramic surface. Thus, the presence of oxygen in alumina provides a driving force for the diffusion of Ti to the alumina surface.

In order to prevent Ti from the braze alloy oxidising, by reacting with oxygen in the furnace, the vacuum level is theoretically required to be at least $\sim 1.0 \times 10^{-19}$ mbar (Lugscheider and Tillmann, 1993). This is significantly lower than the $\sim 1 \times 10^{-5}$ mbar pressure typically reported in the literature for AMB experiments. Theoretically, the partial pressure of oxygen in a vacuum furnace, therefore, is likely to be insufficient to completely prevent the oxidation of Ti (do Nascimento et al., 2003). Despite this, a pressure of at least $\sim 1 \times 10^{-5}$ mbar has been reported to be one of the best conditions for vacuum furnace brazing (Gauthier, 1995).

Differences in the brazing atmosphere have been shown to affect the wetting behaviour of Ag-Cu-Ti braze alloys on alumina. Voytovych et al. (2006) performed wetting experiments on monocrystalline alumina using Cusil ABA®. The equilibrium contact angle achieved in a vacuum of 1×10^{-5} mbar was found to be 75°. This decreased to 20° when the wetting experiment was performed in a purified He atmosphere (Figure 2.23). In a reduced vacuum of 1×10^{-7} mbar a further decrease in the equilibrium contact angle to just ~10° was observed (Voytovych et al., 2006). While this shows that the brazing atmosphere is an important consideration in the AMB of ceramics, the equilibrium contact angle of 75° which resulted in a vacuum of 1×10^{-5} mbar is not in agreement with the literature.

The brazing atmosphere has been suggested to affect the composition of the reaction layer at the alumina/Ag-Cu-Ti joint interface. Lin et al. (2014) formed alumina-to-alumina brazed joints in a vacuum of 1.3×10^{-6} mbar. The brazing conditions used were similar to those used by Voytovych et al. (2006); however, Voytovych et al. (2006) performed wetting trials on alumina in a He atmosphere. Both studies used similar Ag-Cu-Ti braze alloy compositions. As the Ti concentration was increased from ~1.75 to ~4.5 wt.% Ti, Lin et al. (2014) observed the M_6O-type layer to be composed of Ti_3Cu_3O whereas Voytovych et al. (2006) observed this layer to include Ti_4Cu_2O as well as Ti_3Cu_3O. Lin et al. (2014) suggested that this discrepancy could be due to differences in the brazing atmospheres selected, in the two studies.

Literature Review

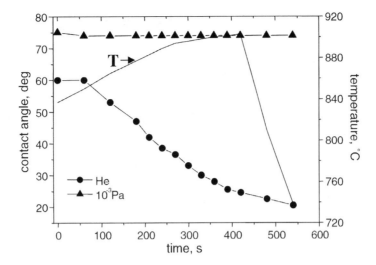

FIGURE 2.23 The effect of brazing atmosphere on the wettability of alumina using Cusil ABA® braze paste (Voytovych et al., 2006).

Ali et al. (2015) formed alumina-to-alumina brazed joints using Cusil ABA® in a conventional vacuum furnace and in an Ar-filled horizontal electric furnace; a pressure of ~1 × 10^{-5} mbar was measured in both cases. The reaction layers were found to be similar in both sets of brazed joints.

These conflicting views suggest that the brazing atmosphere could perhaps affect the reliability of joint formation in the alumina/Ag-Cu-Ti system as well as in the AMB of ceramics in general. Further investigation is required, however, to determine the conditions at which the brazing atmosphere begins to adversely affect joint formation.

2.4 VARIABLES INFLUENCING CERAMIC PROPERTIES

The monolithic strength of an alumina ceramic selected for brazing is critical as this can subsequently limit the maximum strength of a brazed joint. The properties of a ceramic which may influence joint strength, therefore, include its chemical composition, geometry, surface condition, and the presence of microstructural defects such as interparticle porosity. These properties are typical considerations during ceramic processing and manufacturing. Details of the various types of ceramic processing techniques and the steps involved are provided in Richerson (1992). Of these properties, the effects of composition or alumina purity, surface condition and surface roughness in the formation of alumina/Ag-Cu-Ti brazed joints are discussed in the following sections.

2.4.1 ALUMINA PURITY

Alumina purity refers to the concentration of alumina (wt.% Al_2O_3) in an alumina ceramic which may otherwise contain secondary phases such as magnesia (MgO), silica (SiO_2) and calcia (CaO). Typically, these secondary phases are sintering aids which facilitate the sintering process.

The mechanical properties of alumina ceramics are strongly correlated to the alumina purity. Morrell (1987) has classified the properties of alumina ceramics based on their relative purities, and this has recently been summarised in Auerkari (1996).

Sintering is the process by which green body powder compacts that are shaped by forming processes such as die pressing or otherwise, are fired at elevated temperatures. Sintering imparts mechanical strength to polycrystalline ceramics through densification of the green body compact, and the elimination of interparticle porosity.

The mass transport of alumina during densification occurs through diffusion mechanisms that depend on the type of sintering method employed. The sintering method depends on the alumina composition or purity, thus, alumina ceramics are either typically solid-state sintered or liquid phase sintered.

Alumina ceramics that are composed of at least 99.0 wt.% Al_2O_3 alumina (Morrell, 1987) or at least 99.7 wt.% Al_2O_3 alumina (Lee and Rainforth, 1994) are solid-state sintered between 1,500°C and 1,900°C. These relatively high sintering temperatures enable densification via solid-state diffusion mechanisms. The remaining < 1.0 wt.% usually consists of MgO which rapidly distributes itself at the alumina grain boundaries thereby inhibiting discontinuous grain growth to provide a uniform grain size distribution (Coble, 1961; Jorgensen and Westbrook, 1964).

The presence of MgO counters discontinuous grain growth during particle coarsening and thereby increases the time available for densification by prolonging the intermediate stage of sintering. This can help to achieve near theoretical densities; hence, solid-state sintered alumina ceramics are typically used in demanding applications such as those requiring excellent mechanical strength at high temperatures, e.g. sodium vapour lamp envelopes.

The greater stringency in the purity of starting powders and the higher sintering temperatures required can make the production of solid-state sintered alumina ceramics relatively costly as compared with liquid phase sintered alumina ceramics.

Alumina ceramics that are composed of between 80.0 and 99.7 wt.% Al_2O_3 alumina are typically sintered at 1,400°C to 1,600°C. These relatively lower sintering temperatures enable densification via the formation of a liquid phase, which typically consists of the sintering aids MgO, SiO_2 and CaO. In the Al_2O_3-CaO-SiO_2 system, the eutectic temperature ranges between 1,140°C and 1,400°C (Galusek et al., 2002). The distribution of a liquid phase at the alumina grain boundaries facilitates sintering by enabling particle arrangement through viscous flow and mass transport of alumina through solution/re-precipitation mechanisms (Auerkari, 1996).

Liquid phase sintered alumina ceramics are often considered cheaper alternatives to solid-state sintered alumina ceramics owing to the less stringent requirement in the purity of starting powders and the use of relatively lower sintering temperatures. Furthermore, the properties of liquid phase sintered alumina ceramics are generally inferior as compared with those of solid-state sintered alumina ceramics (Morrell, 1987). This can vary, however, with some liquid phase sintered alumina ceramics exhibiting higher flexural strengths than their solid-state sintered counterparts.

Ceramic processing methods can significantly influence the final properties of a ceramic part. Differences in microstructural features such as grain size distribution and interparticle porosity, or bulk features such as surface defects and complex geometries, are examples of features introduced during ceramic processing that may alter the final properties of an alumina part in addition to its composition or purity. For example, in liquid phase sintered alumina ceramics, if the CTE of the grain boundary phase is higher than that of the adjacent alumina grains, a clamping effect can provide a strengthening effect. However, a large CTE mismatch may produce a highly strained alumina microstructure which may have an adverse effect on its strength (Figure 2.24) (Galusek et al., 2002).

Most studies reviewed in this section have opted to use solid-state sintered alumina ceramics composed of at least 99.0 wt.% Al_2O_3 alumina, when performing brazing experiments using Ag-Cu-Ti braze alloys. Less than 20% of studies found in the literature have selected to use liquid phase sintered alumina ceramics, while in other studies the alumina purity has not been reported (Figure 2.25). The superior mechanical properties of solid-state sintered alumina ceramics in comparison to liquid phase sintered alumina ceramics may have influenced these selections.

2.4.1.1 Alumina Purity and Wetting

The effect of alumina purity on the wetting behaviour of Ag-Cu-Ti braze alloys is not widely reported in the literature. Typically, slightly higher equilibrium contact angles are formed following wetting experiments performed on liquid phase sintered alumina ceramics as compared with solid-state

Literature Review

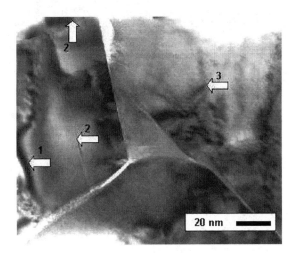

FIGURE 2.24 TEM image of a triple pocket grain boundary region in a liquid phase sintered alumina ceramic. Arrows indicate the strains caused by a CTE mismatch between the secondary phase (MgSiO$_3$) and adjacent alumina grains (Galusek et al., 2002).

sintered alumina ceramics. These findings suggest that the presence of a secondary phase on the alumina surface may interact with the Ag-Cu-Ti braze alloy, adversely affecting its wetting behaviour.

Loehman and Tomsia (1992) performed wetting experiments on 95.0 and 99.8 wt.% Al$_2$O$_3$ alumina using 59.1Ag-39.0Cu-1.9Ti and 55.1Ag-36.3Cu-8.6Ti at.% braze alloys. Brazing was conducted at 1,000°C for 60 min. The equilibrium contact angles achieved on 95.0 wt.% Al$_2$O$_3$ alumina using the 1.9 and 8.6 at.% Ti braze alloys were found to be 42° and 6°, respectively. The corresponding equilibrium contact angles achieved on 99.8 wt.% Al$_2$O$_3$ alumina were found to be 20° and 5°, respectively. The error in contact angle measurements was reported to be ± 3°. The ~20° difference in the equilibrium contact angles observed on the two grades of alumina by the 1.9 at.% Ti braze alloy, could not be explained.

Voytovych et al. (2006) performed wetting experiments on 96.0 wt.% Al$_2$O$_3$ alumina, 99.5 wt.% Al$_2$O$_3$ alumina and monocrystalline alumina using an Ag-Cu-2.9Ti at.% braze alloy. Brazing was conducted at 900°C for 30 min. The highest equilibrium contact angle observed was 19°, and this formed on 96.0 wt.% Al$_2$O$_3$ alumina. EPMA was used to detect ~0.3 at.% Si in the Ti$_3$Cu$_3$O layer which formed at the joint interface. Slightly lower equilibrium contact angles were achieved on both the 99.5 wt.% Al$_2$O$_3$ alumina and monocrystalline alumina ceramics. This improvement in wetting behaviour may have been associated to the lack of any Si-rich secondary phase in the higher purity alumina ceramics (Table 2.16).

2.4.1.2 Alumina Purity and the Reaction Layer

Several studies have observed the presence of secondary phase elements at the alumina/Ag-Cu-Ti joint interface. Asthana and Singh (2008) formed alumina-to-alumina brazed joints using alumina ceramics that were composed of 99.95% pure alumina powder and 0.05% MgO. Several grades of alumina were produced, sintered between 1,200°C and 1,600°C for 0.5 to 4.0 h. Brazing was conducted at 830°C for 20 min for joints made using Cusil ABA®, and at 915°C for 20 min for joints made using TICUSIL®. EDX line profiles performed at the interfaces of the brazed joints showed the diffusion of Mg across the joint interface, from the alumina into the braze interlayer. A maximum of ~3.25 at.% Mg was observed at the interfaces of joints made using TICUSIL® (Figure 2.26).

Vianco et al. (2003) formed alumina-to-alumina brazed joints using 94.0 wt.% Al$_2$O$_3$ alumina and 54-μm-thick Cusil ABA® braze foils. It was reported that when characterising the joint microstructures using EPMA techniques, a small amount of Si from the secondary phase in alumina was detected in the reaction layer.

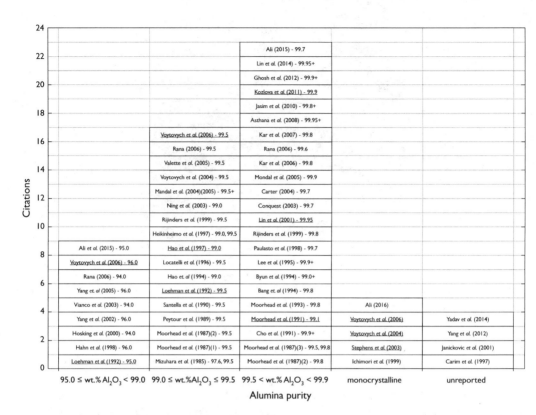

FIGURE 2.25 A summary of the grades of alumina selected in wetting experiments and in the formation of alumina-to-alumina brazed joints, using Ag-Cu-Ti braze alloys. Underline denotes wetting experiments as opposed to the formation of alumina-to-alumina brazed joints. + denotes alumina ceramics were processed in-house, according to the following: Cho et al. (1991) – 99.9 % pure alumina powder with 1500 ppm MgO, sintered at 1600°C for 2.0 h. Byun et al. (1994) – polycrystalline alumina containing 1% silicate. Lee et al. (1995) – 99.9 % pure alumina powder with 1000–1500 ppm MgO addition. Mandal et al. (2004)(2005) – 99.5 % pure alumina powder sintered at 1600°C for 0.5 h. Asthana et al. (2008) – 99.95 % pure alumina powder with 0.05 % MgO, sintered between 1200 and 1600°C for 0.5 to 4.0 h. Jasim et al. (2010) – 99.8 % pure alumina powder sintered at 1600°C. Ghosh et al. (2012) – 99.99 % pure alumina powder sintered at 1600°C for 2.0 h. Lin et al. (2014) – 99.95 % pure alumina powder with 0.05 % MgO, sintered between 1200 and 1600°C for 0.5 to 4.0 h.

TABLE 2.16
The Effect of Alumina Purity on the Wetting Behaviour of an Ag-Cu-2.9Ti at.% Braze Alloy, Following Wetting Experiments Performed at 900°C for 30 min

Alumina purity, (wt.% Al$_2$O$_3$)	Surface roughness, Ra (µm)	Equilibrium contact angle, θ_f (°)
96.0	0.15	19.0 ± 3
99.5	0.075	17.0 ± 3
99.5	1.3	13.5 ± 3
Monocrystalline	0.004	10.0 ± 3

Source: Voytovych et al. (2006).

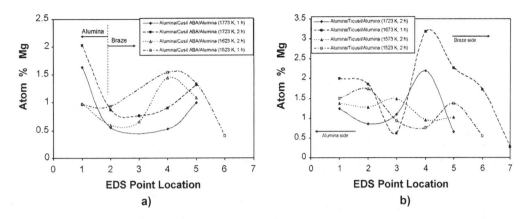

FIGURE 2.26 SEM-EDX line profiles showing the variation of Mg at the interfaces of alumina-to-alumina brazed joints made using several different grades of alumina (sintered at 1,200°C to 1,600°C for 0.5–4.0 h) using ~110 μm thick (a) Cusil ABA® and (b) TICUSIL braze foils (Asthana and Singh, 2008).

Ali et al. (2015) formed alumina-to-alumina brazed joints using 95.0 and 99.7 wt.% Al_2O_3 alumina and 50-μm-thick Cusil ABA® braze foils. Using TEM and EDX techniques, solid solutions of Ca and Si from the secondary phase in alumina were observed in the vicinity of both the γ-TiO (layer I) and Ti_3Cu_3O layers (layer II). It was reported that up to 5 at.% Si was present in the Ti_3Cu_3O layer (Figure 2.27).

Kang and Selverian (1992) deposited a 3-μm-thick Ti coating, using electron beam evaporation, onto the surfaces of 95.0 and 99.5 wt.% Al_2O_3 alumina ceramics. The coated specimens were heated to 980°C for 10 min in a vacuum furnace at a pressure of 1×10^{-5} mbar simulating a typical brazing experiment. SEM and TEM techniques were used to characterise the reaction layer as a Ti_3Al-type phase, in both sets of joints. At the interfaces of joints made using 95.0 wt.% Al_2O_3 alumina, the secondary phase at the alumina grain boundaries had reacted with Ti in the reaction layer to form Ti_5Si_3 compounds.

These studies show that secondary phases in alumina ceramics can lead to additional reactions at the alumina/Ag-Cu-Ti joint interface to those described in Section 2.1.7. Further investigation

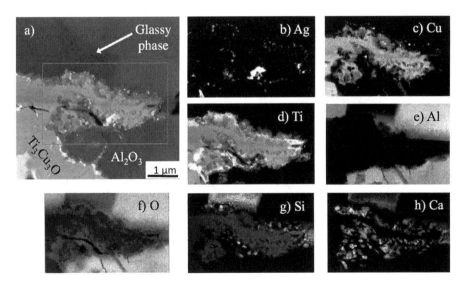

FIGURE 2.27 (a) HAADF STEM images showing the presence of secondary phase elements at the alumina/Ag-Cu-Ti joint interface, and EDX maps showing distributions of (b) Ag, (c) Cu, (d) Ti, (e) Al, (f) O, (g) Si and (h) Ca, corresponding to the selected area shown in (a) (Ali et al., 2015).

is required, therefore, to establish how the secondary phase in alumina, through interactions at the joint interface, may affect interfacial chemistry, reaction layer formation and joint strength.

2.4.1.3 Alumina Purity and Joint Strength

The effect of secondary phases in alumina ceramics on the resulting strength of joints made using Ag-Cu-Ti braze alloys has not been widely reported in the literature. However, brazed joints made using higher purity alumina ceramics are suggested to provide superior joint strength.

Rana (2006) formed alumina-to-alumina brazed joints using 94.0, 99.5 and 99.6 wt.% Al_2O_3 alumina and 60-μm-thick TICUSIL® braze foils. Brazing was conducted at 950°C and joint strength was evaluated using four-point bend testing. The average strength of brazed joints made using 94.0 wt.% Al_2O_3 alumina was 223 MPa. Brazed joints made using 99.5 and 99.6 wt.% Al_2O_3 alumina achieved 236 and 205 MPa, respectively. It was suggested that alumina purity could not be correlated to joint strength. The grades of alumina selected in the study by Rana (2006) were manufactured by different suppliers. The surface conditions of each alumina grade may have differed due to a variation in the grinding methods used. The alumina purity may not have been the only difference between the alumina grades, and these differences may have also affected the resulting joint strengths achieved.

Hahn et al. (1998) formed alumina-to-alumina brazed joints using 96.0 wt.% Al_2O_3 alumina and 50-μm-thick Cusil ABA® braze foils. Brazing was conducted at 830°C for 10 min and joint strength was evaluated using four-point bend testing. Brazed joints tested in an as-brazed condition achieved 294 MPa. Some of the brazed joints were heat treated to 400°C or 600°C for 100 h prior to mechanical testing. These joints achieved 266 and 141 MPa, respectively. The degradation in joint strength observed was suggested to be due to excessive reactions at the joint interfaces involving the crystallisation of grain boundary phases in alumina, activated by the presence of Ti, which weakened the alumina side of the joint interface.

These studies suggest that the secondary phase in alumina ceramics can influence joint strength; however, there is a lack of consistent evidence to support this. Systematic experiments, whereby other variables in the AMB process are controlled, e.g. surface condition and grinding methods, are required to better evaluate the effect of alumina purity on joint strength.

2.4.2 Alumina Surface Condition

The surface condition of alumina is commonly reported to be one of the major variables in obtaining high-strength alumina-to-alumina and alumina-to-metal brazed joints (Mizuhara and Mally, 1985). The structural integrity of the alumina surface is crucial for achieving high-strength joints due to its inherent brittleness; hence, the ceramic/braze region of a brazed joint is often regarded the more critical side of the joint interface.

Surface condition hereby refers to the defect population within ~20 μm of the ceramic surface. This region typically comprises internal defects, e.g. entrapped porosity during sintering, and surface defects, e.g. grinding-induced defects. Other stress concentrators such as sharp corners in ceramic parts with complex geometries or otherwise, unless chamfered, are also considered as surface defects.

Failure of a ceramic under applied load typically depends on the presence of a crack-producing flaw in a given volume – based on probability. A ceramic surface that comprises fewer defects, therefore, ensures lower likelihood of failure and thereby a higher likelihood in the formation of high-strength brazed joints.

Grinding of a ceramic can affect its surface roughness and may induce surface defects, both of which can adversely affect the flexural strength of alumina ceramics (Belenky and Rittel, 2012; Liao et al., 1997). However, grinding is a necessary procedure which imparts dimensional accuracy to an otherwise defect-free but uneven as-sintered ceramic surface. Polishing and lapping methods, commonly used to recover grinding damage, are also abrasive techniques which may induce new defects into the ceramic surface.

Literature Review

In literature studies, the parent alumina ceramics are commonly characterised prior to brazing in terms of geometry, composition and surface roughness. Their microstructures and mechanical properties are less commonly studied particularly since these details may be provided in materials data sheets. Unless the manufacturing steps involved are known, e.g. grinding history, it may be difficult to accurately characterise the surface condition of an alumina part. For example, despite similar surface roughnesses, the surface conditions of alumina ceramics that are as-sintered or as-ground or ground-lapped may differ, such that once brazed, joint strengths may also differ significantly.

Most literature studies have incorporated polishing as a preparation method, prior to the brazing of alumina. A comparison of the procedures applied to alumina ceramics, reported in studies which have selected Ag-Cu-Ti braze alloys is shown in Table 2.17.

TABLE 2.17
A Comparison of the Reported Methods Used in the Preparation of Alumina Prior to Active Metal Brazing Experiments

Publication	Alumina (wt.% Al_2O_3)	Surface Condition	Treatment Procedures	Surface Roughness, Ra (µm)
Ali et al. (2015)	95.0	Ground	–	2.10
	99.7			1.30
	99.7			0.60
Asthana and Singh (2008)	99.5+	Sintered	–	–
Bang and Liu (1994)	99.8	Finely ground	1,200 SiC paper	–
Byun and Kim (1994)	99.0	Heat treated	1,000°C, 0.5 h	
Carim (1991)	–	Polished	10 µm diamond paste	–
Carter (2004)	99.7	As-received	–	–
Cho et al. (1992)	99.9+	Heat treated	1,500°C, 1.0 h	–
Conquest (2003)	99.7	As-received	–	–
Ghosh et al. (2012)	99.9+	Polished	0.25 µm diamond paste	–
		Heat treated	1,600°C, 2.0 h	
Hahn et al. (1998)	96.0	Polished	0.25 µm diamond paste	0.25
Hao et al. (1994)	99.0	Polished	–	–
Hao et al. (1997)	99.0	–	–	–
Heikinheimo et al. (1997)	99.0	Ground	800 SiC paper	–
	99.5			
Hosking et al. (2000)	94.0	Heat treated	1,575°C, 2.0 h	–
	99.8			
Janickovic et al. (2001)	–	–	–	–
Jasim et al. (2010)	99.8+	Polished	0.5 µm diamond paste	0.005
		Heat treated	1,600°C	
Kar et al. (2006)	99.8	Polished	–	–
		Heat treated	1,600°C, 0.5 h	
Kar et al. (2007)	99.8	Polished	0.5 µm diamond paste	–
		Heat treated	1,600°C, 0.5 h	
Kozlova et al. (2011)	99.9	Polished	–	'Few nm'
Ichimori et al. (1999)	99.9	–	–	–
Lin et al. (2014)	99.95+	As-sintered	–	–
Lin et al. (2001)	99.95	–	–	–

(Continued)

TABLE 2.17 (*Continued*)
A Comparison of the Reported Methods Used in the Preparation of Alumina Prior to Active Metal Brazing Experiments

Publication	Alumina (wt.% Al_2O_3)	Surface Condition	Treatment Procedures	Surface Roughness, Ra (μm)
Lee and Kwon (1995)	99.9+	Polished	2.5 μm diamond paste	–
Locatelli et al. (1997)	99.5	Polished	1.0 μm diamond paste	–
Mandal et al. (2004a)	99.5+	Polished Heat treated	0.5 μm diamond paste 1,600°C, 0.5 h	–
Mandal et al. (2004b)	99.5+	Polished Heat treated	0.5 μm diamond paste 1,600°C, 0.5 h	–
Mizuhara and Mally (1985)	99.5 99.5 97.6	As-sintered Ground, Polished Ground, Heat treated	– – –, 1,600°C, 0.5 h	–
Mondal et al. (2002)	99.9+	Heat treated	1,600°C, 0.5 h	–
Moorhead et al. (1987)(1)	99.5	Ground Heat treated	– 1,000°C, 0.25 h	4.46
Moorhead et al. (1987)(2)	99.5 99.8	–	–	–
Moorhead et al. (1987)(3)	99.5 99.8	Ground Heat treated	– 1,000°C, 0.25 h	2.65 4.46
Moorhead et al. (1991)	99.8	Heat treated	800°C, 0.25 h	–
Moorhead et al. (1993)	99.8	Ground Polished	–	0.49 0.11
Ning et al. (2003)	99.0	Ground Polished	150 to 1200 SiC paper	0.16 to 4.40
Paulasto and Kivilahti (1998)	99.7	Polished	–	–
Peytour et al. (1990)	99.5	Polished	2.0 μm diamond paste	0.2
Rana (2006)	94.0 99.5 99.6	As-received	–	–
Rijinders and Peteves (1999)	99.5 99.8	Polished Heat treated	1.0 μm diamond paste 1,100°C, 1.0 h	0.45 0.25
Santella et al. (1990)	99.5	As-sintered Heat treated	– 1,000°C, 0.25 h	4.46
Stephens et al. (2003)	Monocrystalline			
Valette et al. (2005)	99.5	Polished	–	0.6
Vianco et al. (2003)	94.0	Heat treated	1,575°C, 2.0 h	–
Voytovych et al. (2006)	99.5 Monocrystalline	Polished	–	0.5 0.002
Voytovych et al. (2004)	96.0 99.5 99.5 Monocrystalline	Polished		0.15 0.075 1.3 0.004
Yadav et al. (2014)	–	–	–	–
Yang et al. (2002)	96.0	Polished	1000 SiC paper	–
Yang et al. (2005)	96.0	Polished	–	1.0
Yang et al. (2012)	–	–	–	–

2.4.2.1 Surface Roughness

The surface roughness of alumina has been found to correlate to its monolithic flexural strength. Belenky and Rittel (2012) prepared several 99.5 wt.% Al_2O_3 alumina test bars according to the the ASTM C1161 standard which includes details of geometries and grinding procedures. An interferometer was used to measure the surface roughness (Ra values) at the mid-points of the tensile faces of each test bar. Some of the bars were in as-ground condition while others were polished to finer Ra values before mechanical testing. The flexural strengths of the test bars were determined using three-point bend testing and the results were correlated to surface roughness (Figure 2.28).

These results showed that the surface roughness (Ra values) of the 99.5 wt.% Al_2O_3 alumina test bars decreased from 0.9 to 0.02 μm as a result of polishing. As a result, this led to ~10% increase in the flexural strength of the test bars. While polishing led to a decrease in the surface roughness of the as-ground test bars, the defect population in their surfaces may have also been reduced. An improvement in surface condition, therefore, may better explain the increase in flexural strength observed, rather than a direct correlation between surface roughness and flexural strength.

2.4.2.1.1 Surface Roughness and Wetting

Conventionally, the effect of surface roughness on the equilibrium contact angle during wetting depends on the intrinsic wettability of a system (Kubiak et al., 2011; Wenzel, 1936). If a given system is non-wettable ($\theta > 90°$) then an increase in surface roughness which provides an increased surface area may lead to higher equilibrium contact angles. Similarly, if a given system is wettable ($\theta > 90°$) then an increase in surface roughness may lead to a decrease in the equilibrium contact angle.

This relationship is given by Wenzel's theory:

$$\cos(\theta_m) = r\cos(\theta_i) \qquad (2.23)$$

$$r = \frac{\text{Actual area of rough surface}}{\text{Plan area}} \qquad (2.24)$$

where θ_i is the intrinsic contact angle, r represents an increase in the surface area and θ_m is the macroscopic contact angle.

In AMB, a fine surface with a lower Ra value has commonly been reported to be more beneficial for reactive wetting than a rougher surface, which may otherwise pin the movement of the triple point. The grooves, ridges and other surface heterogeneities of a rough surface, therefore, are imperfections that the moving triple line has to overcome (Passerone et al., 2013). This is consistent with the Young–Dupré equation (Equation 2.1), since a rougher surface increases the solid–liquid contact area.

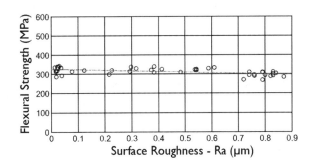

FIGURE 2.28 The effect of surface roughness, induced by grinding and polishing procedures, on the flexural strength of 99.5 wt.% Al_2O_3 alumina test bars (Belenky and Rittel, 2012).

The definition of a rough surface whereby wetting becomes adversely affected needs further clarification. Wetting studies performed using Ag-Cu-Ti braze alloys have found that wetting behaviour is not significantly affected by an increase in the surface roughness of alumina to Ra values that are reasonably induced through abrasive methods such as grinding or polishing. Voytovych et al. (2006) found that despite an increase in the Ra values of a variety of alumina ceramics from 0.004 to 1.3 µm, the equilibrium contact angles varied by just ± 5° (Table 2.16).

Asthana and Singh (2008) found that spreading of an Ag-Cu-Ti braze alloy becomes adversely affected when the alumina surface consists of voids or porosity due to partial sintering or missing grains, e.g. due to grain pull-out during polishing, or otherwise. These surface asperities which produce significantly rougher alumina surfaces than those measured by Voytovych et al. (2006), therefore, can adversely affect wetting behaviour.

Ning et al. (2003) found that by polishing a 99.0 wt.% Al_2O_3 alumina ceramic to reduce its surface roughness, the wetting behaviour of a 70.5Ag-27.5Cu-2.0Ti wt.% braze alloy significantly improved. With an Ra value of 4.40 µm prior to polishing, the alumina ceramic was reported to be 'non-wettable' whereas with an Ra value of 0.16 µm which resulted after polishing, an equilibrium contact angle of just ~7° was achieved.

These results show the importance of surface roughness, which has been shown to influence reactive wetting significantly. Generally, a finely polished alumina surface enhances the wetting behaviour and flow characteristics of an Ag-Cu-Ti braze alloy (Peytour et al., 1990). The extent of this varies, however, as wetting only becomes adversely affected when surfaces become rough beyond a given limit i.e. Ra > ~1.3 µm (Voytovych et al., 2006).

2.4.2.1.2 Surface Roughness and the Reaction Layer

In the reactive wetting model, lateral growth of the reaction layer is required for further spreading of an Ag-Cu-Ti droplet in the quasi-steady state, on an alumina surface (see Figure 2.2). The surface profile of the reaction layer typically follows that of the alumina surface it wets provided that the moving triple line is able to overcome surface asperities. A relatively rough alumina surface may inhibit spreading but exhibits a larger surface area for reactions to occur - a greater degree of chemical bonding, providing interlocking strength. Enhanced reactive wetting occurs on polishing surfaces with lower Ra values (see Section 2.4.2.1.1). On these surfaces, lateral growth of the reaction layer may be more easily achieved.

The effect of surface roughness of the properties of the reaction layer in the alumina/Ag-Cu-Ti system has not been widely reported in the literature. However, Ichimori et al. (1999) observed the thickness of the TiO layer (layer I) on the alumina side of the joint interface to vary consistently according to the roughness of the alumina surface.

Ichimori et al. (1999) formed alumina-to-alumina brazed joints using monocrystalline alumina and 100-µm-thick 66.7Ag-28.4Cu-4.9Ti wt.% braze foils. TEM and EDX techniques were used to characterise a 10- to 50-µm-thick TiO layer, with isolated Ti_2O particles, on the alumina side of the joint interface (layer I) and a 1- to 2-µm-thick Ti_3Cu_3O layer on the braze side of the joint interface (layer II). A thicker TiO layer was observed in rougher parts of the alumina surface as compared with flatter parts. A rougher surface may have provided a greater surface area, therefore, for reactions at the joint interface to occur (Figure 2.29).

2.4.2.1.3 Surface Roughness and Joint Strength

The surface roughness of an alumina part, once brazed, has been found to influence joint strength. In several studies, an improvement in joint strength has been achieved via a reduction in the surface roughness of an alumina part prior to brazing - this has been induced via polishing which may provide an additional benefit to its surface condition i.e. through the recovery of grinding damage (see Section 2.4.2.2). Therefore, joint strength may not be simply attributable to surface roughness.

Ning et al. (2003) formed alumina-to-alumina brazed joints using 99.0 wt.% Al_2O_3 alumina and 0.5 mm thick 70.5Ag-27.5Cu-2.0Ti wt.% braze foils. Joint strength was evaluated using tensile

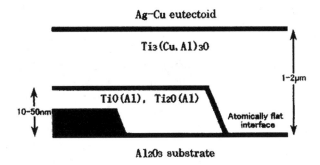

FIGURE 2.29 The effect of small variations in the roughness of monocrystalline alumina on the thicknesses of the TiO(Al)/Ti$_2$O(Al) (layer I) and Ti$_3$(Cu, Al)$_3$O (layer II) at the alumina/Ag-Cu-Ti joint interface (Ichimori et al., 1999).

testing and the hermeticity of the joints was evaluated using helium leak testing. The alumina specimens were polished using SiC papers, with progressively finer grit designations, followed by use of a diamond suspension. This led to a significant decrease in surface roughness; Ra values decreased from 4.40 to 0.16 μm. This coincided with an increase in joint strength, from 20 to 63 MPa, respectively (Figure 2.30). The hermeticity of the joints also improved from ~2 × 10^{-5} Pa m^3/s to 5.00 × 10^{-13} Pa m^3/s, respectively.

While a decrease in the surface roughness of alumina, induced by polishing, can be correlated to an increase in joint strength, the underlying mechanism of an improvement in the ceramic surface condition through the recovery of grinding damage cannot be ignored. Understandably, variation in the surface roughness of alumina ceramics, following sintering, is induced through abrasive methods (see Section 2.4.2.2). However, variation in the surface condition of alumina ceramics may not necessarily be restricted in the same way, e.g. post-grinding heat treatment, which is also used to recover grinding damage in alumina ceramics (see Section 2.4.2.3).

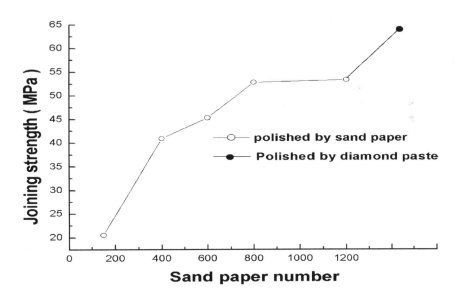

FIGURE 2.30 The effect of surface roughness, induced by polishing, on the strength and hermeticity of alumina-to-alumina brazed joints (Ning et al., 2003).

2.4.2.2 Grinding and Polishing

Grinding is a necessary procedure which imparts dimensional accuracy to an otherwise defect-free but uneven as-sintered ceramic surface. Grinding of a ceramic typically produces a relatively rough surface and induces surface defects, both of which adversely affect the flexural strength of alumina. Polishing, or lapping, of a ceramic is a technique used to recover grinding damage and typically produces a relatively fine ceramic surface (Belenky and Rittel, 2012; Liao et al., 1997; Mizuhara and Mally, 1985; Mizuhara et al., 1989).

Two alumina parts may be polished to the same surface roughness but due to differences in their grinding histories, may achieve different strengths despite similar brazing conditions - surface roughness may not be directly correlated to joint strength. According to a study where fractography was combined with confocal scanning laser microscopy and dye penetrant inspection, 85% of failures in the flexural testing of alumina ceramics originated from defects in their surfaces (Nakamura et al., 2009). Therefore, it is necessary to consider the surface of the alumina parts including defects within at least ~20 μm of their surfaces, e.g. sub-surface defects, grain pull-out, microcracks, etc. which may all act as failure initiation sites.

Alumina ceramics that have been ground and polished, once brazed, have been reported to achieve joint strengths that are significantly higher than those made using alumina ceramics in as-ground condition.

Mizuhara and Mally (1985) formed two sets of alumina-to-alumina brazed joints using 99.5 wt.% Al_2O_3 alumina and 50-μm-thick Cusil ABA® braze foils. The first set of these joints were made using alumina ceramics that were in as-ground condition whereas in the second set, the as-ground alumina ceramics had been lapped prior to brazing. Brazing was conducted at 840°C for 10 min and joint strength was evaluated using tensile testing (ASTM F-19). While the first set of joints achieved an average strength of 49.6 MPa and failed at the joint interfaces, the second set of joints failed in the ceramic, away from the joint interfaces with an average strength of 65.5 MPa. The increase in joint strength was attributed to lapping, which had improved the surface condition of the alumina ceramics by removing grinding damage, prior to brazing.

Moorhead and Simpson (1993) formed two sets of alumina-to-alumina brazed joints using 99.8 wt.% Al_2O_3 alumina and 76 μm-thick 41Ag-48Cu-4Sn-7Ti at.% braze foils. The first set of these joints were made using alumina ceramics that were in an as-ground condition (Ra = 0.49 μm) whereas in the second set, the as-ground alumina ceramics had been lapped (Ra = 0.11 μm). Brazing was conducted at 800°C for 10 min and joint strength was evaluated using three-point bend testing. The first set of joints achieved an average strength of 323 ± 24 MPa, however, mixed-mode failures were observed. Half of these joints failed in the ceramic and achieved 301 ± 8 MPa, whereas the other half failed at the joint interface and achieved 345 ± 6 MPa. The second set of joints failed consistently at the joint interface and achieved 341 ± 18 MPa. Lapping had led to a marginal ~6% increase in the average strength of the joints, markedly lower than the improvement observed as a result of lapping by Mizuhara and Mally (1985). New cracks induced by polishing were observed by Moorhead and Simpson (1993) in the surfaces of the alumina ceramics using ultrasonic testing with high frequency (50 MHz) radially propagating surface waves. These new cracks were suggested to have inhibited any significant increase in joint strength.

In Mizuhara and Mally (1985), lapping was performed on the faying alumina surfaces prior to brazing while in Moorhead and Simpson (1993), the lapping step involved the polishing of brazed joints prior to mechanical testing (see Figure 2.31). These differences led to notable changes in the failures observed which corresponded to the increase in joint strength. In Mizuhara and Mally (1985), the failure locations shifted from the joint interface to the ceramic, whereas, in Moorhead and Simpson (1993) failure locations shifted from the ceramic to the joint interface. Mizuhara and Mally (1985) found that by polishing the faying alumina surfaces, an improvement in the continuity of the bond line at the joint interface led to the increase in joint strength. Moorhead and Simpson

Literature Review

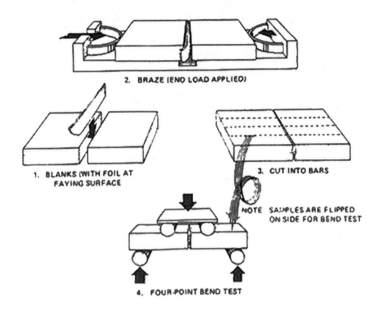

FIGURE 2.31 Steps in fabrication of brazed alumina-to-alumina flexural test bars (Moorhead and Simpson, 1993).

(1993) found that by polishing the outer surfaces of the brazed joints, the ceramic portions of the brazed joints could withstand relatively higher loads.

The method used by Moorhead and Simpson (1993) in the preparation of brazed joints for four-point bend testing involved a post-brazing machining step (Figure 2.31, step 3) – this has also been employed elsewhere in the literature (Kim et al., 2006). The severity of the damage induced by this step may not have been fully recoverable by subsequent lapping; hence, the marginal ~6% increase in the average strength observed. An alternative method in the preparation of brazed joints for bend testing, which eliminates any post-brazing machining step, involves the use of a brazing fixture to support the butt-joint assemblies during brazing, as shown in Xiong et al. (1990), (see Figure 2.22).

2.4.2.3 Post-Grinding Heat Treatment

Post-grinding heat treatment is a non-abrasive technique used for the recovery of grinding damage in alumina ceramics. This has been found to lead to an increase in the monolithic flexural strength of as-ground 96.0 wt.% Al_2O_3 alumina ceramics, to test temperatures of up to ~550°C.

Kirchner et al. (1970) studied the effect of post-grinding heat treatment, which was conducted at 1,600°C for 15 min, on the elevated temperature flexural strength of 96.0 wt.% Al_2O_3 alumina rods. Joint strength was evaluated using four-point bend testing. The average room temperature flexural strength of ~300 MPa increased to ~430 MPa following the heat treatment (Figure 2.32). This corresponded to an increase in scatter of the measured strength values. It was concluded that the post-grinding heat treatment had healed at least one type of flaw in the alumina microstructure whereby an increase in the average flexural strength was observed.

The heat treatment temperature of ~1,600°C in Kirchner et al. (1970) is similar to the typical temperatures at which alumina is sintered. Therefore, liquid phase formation in the 96.0 wt.% Al_2O_3 alumina rods is likely to have occurred. The mass transport of alumina through solution/re-precipitation in the liquid phase may have effectively re-sintered the alumina rods, imparting strength through densification.

The re-sintering of alumina during post-grinding heat treatment may depend largely on the original sintering regime. The healing of defects, such as grinding damage or otherwise, is likely to

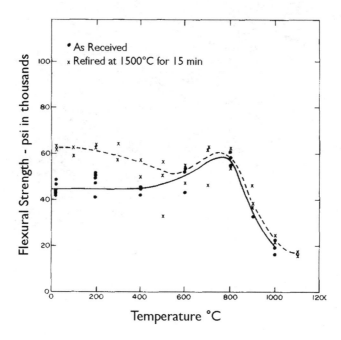

FIGURE 2.32 The effect of post-grinding heat treatment, conducted at 1,600°C for 15 min, on the elevated temperature flexural strength of liquid phase sintered 96.0 wt.% Al_2O_3 alumina rods (Kirchner et al., 1970).

depend therefore on the alumina purity since the solid-state diffusion mechanisms in higher purity alumina ceramics differ from those in relatively lower purity alumina ceramics whereby sintering is facilitated through liquid phase formation. However, the relatively short heat treatment times may make re-sintering an altogether unlikely mechanism.

The near-sintering heat treatment temperatures may or may not benefit solid-state sintered alumina ceramics that were originally fired to near-theoretical densities. In liquid phase sintered alumina ceramics, however, additional mechanisms may occur. This includes the re-distribution of the liquid phase as it flows into crevices occupying these regions in the same way as in the original sinter, whereby it had accommodated the intergranular and triple pocket grain boundary regions of the alumina microstructure.

In the literature, several studies have employed a post-grinding heat treatment step, prior to brazing alumina, as a means to achieve high-strength joints by first improving its surface condition (Cho and Yu, 1992; Oyama and Stribe, 1998; Stephens et al., 2003; Vianco et al., 2003).

In studies which have opted to employ the post-grinding heat treatment of alumina prior to brazing, a variety of alumina ceramics have been selected; alumina purity has not therefore been considered (Table 2.18). This shows that post-grinding heat treatment is thought to be beneficial in improving the surfaces of all alumina ceramics irrespective of whether they are either liquid phase, or solid-state sintered.

The effect of post-grinding heat treatment of alumina, prior to brazing, on the strength of alumina-to-Kovar® brazed joints was found to affect joints made using liquid phase sintered alumina ceramics differently to those made using solid-state sintered alumina ceramics.

Mizuhara and Mally (1985) studied the effect of post-grinding heat treatment, conducted at 1,650°C for 1 h, on the strength of 97.6 and 99.5 wt.% Al_2O_3 alumina-to-Kovar® brazed joints. Three different 50-μm-thick Ag-Cu-Ti braze foils with increasing Ti concentrations were used (Table 2.4). The strengths of 97.6 and 99.5 wt.% Al_2O_3 alumina-to-Kovar® brazed joints whereby the alumina ceramics were heat treated prior to brazing were compared with the strengths of

TABLE 2.18
A Comparison of Studies Which Have Used Post-Grinding Heat Treatment to Improve the Surface Condition of Alumina Prior to Active Metal Brazing

References	wt.% Al_2O_3	Heat Treatment Temp (°C)	Time (min)
Mizuhara and Mally (1985)	99.5	1,650	60
	97.6		
Mizuhara et al. (1989)	99.5	1,650	60
Cho et al. (1991)	99.9	1,500	30
Oyama and Stribe (1998)	97.6	1,500	180
Hosking et al. (2000)	94.0	1,575	120
	99.8		
Vianco et al. (2003)	94.0	1,575	120
Stephens et al. (2003)	Monocrystalline	1,575	120

joints made using the same alumina ceramics in as-sintered, as-ground and in ground/lapped conditions.

For joints made using 97.6 wt.% Al_2O_3 alumina, the weakest joints were those with alumina in as-ground condition and the highest strength joints were those whereby alumina was heat treated prior to brazing. For joints made using 99.5 wt.% Al_2O_3 alumina, the weakest joints were also those with alumina in as-ground condition; however, the highest strength joints were those with alumina in as-sintered condition. Brazed joints made using 99.5 wt.% Al_2O_3 alumina in heat treated condition achieved strengths that were only similar to their ground/lapped counterparts.

While post-grinding heat treatment was found to be beneficial for joints made using 97.6 wt.% Al_2O_3 alumina, it had not provided the same benefit to joints made using 99.5 wt.% Al_2O_3 alumina (Figure 2.33). Thus, these results show that the effect of post-grinding heat treatment on the strength of brazed joints is strongly correlated to alumina purity.

Further analysis of the results obtained by Mizuhara and Mally (1985) shows that joints brazed in ground/lapped condition achieved higher strengths than those joints brazed in as-ground condition, for both grades of alumina. Therefore, while grinding of alumina reduces brazed joint strength, polishing or lapping can somewhat recover this. As expected, abrasive processes introduce defects into the alumina surface leading to joint strengths that are relatively lower than those achieved when alumina is brazed in as-sintered condition. Post-grinding heat treatment of alumina, a non-abrasive process, can potentially lead to an improvement in joint strength as compared with joints made using alumina in as-ground or in ground/lapped conditions; this depends, however, since for liquid phase sintered alumina ceramics, this provides joints that are stronger than even their as-sintered counterparts, while for solid-state sintered alumina ceramics, this does not provide the same benefit. Therefore, the mechanism by which post-grinding heat treatment of alumina enables the formation of high-strength joints depends on the presence of a liquid phase, which contradicts the application of heat treatment prior to brazing applied to a range of alumina ceramics, as found in the literature.

2.5 MECHANICAL TESTING

Joint design and testing methods can influence the strength of ceramic-to-ceramic and ceramic-to-metal brazed joints (Figure 1.1). Standard testing methods often define specimen geometries and joint configurations. However, while these test methods are adequate for mechanical testing purposes, their suitability may depend on the types of stresses, temperatures, atmospheres and

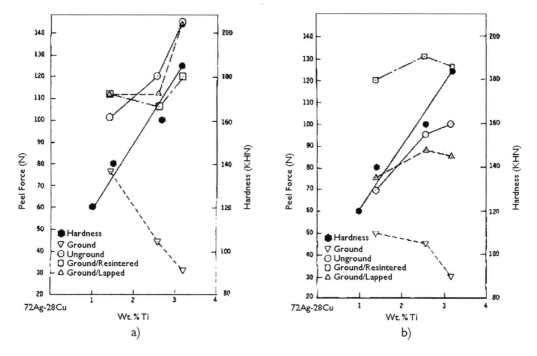

FIGURE 2.33 The strength of alumina-to-Kovar® brazed joints made using (a) 99.5 wt.% Al_2O_3 and (b) 97.6 wt.% Al_2O_3, alumina in as-sintered, as-ground, ground/lapped and ground/re-sintered conditions correlated to Ti concentration in an Ag-Cu-Ti braze alloy (Mizuhara and Mally, 1985).

other conditions which brazed components are later subjected to in-service (Moorhead and Santella, 1987). While test methods are often performed to derive meaningful strength data, this can be challenging considering the inherent brittleness of ceramic materials. In the literature, the most commonly used test methods for evaluating the strengths of brazed joints comprise tensile testing, shear testing and bend testing according to the configurations shown in Figure 2.34.

2.5.1 Typical Testing Methods

The most commonly used standardised test methods for evaluating the strength of alumina-to-alumina brazed joints as reported in the literature are tensile testing - based on ASTM F-19 (ASTM Standards, 2016; Ali et al., 2015) (Figure 2.34a), and four-point bend testing - based on AWS C3.2M/C3.2:2008 (AWS Standards, 2008) and an adaptation from ASTM C1161 (ASTM Standards, 2013) (Conquest, 2003) (Figure 2.34b). While shear testing has also been commonly used, bespoke procedures have usually been adopted (Hao et al., 1994) (Figure 2.34c).

Other test methods include single-edge notched beam (SENB) testing (Cho and Yu, 1992), double cantilever beam (DCB) testing (Heikinheimo et al., 1997), three-point bend (3PB) testing (Moorhead and Simpson 1993) and double-bonded shear (DBS) testing (Mohammed Jasim et al., 2010).

Test methods used to evaluate the mechanical properties of the individual constituents of brazed joints include microhardness testing (Asthana and Singh, 2008) and nanoindentation testing (Ghosh et al., 2012).

Joint strength values obtained from different test methods are not typically comparable. The failure location in a brazed joint, following mechanical testing, is often used as an indication to qualitatively compare joint strengths. Failure at the joint interface is often correlated to a weak bond requiring further optimisation while failure away from the joint interface, in the parent ceramic material, is often associated to a strong bond that is limited only by the strength of the ceramic.

Literature Review

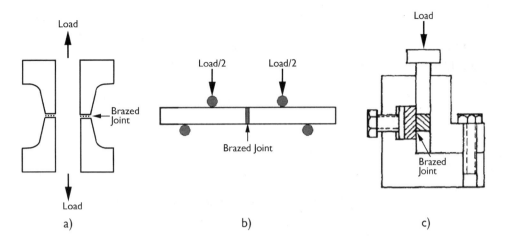

FIGURE 2.34 The most commonly used test methods for evaluating the strength of alumina-to-alumina brazed joints: (a) tensile testing, (b) four-point bend testing and (c) shear testing (Hao et al., 1994).

These comparisons between failure locations without consideration of joint strength values or to types of tests employed, by which joint strength data is derived, may result in somewhat misleading conclusions.

Failure in the ceramic may not necessarily infer a high-strength brazed joint for the following reasons:

a. Defects in the ceramic surface may limit joint strength; hence, failure is more likely to occur away from the joint interface when the ceramic surface is defective or structurally weak. If the appropriate surface preparation techniques have not been employed, then failure may inevitably occur in the parent ceramic.
b. If failure location is to be regarded as a measure of joint performance, then for a joint that fails in the ceramic the failure load should be similar to that of its monolithic strength under the same test conditions. Brazed joints that fail in the ceramic at strengths significantly lower than monolithic strength of the ceramic, may not necessarily infer a strong joint – of course, some allowance is to be made for the effects of thermally induced residual stresses.
c. The test method employed may be biased towards failure in the ceramic, particularly since ceramics are significantly weaker in tension than in compression. For example, in ASTM F-19 where the brazed is placed in tension, failure is almost always observed in the ceramic. For brazed components which do not exhibit tensile loading in-service, this test method cannot be used to fairly derive joint performance largely since load application during mechanical testing is biased towards inducing failure in the ceramic.

Flexural testing, conventionally used for the mechanical testing of ceramics, limits the volume of the ceramic material that is exposed to tensile stress; this enables more meaningful strength data to be derived.

Yadav et al. (2014) compared strength data and failure locations of brazed joints made using alumina ceramics following both tensile testing (ASTM F-19) and four-point bend testing (based on ASTM C1161). Yadav et al. (2014) formed alumina-to-alumina brazed joints using 50-µm-thick Cusil ABA® braze foils. Brazing was conducted at 860°C for 3 min. SEM and EDX techniques were used to characterise a 2- to 3-µm-thick reaction layer composed of 46.2Cu-36.1Ti-15.9Al-1.8Ag wt.% at the joint interface in both sets of joints, which had identical microstructures. The average joint strength achieved following tensile testing was 62 MPa (Table 2.19) and failures consistently

TABLE 2.19
Strength of Alumina-to-Alumina Brazed Joints Evaluated Using Tensile Testing (ASTM F-19) and Four-Point Bend Testing (ASTM C1161)

Test Method	Strength (MPa)	Failure Location
Tensile (ASTM F-19)	62	Ceramic
Four-point bend (based on ASTM C1161)	110	Interface

Source: Based on ASTM C1161 and Yadav et al. (2014).

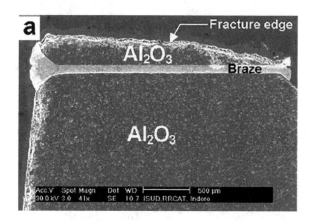

FIGURE 2.35 Failure in the parent alumina ceramic, away from the joint interface, in an alumina-to-alumina brazed joint following tensile testing according to ASTM F-19 (Yadav et al., 2014).

occurred in the ceramic, away from the joint interface (Figure 2.35). The average joint strength achieved following four-point bend testing was 110 MPa, and failures consistently occurred at the joint interface. Therefore, four-point bend testing had enabled the brazed joints to be tested to a greater extent than tensile testing.

These results show that the mechanical testing method selected affects both the failure location and the failure load of a brazed joint. Failure location alone is insufficient to assess joint performance unless the failure load is considered relative to the parent material strength under the same test or in-service conditions.

2.6 GAPS IDENTIFIED IN THE LITERATURE

A review of the some of the variables in the AMB process which affect interfacial chemistry and joint strength in the alumina/Ag-Cu-Ti system has helped to identify gaps requiring further investigation. Three variables, discussed below, have been selected to form the basis of the work discussed in the onward sections of this book.

2.6.1 Ag-Cu-Ti Braze Preform Thickness

The Ag-Cu-Ti braze volume selected or braze preform dimensions have seldom been reported in the literature concerning the formation of alumina-to-alumina and alumina-to-metal brazed joints. In most studies, the starting braze foil thickness is typically reported, and has been found to vary between 50 and 200 μm (see Figure 2.9).

Literature Review

The Ag-Cu-Ti braze preform dimensions can determine the amount of Ti which is available to diffuse to the alumina surface – dimensions of which are also significant. This leads to the formation of a reaction layer, commonly reported as a nm-thick Ti-O layer on the alumina side of the joint interface and a μm-thick Ti-Cu-O layer on the braze side of the joint interface. The continuity, composition and thickness of this reaction layer have been correlated to joint strength and can be controlled by the brazing temperature and brazing time. As the Ti concentration in the Ag-Cu-Ti braze alloy increases, reaction kinetics at the joint interface and the wetting of alumina improve. Therefore, the Ag-Cu-Ti braze preform dimensions may also affect the reaction layer properties and joint strength.

The complete diffusion of Ti to the joint interface leads to the formation of a ductile Ag-Cu braze interlayer, which can plastically deform to accommodate thermally induced residual stresses. Therefore, the dimensions of an Ag-Cu-Ti braze preform may alter the braze interlayer thickness and thereby the resulting joint strength.

The braze interlayer is the only ductile constituent in the alumina/Ag-Cu-Ti system. While most studies have shown how brazing parameters can influence the reaction layer thickness, the effect of Ti retained in the braze interlayer on microstructural changes and joint strength have not been widely reported.

The incomplete diffusion of Ti to the joint interface may result from relatively low Ag concentrations or relatively high Cu concentrations in the Ag-Cu-Ti braze alloy. Reduced Ti activity levels, particularly when brazing to limited alumina surfaces, may arise also from excessive braze volumes selected or insufficiently low brazing temperatures. Therefore, the braze preform dimensions, which can determine the amount of Ti that is available to diffuse to the joint interfaces, if excessive, may alter the microstructure of the braze interlayer and thereby, the resulting joint strength.

2.6.2 Secondary Phase Interaction

Polycrystalline alumina ceramics selected in joining studies using Ag-Cu-Ti braze alloys have ranged in composition, comprising between 95.0 and 99.9 wt.% Al_2O_3 alumina (see Figure 2.25). Due the superior mechanical properties of solid-state sintered alumina ceramics in comparison to liquid phase sintered alumina ceramics, the former is more commonly selected.

Several studies have shown that the secondary phase in liquid phase sintered alumina ceramics can lead to additional reactions at the alumina/Ag-Cu-Ti joint interface. This may affect interfacial chemistry, reaction layer formation and joint strength. Therefore, while solid-state sintered alumina ceramics may exhibit superior mechanical properties, liquid phase sintered alumina ceramics may provide superior joint performances.

2.6.3 Post-Grinding Heat Treatment

Post-grinding heat treatment has been used as a non-abrasive technique for recovering grinding damage in alumina ceramics prior to brazing with Ag-Cu-Ti braze alloys (Table 2.18). Several studies, in which both liquid phase and solid-state sintered alumina ceramics are selected, have employed a post-grinding heat treatment prior to brazing as a means to achieve high-strength joints by first improving the surface condition of alumina (Table 2.18).

The likely mechanism by which post-grinding heat treatment enables an improvement in the mechanical properties of alumina has been shown to require the presence of a liquid phase. Therefore, while post-grinding heat treatment has been shown to improve the monolithic strength of as-ground liquid phase sintered alumina ceramics, there is insufficient evidence to suggest solid-state sintered alumina ceramics can benefit in the same way.

The effect of post-grinding heat treatment on the monolithic strength of alumina ceramics is better understood than its effect on the strength of brazed joints. Since in liquid phase sintered alumina ceramics the heat treatment induces liquid phase formation then this may also affect interfacial chemistry, reaction layer formation and joint strength. Therefore, post-grinding heat treatment may affect the monolithic and brazed joint strengths differently.

2.7 SUMMARY OF OBJECTIVES

The objectives of this work identified from gaps in the literature are as follows:

1. To study the sub-surface conditions and microstructures of a liquid phase sintered and a solid-state sintered grades of alumina, prior to brazing.
2. To test and compare the monolithic strengths of both of these grades of alumina with the strengths of brazed joints, using four-point bend testing.
3. To investigate the effect of braze preform thickness on the microstructure and mechanical properties of alumina-to-alumina brazed joints.
4. To characterise the reaction layer and braze interlayer of all joints using SEM and TEM techniques, and to investigate any secondary phase interaction.
5. To investigate the effects of post-grinding heat treatment on the monolithic strengths of both grades of alumina and on the strengths of brazed joints made using different braze preform thicknesses.
6. To evaluate the mechanical properties of all phases in the microstructures of the brazed joints using nanoindentation and to include in this, microstructural changes which may occur due to retained Ti in the braze interlayer as a result of an increase in the braze preform thickness.

The experimental considerations required to achieve the above objectives are as follows:

1. To select a liquid phase sintered and a solid-state sintered alumina ceramics and to prepare monolithic alumina specimens and halved specimens (for joining purposes) according to standardised procedures (ASTM C1161-13).
2. To conduct surface roughness measurements to ensure that both grades of alumina have been machined and ground in the same way such that they only differ in their respective wt.% Al_2O_3 alumina concentrations and processing routes i.e. sintering.
3. To design a brazing fixture, in which at least five joints can be produced per brazing cycle, and to avoid any machining prior to mechanical testing.
4. To select, in a range of thicknesses, an Ag-Cu-Ti braze foil with sufficient Ti concentration, and to adopt a fixed recommended brazing procedure.
5. To capture, using photographs, the quality of each set of brazed joints, and the failure locations in each joint after mechanical testing.

3 Experimental Methods

3.1 ALUMINA MATERIALS SELECTION AND DESIGN

Two commercially available grades of polycrystalline alumina, Dynallox 96 (D-96, 96.0 wt.% Al_2O_3 alumina) and Dynallox 100 (D-100, 99.7 wt.% Al_2O_3 alumina), manufactured by CoorsTek Ltd, Crewe, UK, were used to produce test bars of two different geometries. Standard test bars of D-96 and D-100 alumina in as-ground condition, hereby designated as 'D-96 AG' and 'D-100 AG', respectively, were manufactured to the dimensions 90 mm × 8 mm × 6 mm, according to configuration c of ASTM-C1161-13 – 'Standard Test Method for Flexural Strength of Advanced Ceramics at Ambient Temperature' (Figure 3.1). Short test bars of both grades of alumina with lengths half that of the standard test bars were manufactured to the dimensions 45 mm × 8 mm × 6 mm (Figure 3.2). These short test bars were prepared for brazing trials. Once brazed, butt-joints formed using the short test bars closely matched the dimensions of the standard test bars (see Sections 3.2 and 3.5). In this way, the strength of brazed joints could be compared with that of the standard test bars. All test bars were ground and chamfered in the same way, according to ASTM C1161-13 (ASTM Standards, 2013).

According to the materials data sheets supplied by CoorsTek Ltd, the average flexural strengths of D-96 and D-100 alumina ceramics, measured at 20°C, are 360 MPa and 350 MPa, respectively. Despite differences in the purity of these alumina ceramics, both materials provide similar flexural

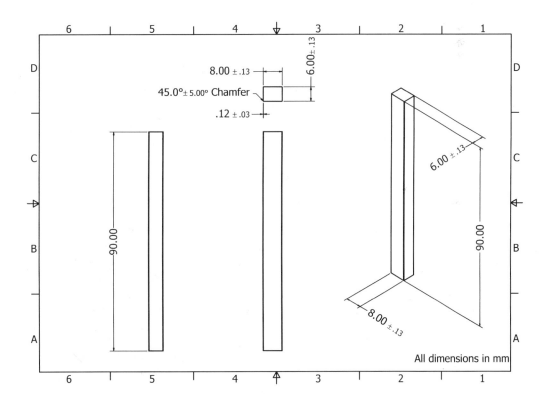

FIGURE 3.1 Dimensions of D-96 and D-100 standard alumina test bars, manufactured according to configuration c of ASTM C1161-13 (all dimensions in mm).

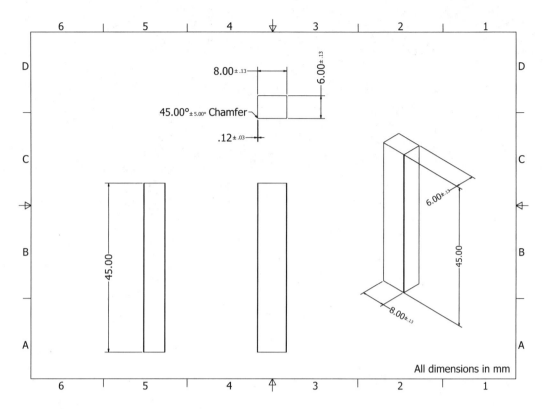

FIGURE 3.2 Dimensions of D-96 and D-100 short alumina test bars, manufactured in the same way as the standard alumina test bars, except with lengths halved (all dimensions in mm).

strengths. Table 3.1 shows a comparison between the properties of the D-96 and D-100 alumina ceramics. The main differences between the liquid phase sintered D-96 and the solid-state sintered D-100 alumina ceramics include that the former is less dense and exhibits a relatively lower CTE while the latter is slightly harder and comprises significantly higher elastic modulus and fracture toughness.

3.2 SURFACE ROUGHNESS MEASUREMENTS

Arithmetic mean surface roughness (Ra) measurements were made using a Zeiss Surfcom 130A stylus contact profilometer (Figure 3.3). Three scans in both longitudanal (*L*) and transverse (*T*) directions were made at the mid-points of the outer surfaces of each standard test bar (Rst_{outer}) and short test bar (Rsh_{outer}), as shown in Figure 3.4. The surfaces on which the measurements were made were maintained as the tensile surfaces during four-point bend testing (see Section 3.6). In this way, any variation in surface roughness could be correlated to flexural strength.

Surface roughness measurements were also performed at the faying surfaces of short test bars (Rsh_{faying}), with three scans in both longitudanal (*L*) and transverse (*T*) directions (Figure 3.4). This was intended to enable differences in surface roughness of the test bars to be correlated to wetting and spreading of the Ag-Cu-Ti braze alloy as well as reaction layer formation. Since all of the test bars were ground in the same way, variation in surface roughness between the test bars was expected to be minimal except as a result of inherent differences between the two grades of alumina.

A roughness sampling length of 0.8 mm and an evaluation length of 4.0 mm were used when performing surface roughness measurements in accordance with the ISO 4288 standard for Ra values between 0.1 and 2.0 μm.

Experimental Methods

TABLE 3.1
Properties of Alumina Ceramics, Manufactured by Coorstek Ltd, Used in This Study

	Dynallox 96 (D-96)	Dynallox 100 (D-100)
Physical properties		
Purity (nominal comp.) (wt.% Al_2O_3)	96.0	99.7
Density (g cm^{-3})	3.7	3.9
Mechanical properties		
Flexural strength @ 20°C (MPa)	360	350
Flexural strength @ 800°C (MPa)	250	250
Compressive strength (MPa)	2,100	2,500
Elastic modulus (GPa)	275	350
Hardness (R45N)	81	84
Hardness ($Hv_{0.3}$)	1,590	1,700
Fracture toughness (K_{IC})	3.5	4.5
Thermal properties		
Max use temperature (°C)	1,700	1,725
Coeff. of thermal expansion (CTE) ($\times 10^{-6}$·°C^{-1})	7.8	8.5
Thermal conductivity (W·m^{-1}·K^{-1})	25	28
Thermal shock resistance (°C)	200	220
Electrical properties		
Resistivity @ 25°C (Ω·cm)	$> 10^{14}$	$> 10^{14}$
Resistivity @ 500°C (Ω·cm)	4.0×10^9	$> 10^{12}$
Resistivity @ 1,000°C (Ω·cm)	1.0×10^6	$> 10^6$

Source: CoorsTek Ltd.

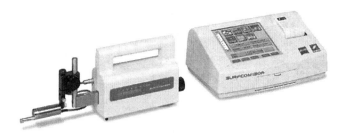

FIGURE 3.3 Zeiss surfcom 130A stylus contact profilometer (Courtesy of Carl Zeiss Ltd).

An Alicona InfiniteFocusSL microscope was also used to measure surface roughness as well as surface condition of the alumina test bars in both as-ground (AG) and in ground-and-heat-treated (GHT) conditions. This included scanning the test bar surfaces of both grades of alumina to visualise any changes in the depths of damaged regions, from grinding or otherwise, following post-grinding heat treatment.

3.3 POST-GRINDING HEAT TREATMENT

A selection of both standard and short test bars were heat treated at 1,550°C for 1 h in a Carbolite HTC 18/8 air furnace (Figure 3.5a). This produced ground-and-heat-treated (GHT) D-96 and

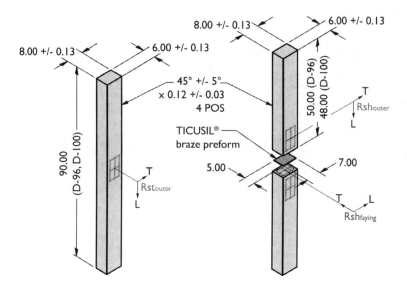

FIGURE 3.4 Test bar geometries and surface roughness measurement positions in (a) a standard alumina test bar and (b) two short test bars in a butt-joint assembly with a TICUSIL® braze preform sandwiched between the faying alumina surfaces (all dimensions in mm).

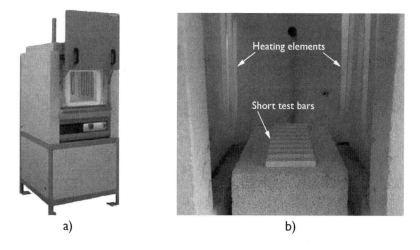

FIGURE 3.5 (a) carbolite HTC 18/8 air furnace in which heat treatments were performed (Courtesy of Carbolite Gero Ltd) and (b) photograph showing the arrangement of short test bars on a brick during a heat treatment.

D-100 alumina test bars, hereby designated as 'D-96 GHT' and 'D-100 GHT', respectively. Test bar surfaces that had been used for surface roughness measurements were maintained as free surfaces during the heat treatment following which surface roughness measurements were repeated. These measurements were performed at the same locations, R_{outer} and R_{faying}, as those measurements performed on the as-ground (AG) standard and short alumina test bars (Figure 3.5b). In this way, the effect of post-grinding heat treatment on surface roughness could be considered when evaluating any changes in strength. The specification of the air furnace used in conducting the post-grinding heat treatment is shown in Table 3.2.

TABLE 3.2
Specifications of the Carbolite HTC 18/8 Furnace, Manufactured by Carbolite Gero Ltd, Used in This Study

	Carbolite HTC 18/8 Furnace
Insulation Type	Refractory Brick
Max Temp °C	1,800
Capacity (litres)	7.7
Chamber dimensions (height × width × depth) (mm)	190 × 150 × 270
External dimensions (height × width × depth) (mm)	1,580 × 690 × 800
Maximum power (kW)	8
Heat up time to maximum temperature less 100°C (min)	150

Source: Carbolite Gero Ltd.

3.4 Ag-Cu-Ti BRAZE ALLOY SELECTION AND DESIGN

The commercially available braze alloy TICUSIL®, manufactured by Wesgo Ceramics GmbH, Erlangen, Germany, was obtained as braze foils with nominal thicknesses of 50, 100, 150 and 250 μm. With a relatively high Ti concentration of 4.5 wt.% Ti, TICUSIL® exhibits good wetting behaviour on alumina. TICUSIL® was selected on the basis that it provides a sufficiently high amount of Ti to enable the effects of variables in the AMB process such as brazing time, brazing temperature and braze preform thickness to be studied. This includes the effects of these variables on the properties of the reaction layer and the braze interlayer in the alumina/Ag-Cu-Ti system. According to the materials data sheet supplied by Morgan Advanced Materials plc, the recommended brazing procedure when using TICUSIL® comprises a brazing temperature of 860°C and a brazing time of 10 min, with heating and cooling rates of 10°C/min and 5°C/min, respectively.

The recommended brazing temperature of 860°C is below the liquidus temperature of TICUSIL®, which is 900°C. Furthermore, the recommended brazing temperature is provided independent of the braze foil thickness, supplied in the range of 50 to 250 μm. The properties of TICUSIL® provided in the materials data sheet are shown in Table 3.3.

3.5 BRAZING PROCEDURE

Alumina-to-alumina brazed joints, whereby short test bars were brazed to themselves, were prepared by first arranging short test bars into butt-joint assemblies (Figure 3.6a). Several vertically aligned joint assemblies were supported in a bespoke stainless steel fixture, which was designed to allow uniform heating of each joint interface during brazing (Figure 3.6b).

For each brazed joint assembly, a braze preform was placed between the faying surfaces of two short test bars and five joints were produced in each brazing cycle. The braze preforms had dimensions of 7 mm × 5 mm and were mechanically punched from 50-, 100-, 150- and 250-μm-thick braze foils.

No additional load was applied other than the self-weight of the upper short test bar in each joint configuration. Prior to each brazing cycle, the short test bars, braze preforms and the brazing fixture were all ultrasonically cleaned in acetone for 15 min.

The brazing fixture was spray coated using the boron nitride-based LOCTITE SF 7900 also known as AERODAG CERAMSHIELD. This was used as a stop-off to protect the brazing fixture and to prevent any undesirable effects such as the alumina test bars reacting with, or becoming clamped in, the stainless steel brazing fixture during the brazing procedure (Figure 3.7).

TABLE 3.3
Properties of TICUSIL® Braze Alloy, Manufactured by Wesgo Ceramics GmbH, Used in This Study

	TICUSIL®
Physical properties	
Composition (wt.%)	68.8Ag-26.7Cu-4.5Ti
Liquidus temperature (°C)	900
Solidus temperature (°C)	780
Density (g cm^{-3})	9.4
Mechanical properties	
Elastic modulus (GPa)	85
Poisson's ratio	0.36
Yield strength (0.2% offset) (MPa)	292
Tensile strength (MPa)	339
Elongation	28%
Thermal properties	
Thermal conductivity (W·m^{-1}·K^{-1})	219
Coeff. of thermal expansion (CTE) ($\times 10^{-6}$·°C^{-1})	18.5
Electrical properties	
Resistivity @ 25°C (Ω·m)	34×10^{-9}
Conductivity (Ω^{-1}·m^{-1})	29×10^{6}

Source: Morgan Advanced Materials plc.

Brazing was performed in a vacuum furnace manufactured by Vacuum Generation Ltd at a pressure of 1×10^{-5} mbar (Figure 3.8a). The vacuum furnace specifications are listed in Table 3.4. Two thermocouples placed in the brazing fixture were used to control the furnace temperature (Figure 3.8b). Alumina crucibles containing Ti granules, known as getters, were used to reduce the oxygen level in the vacuum chamber during brazing.

Each brazing cycle comprised of a peak brazing temperature of 850°C and a brazing time of 10 min (Figure 3.9). This followed a 10-min isothermal soak at 750°C which was incorporated to enable the temperature inside the furnace, including the temperatures of the brazing fixture and joint assemblies, to homogenise before the peak brazing temperature was reached.

3.6 MECHANICAL TESTING

Standard test bars and brazed joints of both D-96 and D-100 alumina ceramics, in AG and in GHT conditions, were mechanically tested using four-point bend testing.

The tests were performed using an articulating four-point bend test fixture mounted in a Hounsfield universal testing machine according to ASTM C1161-13 (ASTM Standards, 2013). The loading rate was controlled at a crosshead speed of 1.0 mm/min. The test fixture had an inner span of 40 mm and an outer span of 80 mm, between 9 mm diameter silver steel loading rollers (Figure 3.10a). The loading rollers were heat treated, quenched and tempered to a Rockwell hardness value of ~60 HRC.

The test bars were positioned such that the Rst_{outer} surface of the standard test bars (Figure 3.10b) and the Rsh_{outer} surface of the brazed joints (Figure 3.10c) were the lower surfaces, exposed to tensile stresses during the bend test.

Experimental Methods

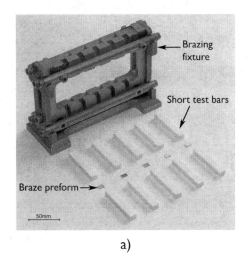

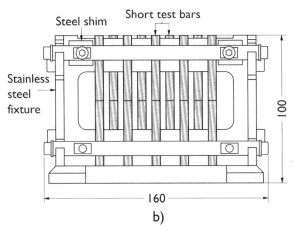

FIGURE 3.6 (a) stainless steel brazing fixture, short alumina test bars and TICUSIL® preforms and (b) schematic of stainless steel fixture showing 18 short test bars in butt-joint assemblies to form nine brazed joints (all dimensions in mm).

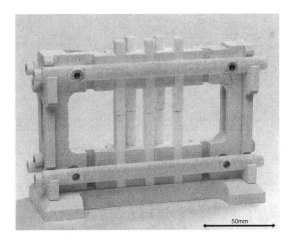

FIGURE 3.7 Brazing fixture coated in AERODAG CERAMSHIELD with 10 short test bars in butt-joint assemblies, each with a TICUSIL® braze preform sandwiched between the faying alumina surfaces.

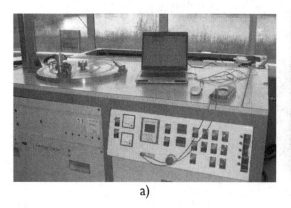
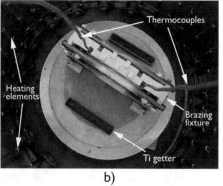

FIGURE 3.8 (a) Vacuum Generation Ltd vacuum furnace connected to a data logger to record thermocouple readings and (b) brazing fixture placed in the vacuum chamber, attached to thermocouples and positioned between Ti getters (Courtesy of TWI Ltd).

TABLE 3.4
Specifications of the Vacuum Furnace, Manufactured by Vacuum Generation Ltd, Used in This Study

	Vacuum Generation Ltd Furnace
Model	DPCVF 300
Chamber dimensions (mm)	320×330 mm
Vacuum level (mbar)	1×10^{-5}
Maximum temperature capability	1,300°C–1,350°C
Controller	Eurotherm

Source: TWI Ltd.

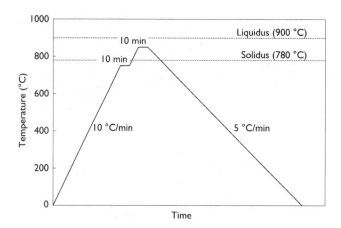

FIGURE 3.9 Heating cycle used in this study for brazing alumina ceramics using TICUSIL® braze alloy.

Experimental Methods

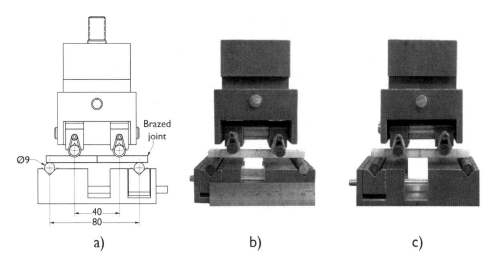

FIGURE 3.10 (a) schematic of four-point bend test fixture (all dimensions in mm), and photographs of bend tests at zero displacement performed on (b) a standard test bar and (c) a brazed joint.

During mechanical testing, load-displacement data was recorded. The maximum load at failure was used to calculate the flexural strengths of the standard test bars and the joint strengths of the brazed joints:

$$\sigma = \frac{3}{4}\frac{PL}{bd^3} \tag{3.1}$$

where σ is the stress at failure (MPa), P is the failure load (N), L is the outer span, b is the specimen width and d is the specimen thickness.

This calculation was simplified given that the outer span of the test fixture was fixed to 80 mm according to configuration c of ASTM C1161-13, and the widths and thicknesses of all test bars were 8 mm and 6 mm, respectively:

$$\sigma = 0.208\dot{3}P \tag{3.2}$$

The failure location along the length of each test bar was measured, and failed specimens were preserved for fractography analysis.

3.7 MACRO IMAGES

For each set of brazed joints, photographs of all of the sides of the joints were taken using a Cannon F-1 macro stand system. This was used to inspect the alignment of the butt-joint assemblies as well as the suitability of the brazing fixture. The joints were also inspected for any braze which had flowed out of the joints during the brazing procedure, hereby termed 'braze outflow'. Following mechanical testing, macro images of the failed joints were also taken to capture the locations and modes of the failure.

3.8 MOUNTING AND POLISHING

The design of experiments in this study comprised 15 brazing cycles (see Section 3.15). For each of the 15 sample sets, five brazed joints were produced. Four of these joints were mechanically tested - the minimum number of joints required to evaluate average joint strength according to

AWS C3.2M/C3.2:2008. The remaining joint in each sample set, retained for further analysis, was machined along the paths shown in Figure 3.11a using an ATM Brilliant 220 precision cutting machine with a wet diamond cutting disc (Figure 3.11b). Joints which failed away from the interface and in the ceramic, following mechanical testing, were also sectioned in this way to produce additional specimens for the further analysis, which included scanning electron microscopy (SEM), transmission electron microscopy (TEM) and nanoindentation techniques.

Specimens sectioned from the brazed joints were cold mounted in the commercially available acrylic resin Durocit, manufactured by Struers Ltd. Sections of the standard test bars were machined and mounted in a similar way in order to study the microstructural differences between D-96 and D-100 alumina ceramics.

The mounted specimens were subsequently ground and polished using an ATM Saphir 560 machine with Rubin 520 head (Figure 3.12). Grain pull-out was minimised by introducing the commercially available MD-Allergo and MD-Largo polishing discs during the transition from the fine grinding and diamond polishing steps. Table 3.5 shows the grinding and polishing steps which were employed in the preparation of specimens for the abovementioned analytical techniques.

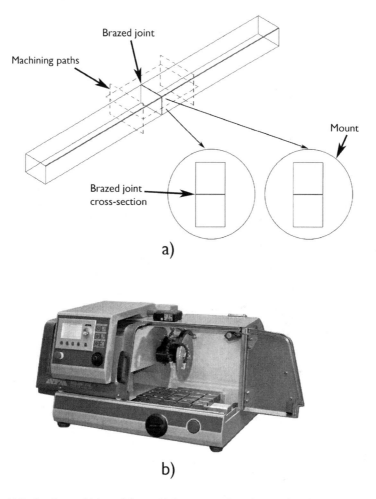

FIGURE 3.11 (a) Paths along which each brazed joint was sectioned to produce two specimens of the brazed cross-section for analysis, and (b) ATM Brilliant 220 precision cut-off machine with wet diamond cutting disc, used for machining the specimens (Courtesy of ATM GmbH).

Experimental Methods

FIGURE 3.12 ATM Saphir 560 with Rubin 520 head automatic grinding and polishing machine (Courtesy of ATM GmbH).

TABLE 3.5
Grinding and Polishing Steps Employed in the Preparation of Mounted Specimens for Analysis

Step	Surface (MD-)	Lubricant Type	Abrasive Type	Force (N)	Time (min)
PG	Piano 220	Water	–	35	1:00
FG1	Piano 600	Water	–	35	2:00
FG2	Piano 1,200	Water	–	35	2:00
FG3	Allergo	Lubricant	9 μm diamond	35	8:00
FG4	Largo	Lubricant	9 μm diamond	35	8:00
P1	Dac	Lubricant	3 μm diamond	35	20:00
P2	Nap	Lubricant	1 μm diamond	35	20:00
OPU	Chem	–	< 0.25 μm silica	35	7:00

PG - Plane grinding; FG - Fine grinding; P - Polishing; OP - Oxide (final) polishing using a standard colloidal silica suspension.

3.9 ETCHING TECHNIQUES

Mounted and polished sections of the standard D-96 and D-100 test bars were etched using two different techniques; thermal etching and chemical etching. These etching techniques were used to reveal different aspects of the alumina microstructures. Both of these etching techniques required the specimens to be broken out of their mounts.

Thermal etching, which can reduce the free energy of a polished alumina surface, was used to generate grain boundary contrast in the alumina microstructures. Thermal etching was conducted at 1,550°C for 15 min in a Carbolite HTC 18/8 air furnace (Figure 3.5a). The polished surfaces of the specimens were maintained as free surfaces during the thermal etching procedure.

Chemical etching was performed using 10 vol.% hydrofluoric acid-aqueous solution for the purpose of revealing the distribution of Si-rich secondary phase elements in the alumina microstructures (Morrell, 1987). Chemical etching was performed on polished specimens of both D-96 and D-100 alumina.

3.10 OPTICAL AND SCANNING ELECTRON MICROSCOPY

Polished sections of both standard alumina test bars and brazed joints were first studied under an Olympus BX41M optical microscope. Subsequently, specimens were gold coated using an Edwards S150B sputter coating machine for SEM analysis performed using a Zeiss ΣIGMA™ scanning electron microscope (SEM) set to an accelerating voltage of 15 kV (Figure 3.13).

SEM images were acquired using secondary electron and the backscattered electron detectors coupled with Zeiss image acquisition software. Energy dispersive X-ray (EDX) chemical analysis and EDX-SEM mapping were performed using an Oxford Instruments detector coupled with Aztec software. EDX chemical analysis included the generation of X-ray spectra at individual points and selected areas of the alumina and brazed joint microstructures. This enabled the compositions of phases to be characterised. SEM-EDX mapping was used to study the distribution of the elements: Al, O, Ag, Cu, Ti, Si, Ca and Mg in each specimen. Line scans were performed to analyse any variation in concentration of these elements across the joint interfaces. In-situ thickness measurements at regular intervals across the bond line in each specimen were used to calculate the average reaction layer and brazed interlayer thicknesses. Joint edges were examined to study the composition of any braze outflow as the braze preform thickness was increased.

These methods were performed for all 15 sample sets; brazed joints comprising TICUSIL® braze preforms of different thicknesses made using D-96 and D-100 alumina ceramics in both AG and in GHT conditions (see Section 3.15).

3.11 ELECTRON PROBE MICROANALYSIS

Electron probe microanalysis (EPMA) was performed using a Cameca SX-100 with five wavelength dispersive spectroscopy detectors. EPMA was carried using a 15-kV beam, 40-nA current and a nominal 1-μm spot size. Each element was calibrated against a metal or oxide standard, and the oxides were calculated stoichiometrically. The trace elements such as Si, Mg and Ca were counted for 120 s and Ag, Cu, Ti and Al were counted for 60 s to satisfy counting statistics over a large range of compositions expected across a traverse line scan from the parent alumina ceramic to the braze interlayer.

3.12 FOCUSSED ION BEAM MILLING

TEM specimens were prepared using an FEI Nova 600 Nanolab Dualbeam focussed ion beam (FIB)/FEG-SEM by performing an in-situ 'lift-out' method at selected regions of the specimens (Figure 3.14).

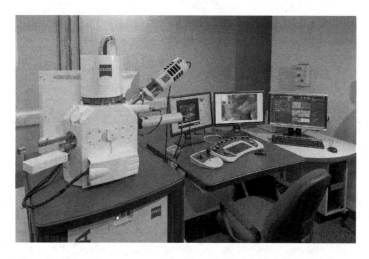

FIGURE 3.13 Zeiss ΣIGMA™ field emission scanning electron microscope used in this study (Courtesy of Carl Zeiss Ltd and TWI Ltd).

Experimental Methods

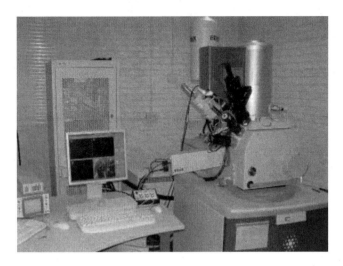

FIGURE 3.14 FEI Nova 600 Nanolab Dualbeam focussed ion beam (FIB/FEG-SEM) used in this study for the preparation of transmission electron microscopy (TEM) specimens (Courtesy of LMCC, Loughborough University).

TEM specimens were prepared from the brazed cross-sections of D-96 AG, D-96 GHT and D-100 AG brazed joints made using 50- and 100-μm-thick TICUSIL® braze preforms. TEM specimens were also prepared from areas of the joint interface where the triple pocket grain boundary regions of the alumina surface intersected with the Ti-rich reaction layers. The in-situ 'lift-out' method employed for the preparation of TEM specimens comprised four main stages.

A platinum (Pt) layer was first sputtered over a region of interest, forming a protective strip. The beam voltage, current and aperture size were 10 kV, 0.5 nA and 30 μm, respectively. Figure 3.15a shows the Pt strip deposited on a brazed specimen across the alumina ceramic, reaction layer and the braze interlayer to produce a TEM specimen of the joint interface.

The ion column current subsequently increased to 20 nA and two trenches were milled, using the ion beam, on either side of the deposited Pt strip (Figure 3.15b). At a tilt of 50.5° to 53.5° and with a reduced ion column current of 3 nA, the lamella formed was milled further to a thickness of ~1.5 μm.

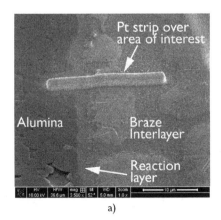
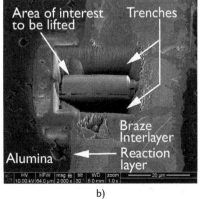

a)   b)

FIGURE 3.15 Ion beam images: (a) Stage 1 – a platinum strip deposited over region of interest, and (b) Stages 2 and 3 – two trenches were milled on either side of the area of interest and a u-shaped cut was performed to free a lamella specimen which could be transferred.

In the sample transfer stage, a u-shaped cut was made at a tilt of 7° by milling around the edges of the lamella. This freed the lamella which was subsequently transferred using a tungsten (W) omniprobe. Using an ion column current of 30 pA, the omniprobe was attached to the lamella by depositing Pt at the point of contact forming a Pt weld. The omniprobe was retracted to complete the lift-out operation (Figure 3.16a).

The lamella was transferred in-situ to a molybdenum (Mo) grid on which it was mounted by depositing Pt in at least two points of contact forming Pt welds. Once attached, the omniprobe was released by milling away the Pt welds between the lamella and the omniprobe. Using an ion column current of 1 nA and at a tilt of 52°, the lamella was milled further at angles of ±1.5° to a thickness of less than 1 µm (Figure 3.16b).

In the final stage of the lift-out method, regions of interest in the lamella were milled to thicknesses of less than 600 nm, forming 'windows' using an ion column current of 0.5 nA (Figure 3.17a). In one specimen, a first window was formed to reveal a region of the reaction layer, and a second window was formed to reveal a region of the braze interlayer.

The thicknesses of the windows in each lamella were carefully reduced in further steps, first to ~200 nm and then to between 80 and 130 nm using ion column currents of 0.3 nA and 100 pA,

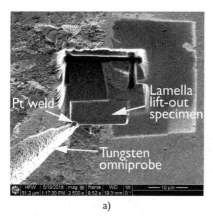
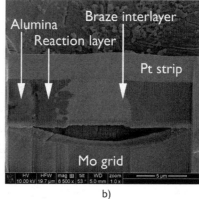

a)   b)

FIGURE 3.16 Ion beam images: (a) Stage 3 (continued) – the lamella was attached to a tungsten omniprobe which when retracted, completed the lift-out operation and (b) Stage 4 – the lamella was mounted and attached to a molybdenum grid where it was thinned further.

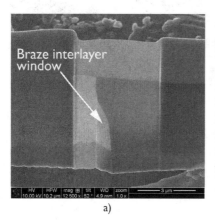
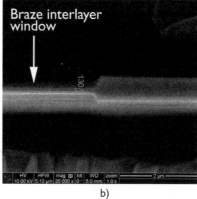

a)   b)

FIGURE 3.17 Ion beam images: (a) Stage 4 (continued) – a window in the lamellar thinned to ~130 nm - phases in the braze interlayer made electron transparent and (b) top-view of the window shown in (a).

Experimental Methods   81

respectively (Figure 3.17b). Thus, both of these windows were made electron transparent as required to perform TEM analysis techniques such as selected area diffraction (SAD) analysis.

3.13 TRANSMISSION ELECTRON MICROSCOPY

TEM was performed using a JEOL 2100 field emission gun TEM with scanning-TEM (STEM) capabilities set to an operating voltage of 200 kV (Figure 3.18).

STEM images were acquired using Jeol bright-field and/or dark-field detectors and Gatan Digiscan image acquisition software. Image processing of micrographs was carried out using Gatan Microscopy Suite. TEM electron micrograph and electron diffraction micrograph images were acquired using a Gatan Orius CCD camera. EDX analysis and STEM-EDX mapping were performed using a Thermo Scientific Noran 7 silicon drift detector EDX system coupled with NSS software.

3.14 NANOINDENTATION

Displacement-controlled nanoindentation tests were performed using a diamond Berkovich indenter mounted in a Hysitron TriboIndenter TI-950 nanomechanical testing system with scanning probe microscopy imaging capabilities (Figure 3.19a). The components of the Hysitron TriboIndenter TI-950 are shown in Figure 3.19b.

A depth-controlled calibration trial was first performed on a standard sample of fused quartz to calculate the contact area between the indenter and the test specimens – the area function. Once calibrated, depth profiles were performed in the alumina ceramics using load-controlled nanoindentation tests to evaluate any variation in nanohardness with contact depth. Targeted displacement-controlled nanoindentation tests were subsequently performed in the alumina grains in polished specimens of the brazed cross-sections. This enabled the hardness and elastic modulus properties of phases in the alumina ceramics to be evaluated.

Load/partial-unload nanoindentation tests were performed in the reaction layer and braze interlayer phases of the brazed joints in order to evaluate any variation in nanohardness with contact

FIGURE 3.18 Jeol 2100 field emission gun transmission electron microscope used in this study (Courtesy of ETC, Brunel University London).

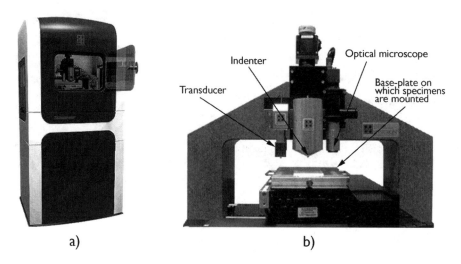

FIGURE 3.19 (a) Hysitron Triboindenter TI-950 nanoindenter used in this study and (b) components of the Hysitron Triboindenter TI-950 (Courtesy of Hysitron, Inc.).

depth. Targeted displacement-controlled nanoindentation tests were subsequently performed to evaluate any variation in the nanomechanical properties of the phases in the reaction layer and braze interlayer of D-96 AG and D-100 AG brazed joints made using 50-, 100- and 150-μm-thick TICUSIL® braze preforms.

Grids of indents were performed in selected areas of the specimens to produce contour maps showing the distribution of nanohardness in the brazed joint microstructures. These contour maps were plotted using OriginLab software. This enabled joint strength to be correlated to the nanohardness distribution in the microstructures of the brazed joints. Further details of the nanoindentation experiments are provided in Section 6.2.

3.15 DESIGN OF EXPERIMENTS

The design of experiments (DoE) in this study comprised 15 brazing cycles. For each of the 15 sample sets, five brazed joints were produced. The Taguchi method was used to optimise the DoE, such that permutations were minimised while the variables: alumina purity, post-grinding heat treatment and braze preform thickness, could be fairly investigated.

Table 3.6 shows the DoE employed for the standard alumina test bars in investigating the effects of alumina purity and surface condition on flexural strength. Table 3.7 shows the DoE employed for the brazed joints in investigating the effects of alumina purity, post-grinding heat treatment and braze preform thickness on joint strength.

TABLE 3.6
Design of Experiments for Standard Alumina Test Bars

Sample Set	Alumina Purity, (wt.% Al_2O_3)	Surface Condition	No. of Specimens
1	96.0	AG	10
2	96.0	GHT	10
3	99.7	AG	10
4	99.7	GHT	10

AG denotes specimens were in as-ground (as-received) condition.
GHT denotes specimens were in ground-and-heat-treated condition.

TABLE 3.7
Design of Experiments for Alumina-to-Alumina Brazed Joints

Sample Set	Alumina purity, (wt.% Al_2O_3)	Surface Condition	Braze Alloy Composition (wt.%)	Braze Preform Thickness (µm)	Vacuum (mbar)	Brazing Temperature (°C)	Brazing time (min)	No. of Specimens
1	96.0	AG		50				5
2	96.0	AG		100				5
3	96.0	AG		150				5
4	96.0	AG		250				5
5	96.0	GHT		50				5
6	96.0	GHT		100				5
7[a]	96.0	GHT	TICUSIL®	100	1.0×10^{-5}	850	10	5
8	96.0	GHT	68.8Ag-26.7Cu-4.5Ti	150				5
9	99.7	AG		50				5
10	99.7	AG		100				5
11	99.7	AG		150				5
12	99.7	AG		250				5
13	99.7	GHT		50				5
14	99.7	GHT		100				5
15	99.7	GHT		150				5

AG denotes specimens were in as-ground (as-received) condition.
GHT denotes specimens were in ground-and-heat-treated condition.
[a] Sample set 6 was repeated in sample set 7 in order to validate the results obtained.

4 Alumina Ceramics

4.1 CHEMICAL COMPOSITION

The average phase compositions of the alumina ceramics were characterised using EPMA. Liquid phase sintered D-96 AG alumina was composed of 96.0 wt.% Al_2O_3 alumina with 3.2 wt.% SiO_2 silica as the main secondary phase, as well as 0.55 wt.% MgO magnesia and 0.06 wt.% CaO calcia. Solid-state sintered D-100 AG alumina was composed of 99.7 wt.% Al_2O_3 alumina with 0.30 wt.% SiO_2 silica, 0.03 wt.% MgO magnesia and 0.02 wt.% CaO calcia (Table 4.1).

4.2 SURFACE ROUGHNESS

The average Ra values of two sets of 20 standard as-ground (AG) alumina test bars at Rst_{outer} in the L-direction were found to be 0.61 μm for D-96 AG alumina and 0.52 μm for D-100 AG alumina using contact profilometry. In the T-direction, the average Ra values were found to be 0.67 and 0.63 μm for the D-96 AG and D-100 AG alumina ceramics, respectively. These findings showed that the average Ra values were higher in the T-direction by 10% and 20% in both D-96 AG and D-100 AG alumina ceramics, respectively, than those in the L-direction. According to ASTM C1161-13, this was due to the outer surfaces of standard test bars having been ground in a direction parallel to their L-axis (Figure 4.1).

Although the same standard grinding procedure had been applied to the standard test bars of both grades of alumina, the average Ra values were found to be higher for D-96 AG alumina than for D-100 AG alumina in both the L- and T-directions, by 17% and 6%, respectively. The rougher surface of D-96 AG alumina may have resulted due to the presence of a secondary phase at the alumina grain boundaries (Table 4.2).

The striation marks resulting from the grinding procedure were clearly observed when analysing the as-ground surfaces of the alumina specimens under the Alicona InfiniteFocusSL microscope (Figure 4.2). The Ra values measured using this non-contact method were slightly lower than those measured using profilometry. However, the striation marks still followed the same grinding direction, in the L-direction of the alumina test bars. The Ra values were again found to be higher for both the D-96 AG and D-100 AG alumina ceramics in the T-direction as compared with those in the L-direction.

The surfaces of the D-96 AG and D-100 AG alumina ceramics were examined using scanning electron microscopy (SEM) (Figure 4.3). The secondary electron images showed the striation marks, as a result of grinding in the L-direction, even more clearly. These striation marks appeared to be more prominent in D-96 AG alumina than in D-100 AG alumina. This was consistent with the relatively higher Ra values that were measured for D-96 AG alumina as compared with those for D-100 AG alumina.

TABLE 4.1
Average Phase Compositions (wt.%) of D-96 AG and D-100 AG of Alumina Ceramics

Alumina	Al_2O_3	SiO_2	MgO	CaO
D-96 AG	96.24 ± 0.81	3.15 ± 0.68	0.55 ± 0.12	0.06 ± 0.01
D-100 AG	99.65 ± 0.08	0.30 ± 0.08	0.03 ± 0.00	0.02 ± 0.00

Analysis performed using EPMA and average values based on ten measurements.

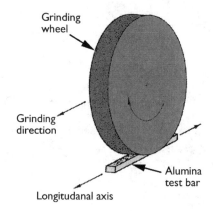

FIGURE 4.1 Alumina test bars that were ground parallel to their longitudanal axes according to ASTM C1161-13 (ASTM Standards, 2013).

TABLE 4.2
Properties of D-96 AG and D-100 AG Alumina Ceramics Used in This Study

Alumina	Density, ρ, (g/cm^3)	Average Grain Size (μm)	Average Surface Roughness Values Using Profilometry, Ra, (μm) (L)	(T)
D-96 AG	3.7	1.5–6.5	0.57 ± 0.0	0.71 ± 0.0
D-100 AG	3.9	2.0–9.0	0.48 ± 0.0	0.63 ± 0.0

The surfaces of the D-96 AG alumina test bars may have suffered greater damage during grinding since they were found to be rougher than the D-100 AG alumina test bars, given that both grades of alumina had been ground in exactly the same way.

The depths of surface asperities in the D-96 AG and D-100 AG alumina test bars were measured using an Alicona InfiniteFocusSL microscope. Figures 4.2c and 4.2d show the positions of line profiles performed over typical asperities on the D-96 AG and D-100 AG alumina surfaces, respectively. The average depths of these surface asperities were found to be 30 μm in the D-96 AG alumina test bars (Figure 4.4a) and 20 μm in D-100 AG alumina test bars (Figure 4.4b).

The average Ra values of two sets of 80 short AG alumina test bars at Rsh_{outer} in the L-direction were found to be 0.62 μm for D-96 AG alumina and 0.56 μm for D-100 AG alumina using contact profilometry. In the T-direction, the average Ra values were found to be 0.69 and 0.72 μm for the D-96 AG and D-100 AG alumina ceramics, respectively. Similarly, the average Ra values of two sets of 80 short AG alumina test bars at Rsh_{faying} in the L-direction were found to be 0.57 μm for D-96 AG alumina and 0.48 μm for D-100 AG alumina. In the T-direction, the average Ra values were found to be 0.71 and 0.63 μm for the D-96 AG and D-100 AG alumina ceramics, respectively (Table 4.3).

These results showed that the average Ra values were generally higher in the T-direction than in the L-direction for both the D-96 AG and D-100 AG short alumina test bars. The surface roughness of the D-96 AG short alumina test bars was consistently found to be higher than that of the D-100 AG alumina test bars. This was consistent with earlier surface roughness measurements performed on the standard test bars. These results also showed that the grinding direction

Alumina Ceramics

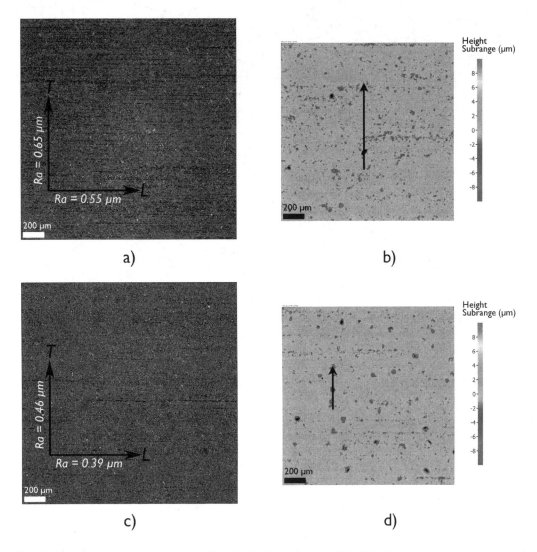

FIGURE 4.2 Surface roughness maps of (a)-(b) D-96 AG and (c)-(d) D-100 AG alumina ceramics measured using an Alicona InfiniteFocusSL microscope. Arrows indicate line profiles used to measure the depths of typical surface asperities.

adopted for the faying surfaces, although not specified in the ASTM C1161-13 standard, was the L-direction.

The variation in surface roughness observed with respect to the grinding direction shows that surface roughness measurements should be made in both the longitudanal and transverse directions, while this has not been commonly practiced in literature studies. This could provide a better means of characterising the wetting behaviour of a braze alloy as it spreads across the alumina surface, with respect to the roughness direction.

These average surface roughness values were found to be statistically significant based on the standard deviation and errors that were calculated. At all measurement positions on both the standard and short alumina test bars, the surface roughness of D-96 AG alumina was found to be relatively higher than that of D-100 AG alumina. These preliminary results indicated that surface roughness, induced by a standard grinding procedure, could be influenced by the composition and purity of an alumina ceramic (Figure 4.5).

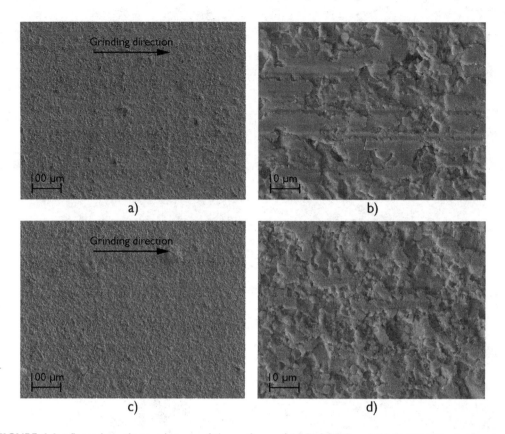

FIGURE 4.3 Secondary electron images of the surfaces of (a) D-96 AG and (b) D-100 AG, alumina test bars and higher magnification secondary electron images of the surfaces of (c) D-96 AG and (d) D-100 AG alumina test bars.

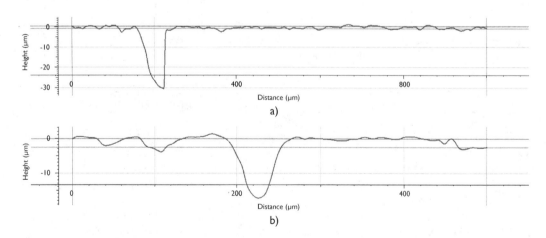

FIGURE 4.4 Depths of typical surface asperities in (a) D-96 AG corresponding to line profile shown in Figure 4.2b and (b) D-100 AG corresponding to line profile shown in Figure 4.2d measured using the Alicona InfiniteFocusSL microscope.

TABLE 4.3
Average Ra Values (μm), along Longitudanal (*L*) and Transverse (*T*) Directions at Rst$_{outer}$ for Sets of 10 D-96 AG and D-100 AG Standard Alumina Test Bars, and at Rsh$_{outer}$ and Rsh$_{faying}$ for Sets of 80 D-96 AG and D-100 AG Short Alumina Test Bars

	D-96				D-100			
	L	Error	*T*	Error	*L*	Error	*T*	Error
Standard test bars								
Rst$_{outer}$ (AG)	0.61	0.01	0.67	0.01	0.52	0.01	0.63	0.01
Short test bars								
Rsh$_{outer}$ (AG)	0.62	0.01	0.69	0.01	0.56	0.01	0.67	0.01
Rsh$_{faying}$ (AG)	0.57	0.01	0.71	0.01	0.48	0.01	0.63	0.01

4.3 MICROSTRUCTURE

The microstructures of polished sections of both D-96 and D-100 alumina ceramics were not easily observed, lacking contrast, when examined using optical microscopy and SEM in the secondary electron detector mode (Figure 4.6). Etching was required to better reveal the alumina microstructures when using these techniques.

4.3.1 D-96 ALUMINA

Thermal etching of D-96 AG alumina revealed a bi-modal grain size distribution. Small and rounded grains had an average grain size of 1.5 μm, while the larger slightly elongated grains had an average grain size of 6.5 μm (Figures 4.7 and 4.8).

Thermal etching of the polished alumina specimens at 1550°C for 15 min provided grain boundary contrast which was found to be suitable for grain size analysis using both optical microscopy and SEM techniques. Thermal etching may have altered the surface chemistry of D-96 AG alumina, however, since the distribution of the secondary phase was only weakly correlated to the grain boundary regions. This was observed in the SEM-EDX maps performed in selected areas of a thermally etched D-96 AG alumina specimen (Figure 4.9). The secondary phase may have evaporated from the surface, therefore, during the thermal etching procedure which was performed for 15 min.

The distribution of the Si-rich secondary phase in D-96 AG alumina was more clearly observed following chemical etching. Chemical etching using 10 vol.% hydrofluoric acid-aqueous solution revealed that the secondary phase in D-96 AG alumina had acicular needle-like morphology, and was distributed at the intergranular and triple pocket grain boundary regions of the D-96 AG alumina microstructure (Figures 4.10 and 4.11).

Although chemical etching did not provide grain boundary contrast suitable for grain size analysis, it did enable the distribution of the secondary phase to be identified using SEM-EDX mapping. Therefore, the Si-rich secondary phase may have been partially dissolved by the hydrofluoric acid-aqueous solution during the chemical etching procedure, which was performed for 10 s.

SEM-EDX maps performed in selected areas of a chemically etched D-96 AG alumina specimen showed high counts of Si and Mg at the grain boundaries (Figure 4.12). While this result showed that the secondary phase consisted of both Si and Mg, relatively lower counts of Ca were observed due to its extremely low concentration in D-96 AG alumina (Table 4.1).

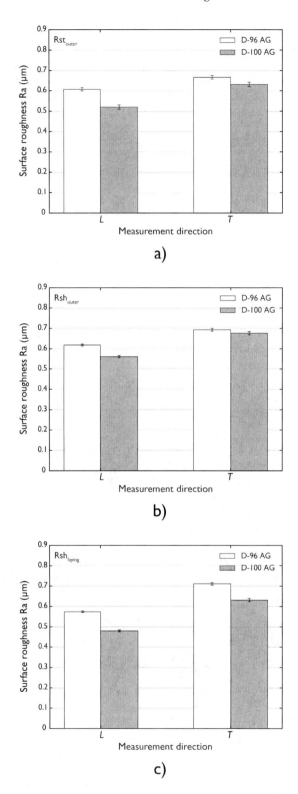

FIGURE 4.5 Average Ra (μm) values along longitudanal (*L*) and transverse (*T*) directions at measurement positions (a) Rst$_{outer}$, (b) Rsh$_{outer}$ and (c) Rsh$_{faying}$, in D-96 AG and D-100 AG standard and short alumina test bars. Measurement positions were as defined in Figure 3.4.

Alumina Ceramics 91

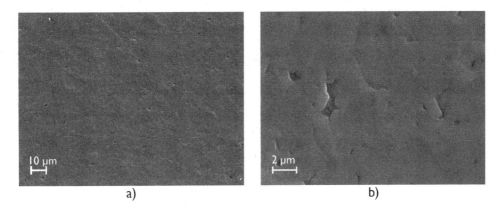

FIGURE 4.6 Secondary electron images of a polished D-100 AG at (a) lower magnification and (b) higher magnification. These images show a relatively smooth alumina surface with poor grain boundary contrast. Some grain pull-out and porosity is observable.

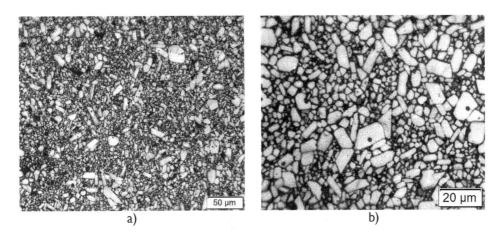

FIGURE 4.7 Optical microscope images of the microstructure of thermally etched D-96 AG alumina at (a) tlower magnification and (b) higher magnification.

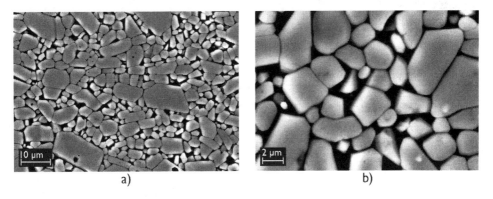

FIGURE 4.8 Backscattered electron images of the microstructure of thermally etched D-96 AG alumina at (a) lower magnification and (b) higher magnification.

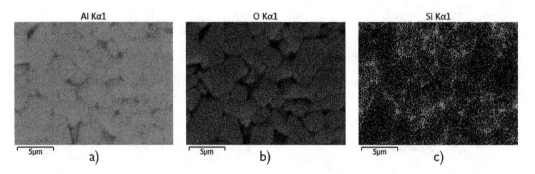

FIGURE 4.9 SEM-EDX maps corresponding to Figure 4.8b, showing distributions of (a) Al, (b) O and (c) Si, in D-96 AG alumina which was thermally etched at 1550°C for 15 min.

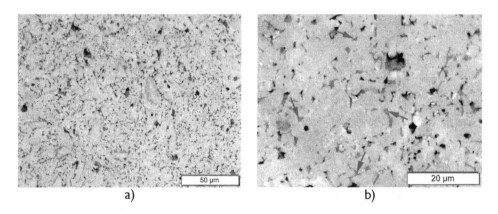

FIGURE 4.10 Optical microscope images of the microstructure of chemically etched D-96 AG alumina at (a) lower magnification and (b) higher magnification (arrows indicate secondary phase).

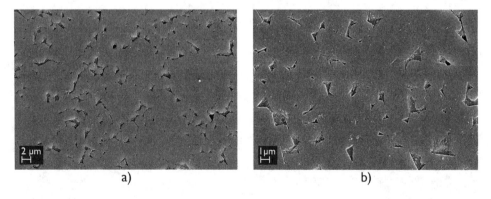

FIGURE 4.11 Secondary electron images of the microstructure of chemically etched D-96 AG alumina at (a) lower magnification and (b) higher magnification.

4.3.2 D-100 Alumina

Thermal etching of D-100 AG alumina revealed a bi-modal grain size distribution. Small and rounded grains had an average grain size of 2.0 μm, while larger slightly elongated grains had an average grain size of 9.0 μm (Figures 4.13 and 4.14). Entrapped porosity was observed both within the grains and at the grain boundaries of D-100 AG alumina, which had been solid-state sintered at ~1650°C.

Alumina Ceramics

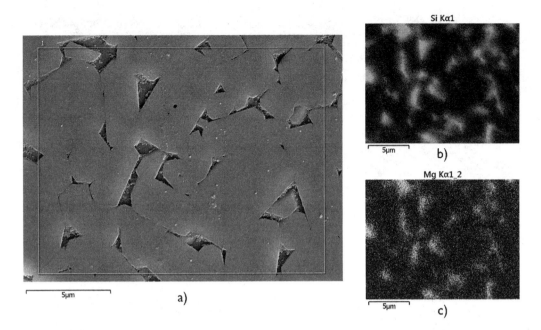

FIGURE 4.12 (a) Secondary electron image of the microstructure of chemically etched D-96 AG alumina and, corresponding SEM-EDX maps showing distributions of (b) Si and (c) Mg.

FIGURE 4.13 Optical microscope images of the microstructure of thermally etched D-100 AG alumina at (a) lower magnification and (b) higher magnification.

Chemical etching of the polished D-100 AG alumina surface using 10 vol.% hydrofluoric acid-aqueous solution did not appear to reveal the presence of any secondary phases. This was consistent with EPMA chemical analysis in which D-100 AG alumina was found to consist of just 0.3 wt.% SiO_2 silica (Table 4.1).

Some grain boundary contrast was observed when D-100 AG alumina, once chemically etched, was analysed using SEM in backscattered detector mode. This contrast was markedly improved, however, when D-100 AG alumina was thermally etched (Figures 4.14 and 4.15).

These results showed that the microstructures of D-96 AG and D-100 AG alumina ceramics were significantly different, although both grades of alumina consisted of bi-modal grain size distributions. The average grain sizes of 1.5 and 6.5 μm in D-96 AG alumina were relatively smaller than those of 2.0 and 9.0 μm in D-100 AG alumina. The liquid-phase sintered D-96 AG alumina comprised an

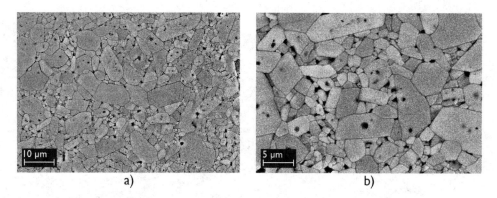

FIGURE 4.14 Backscattered electron images of the microstructure of thermally etched D-100 AG alumina at (a) lower magnification and (b) higher magnification.

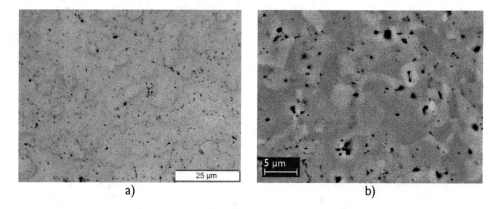

FIGURE 4.15 Images of the microstructure of chemically etched D-100 AG alumina at higher magnification acquired using (a) optical microscopy and, (b) SEM in backscattered electron detector mode.

intergranular secondary phase with an acicular needle-like morphology, which was rich in both Si and Mg. The solid-state sintered D-100 AG alumina, which did not comprise any significant quantities of secondary phase elements, exhibited relatively higher levels of intergranular and entrapped porosity.

These results also showed that thermal etching could suitably reveal grain boundary contrast in alumina ceramics, which enabled grain size measurements to be adequately performed. Thermal etching could not, however, adequately reveal the distribution of the secondary phases. Similarly, while chemical etching revealed the distribution of secondary phase in alumina, it did not yield suitable grain boundary contrast for grain size analysis. Therefore, while thermal etching is a useful technique for grain size analysis in alumina, chemical etching is equally a useful technique for evaluating the presence of Si-rich secondary phases in alumina.

4.4 FLEXURAL STRENGTH

The average flexural strengths of sets of 10 standard alumina test bars were 252 MPa (standard error = 4.8) and 250 MPa (standard error = 3.8) for D-96 AG and D-100 AG, respectively. Weibull statistics were applied to both sets of as-ground standard alumina test bars. These test bars were first ranked in ascending order based on their failure strengths. Relative failure probabilities were subsequently assigned to each standard test bar. The failure probabilities were plotted against the failure strengths achieved (Tables 4.4 and 4.5). These Weibull distribution plots showed the variability in the failure strengths achieved for both grades of alumina.

TABLE 4.4
Weibull Statistics Applied to D-96 AG Standard Alumina Test Bars

Rank	Specimen Number	Flexural Strength, σ, (MPa)	Failure Probability, $P_f = \left(\dfrac{N}{N+1}\right)$	$x = \ln \sigma$	$y = \left(\ln\left(\ln\dfrac{1}{1-P_f}\right)\right)$
1	9	229.8	0.09	5.44	−2.35
2	10	235.0	0.18	5.46	−1.61
3	6	236.5	0.27	5.47	−1.14
4	4	245.8	0.36	5.50	−0.79
5	8	247.5	0.45	5.51	−0.50
6	2	255.6	0.55	5.54	−0.24
7	3	258.3	0.64	5.55	0.01
8	5	267.3	0.73	5.59	0.26
9	1	269.4	0.82	5.60	0.53
10	7	271.5	0.91	5.60	0.87

Average Flexural Strength: σ, (MPa) = 252 MPa

Weibull parameters: $m = 16.2$ $C = 90.3$

Characteristic failure Strength: σ_0, (MPa) = $\exp\left(\dfrac{C}{m}\right)$ = 259 MPa

TABLE 4.5
Weibull Statistics Applied to D-100 AG Standard Alumina Test Bars

Rank	Specimen Number	Flexural Strength, σ, (MPa)	Failure Probability, $P_f = \left(\dfrac{N}{N+1}\right)$	$x = \ln \sigma$	$y = \left(\ln\left(\ln\dfrac{1}{1-P_f}\right)\right)$
1	3	233.5	0.09	5.45	−2.35
2	8	236.9	0.18	5.47	−1.61
3	9	236.9	0.27	5.47	−1.14
4	10	246.3	0.36	5.51	−0.79
5	4	247.3	0.45	5.51	−0.50
6	1	247.5	0.55	5.51	−0.24
7	9	260.8	0.64	5.56	0.01
8	2	254.2	0.73	5.54	0.26
9	5	265.6	0.82	5.58	0.53
10	6	267.7	0.91	5.59	0.87

Average Flexural Strength: σ, (MPa) = 250 MPa

Weibull parameters: $m = 19.4$ $C = 107.7$

Characteristic failure Strength: σ_0, (MPa) = $\exp\left(\dfrac{C}{m}\right)$ = 256 MPa

The Weibull distribution was used to determine the Weibull modulus 'm' (the gradient of the best fit line). This was equal to 16.2 for the Weibull distribution of D-96 AG alumina and 19.4 for that of D-100 AG alumina (Figures 4.16 and 4.17). These values indicated good spread and reliability in the measured failure strengths of both grades of standard alumina test bars, albeit slightly more so for D-100 AG than D-96 AG alumina ceramics. The Weibull distribution plots were used to derive the characteristic strengths, σ_0, as 259 and 256 MPa for the D-96 AG and D-100 AG standard alumina test bars, respectively.

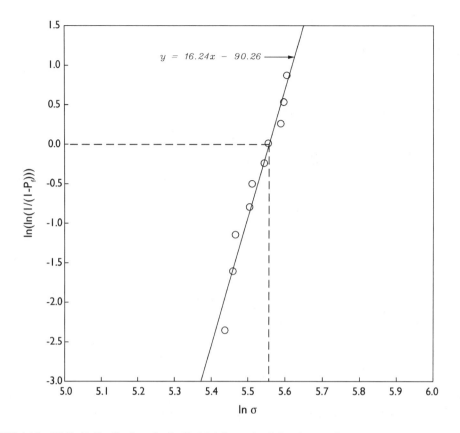

FIGURE 4.16 Weibull distribution plot for D-96 AG standard alumina test bars.

Similar fracture patterns were observed in both the D-96 AG and D-100 AG standard alumina test bars with all failures originating at the tensile surfaces between the inner span of the test fixture. The failure locations for each of the D-96 AG and D-100 AG standard alumina test bars are shown in Figures 4.18a and 4.18b, respectively.

All of the fracture surfaces were perpendicular to the tensile surfaces in both sets of standard alumina tests bars and consisted of either single or double compression curls. No secondary breaks were observed. The failure origins in D-96 AG standard alumina test bars were well within the gage length. The failure origins in D-100 AG standard alumina test bars, however, appeared to occur frequently near to but not directly beneath the loading pins. Since the failures in both sets of standard alumina test bars occurred within the gage length, these tests could be validated. It was likely, therefore, that all tests were free of any misalignment or twisting errors. Such errors may have been more likely, however, in D-100 AG than in D-96 AG standard alumina test bars, based on the failure locations observed.

4.4.1 Flexural Strength and Surface Roughness

The average Ra values of two sets of 10 standard as-ground (AG) alumina test bars at Rst_{outer} were found to be 0.61 μm for D-96 AG alumina and 0.52 μm for D-100 AG alumina in the L-direction. In the T-direction, the average Ra values were found to be 0.67 and 0.63 μm for D-96 AG and D-100 AG alumina ceramics, respectively (Table 4.3). The flexural strength of each standard alumina test bar was compared with their corresponding Ra values, measured in both the L- and T-directions.

Alumina Ceramics

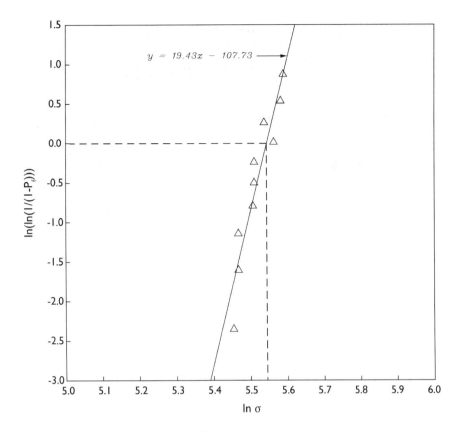

FIGURE 4.17 Weibull distribution plot for D-100 AG standard alumina test bars.

These results showed a relatively weak correlation between flexural strength and surface roughness (Figure 4.19).

Belenky and Rittel (2012) observed a 10% increase in the flexural strength of standard alumina test bars as the Ra values decreased from 0.9 to 0.02 μm (see Section 2.4.2.1). However, it was unclear as to whether this increase in flexural strength may have been influenced by the recovery of grinding damage caused by polishing, the process by which surface roughness was controlled. All standard alumina test bars, in this study, were ground in exactly the same way and the resulting strengths were similar despite some variation in surface roughness. Therefore, the standard alumina test bars, in each set, may have consisted of similar levels of grinding damage. The differences in flexural strengths, thought more likely to be correlated to surface condition than directly to surface roughness, was investigated further by preforming a post-grinding heat treatment as a means to recover the effects of grinding damage in the surfaces of both D-96 and D-100 alumina ceramics.

4.5 POST-GRINDING HEAT TREATMENT

Post-grinding heat treatment was found to significantly affect the flexural strengths of both D-96 AG and D-100 AG standard alumina test bars. However, the effect of post-grinding heat treatment on the flexural strengths of D-96 AG and D-100 AG standard alumina test bars did not occur as a result of a change in surface roughness. In the literature, grinding has been suggested to affect the flexural strength of alumina, and this has been correlated to surface roughness. These results

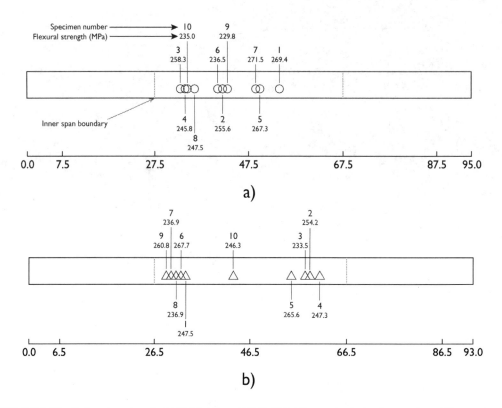

FIGURE 4.18 Failure locations in (a) D-96 AG and (b) D-100 AG, standard alumina test bars. The specimen numbers and failure strengths correspond to Tables 4.4 and 4.5, respectively.

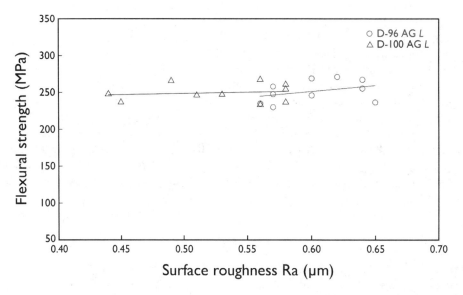

FIGURE 4.19 The effect of surface roughness, Ra (μm), measured in the L-direction of D-96 AG and D-100 AG standard alumina test bars, on flexural strength.

showed that grinding can affect the surface condition of alumina leading to changes in the flexural strength of alumina without affecting surface roughness.

The average flexural strengths of sets of 10 standard alumina test bars were 265 MPa (standard error = 5.3) and 228 MPa (standard error = 3.9) for D-96 ground-and-heat-treated (GHT) and D-100 GHT, respectively. This represented a 5.2% increase in the average flexural strength of the D-96 AG standard alumina test bars and an 8.4% decrease in the average flexural strength of the D-100 AG standard alumina test bars, as a result of the post-grinding heat treatment.

Following the post-grinding heat treatment, the average Ra values of D-96 GHT and D-100 GHT standard alumina test bars were found to be unaffected, with a net change of just 0.2% as compared with the average Ra values of D-96 AG and D-100 AG standard alumina test bars (Table 4.6 and Figure 4.20). Such a small change was within standard experimental error and as such no significant change in the surface roughness of either grade of alumina was observed. This shows that surface roughness is a property induced onto ceramic surfaces by abrasive methods such as grinding or polishing and has an indirect correlation to flexural strength. The effects of these abrasive methods along with other methods such as post-grinding heat treatment can directly affect flexural strength through their effects on the surface condition of an alumina ceramic.

The depths of surface asperities in the ground-and-heat-treated D-96 GHT and D-100 GHT alumina surfaces were measured using an Alicona InfiniteFocusSL microscope. Figures 4.21a and 4.21b show the positions of line profiles performed over typical asperities on the surfaces of D-96 GHT and D-100 GHT alumina ceramics, respectively. The average depths of these surface asperities were found to be 15 and 20 µm in D-96 GHT (Figure 4.22a) and D-100 GHT (Figure 4.22b) alumina ceramics, respectively.

The surface condition of the D-100 AG standard alumina test bars was not affected by the post-grinding heat treatment. However, some of the grinding damage on the surfaces of the D-96 AG standard alumina test bars may have been recovered. This result indicates that the effect of post-grinding heat treatment on the surface condition of alumina depends on the alumina purity, consistent with conclusions derived earlier in Section 2.4.2.3. This opposes several studies in the literature, listed in Table 2.18, in which post-grinding heat treatment has been employed to recover grinding damage, in a range of alumina ceramics, including in higher purity solid-state sintered alumina ceramics, prior to brazing.

TABLE 4.6
Average Ra Values (µm) along Longitudanal (*L*) and Transverse (*T*) Directions at in D-96 AG, D-96 GHT, D-100 AG and D-100 GHT Standard and Short Alumina Test Bars, Selected for Post-Grinding Heat Treatment, at Rst_{outer}, Rsh_{outer} and Rsh_{faying}

	D-96				D-100			
	L	Error	*T*	Error	*L*	Error	*T*	Error
Standard test bars								
Rst_{outer} (AG)	0.61	0.01	0.67	0.01	0.51	0.01	0.62	0.01
Rst_{outer} (GHT)	0.61	0.01	0.66	0.01	0.50	0.01	0.62	0.01
Short test bars								
Rsh_{outer} (AG)	0.61	0.02	0.71	0.02	0.56	0.02	0.68	0.02
Rsh_{outer} (GHT)	0.64	0.01	0.70	0.02	0.57	0.01	0.68	0.02
Rsh_{faying} (AG)	0.59	0.01	0.70	0.02	0.48	0.01	0.64	0.03
Rsh_{faying} (GHT)	0.59	0.01	0.71	0.02	0.52	0.02	0.66	0.04

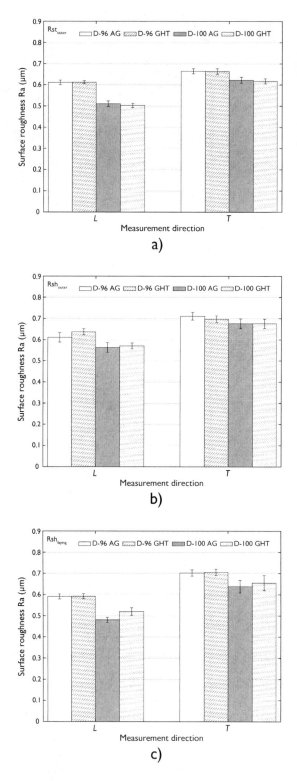

FIGURE 4.20 Average Ra (μm) values along longitudanal (*L*) and transverse (*T*) directions at measurement positions (a) Rst$_{outer}$, (b) Rsh$_{outer}$ and (c) Rsh$_{faying}$, in D-96 AG, D-96 GHT, D-100 AG and D-100 GHT standard and short alumina test bars. Measurement positions were as defined in Figure 3.4.

Alumina Ceramics

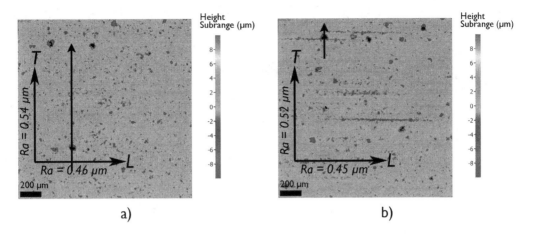

FIGURE 4.21 Roughness maps of (a) D-96 GHT and (b) D-100 GHT produced using the Alicona InfiniteFocusSL microscope. Arrows indicate line profiles, performed to measure the depths of typical surface asperities.

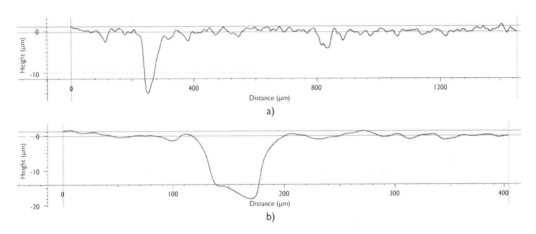

FIGURE 4.22 Depth of typical surface asperities in (a) D-96 GHT corresponding to line profile shown in Figure 4.21a, and (b) D-100 GHT corresponding to line profile shown in Figure 4.21b measured using the Alicona InfiniteFocusSL microscope.

Weibull statistics were applied to both sets of ground-and-heat-treated standard alumina test bars and for each set, failure probabilities were plotted against the failure strengths achieved (Tables 4.7 and 4.8).

The characteristic strength in the Weibull distribution of the D-96 GHT standard alumina test bars increased by ~5%, from 259 to 273 MPa as compared with that of the D-96 AG standard alumina test bars. However, the Weibull modulus in the D-96 GHT alumina distribution decreased by 6.3%, from 16.2 to 15.2 as compared with that of the D-96 AG alumina distribution (Figure 4.23).

The characteristic strength in the Weibull distribution of the D-100 GHT standard alumina test bars decreased by ~6%, from 256 to 241 MPa as compared with that of the D-100 AG standard alumina test bars. In addition, the Weibull modulus in the D-100 GHT alumina distribution decreased by ~59%, from 19.4 to 8.1 as compared with that of the D-100 AG alumina distribution (Figure 4.24).

These results showed that the post-grinding heat treatment had adversely affected the spread and reliability in the measured failure strengths of both sets of standard alumina test bars. Thus, the

TABLE 4.7
Weibull Statistics Applied to D-96 GHT Standard Alumina Test Bars

Rank	Specimen Number	Flexural Strength, σ, (MPa)	Failure Probability, $P_f = \left(\dfrac{N}{N+1}\right)$	$x = \ln \sigma$	$y = \left(\ln\left(\ln\dfrac{1}{1-P_f}\right)\right)$
1	8	241.3	0.09	5.49	−2.35
2	1	245.6	0.18	5.50	−1.61
3	5	246.7	0.27	5.51	−1.14
4	3	251.9	0.36	5.53	−0.79
5	7	267.5	0.45	5.59	−0.50
6	6	271.5	0.55	5.60	−0.24
7	4	274.4	0.64	5.61	0.01
8	2	274.6	0.73	5.62	0.26
9	10	283.8	0.82	5.65	0.53
10	9	286.5	0.91	5.66	0.87

Average Flexural Strength: σ, (MPa) = 264 MPa

Weibull parameters: $m = 15.2$ $C = 85.4$

Characteristic failure Strength: σ_0, (MPa) = $\exp\left(\dfrac{C}{m}\right)$ = 273 MPa

TABLE 4.8
Weibull Statistics Applied to D-100 GHT Standard Alumina Test Bars

Rank	Specimen Number	Flexural Strength, σ, (MPa)	Failure Probability, $P_f = \left(\dfrac{N}{N+1}\right)$	$x = \ln \sigma$	$y = \left(\ln\left(\ln\dfrac{1}{1-P_f}\right)\right)$
1	2	171.9	0.09	5.15	−2.35
2	9	213.8	0.18	5.36	−1.61
3	10	220.8	0.27	5.40	−1.14
4	4	224.6	0.36	5.41	−0.79
5	7	225.4	0.45	5.42	−0.50
6	8	235.8	0.55	5.46	−0.24
7	5	239.0	0.64	5.48	0.01
8	3	239.2	0.73	5.48	0.26
9	1	259.0	0.82	5.56	0.53
10	6	259.0	0.91	5.56	0.87

Average Flexural Strength: σ, (MPa) = 229 MPa

Weibull parameters: $m = 8.1$ $C = 44.3$

Characteristic failure Strength: σ_0, (MPa) = $\exp\left(\dfrac{C}{m}\right)$ = 241 MPa

findings obtained for D-96 GHT standard alumina test bars were consistent with those obtained by Kirchner et al. (1970), in which it was reported that the post-grinding heat treatment of 96.0 wt.% Al_2O_3 alumina rods may attribute to the healing of at least one type of flaw. This led to an increase in the scatter of data points (see Section 2.4.2.3).

The post-grinding heat treatment of D-100 GHT standard alumina test bars led to a decrease in the average flexural strength. Therefore, the post-grinding heat treatment as applied to the higher purity solid-state sintered alumina ceramics may have resulted in a new type of flaw.

Alumina Ceramics

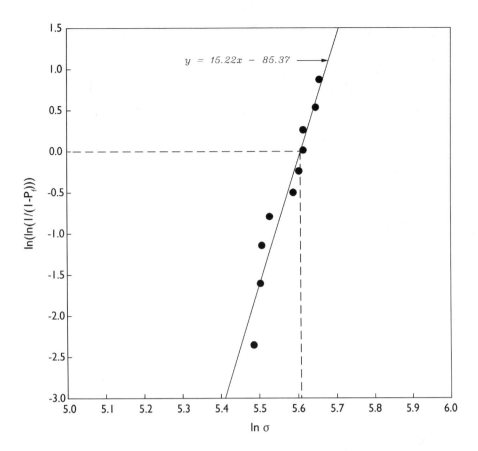

FIGURE 4.23 Weibull distribution plot of D-96 GHT standard alumina test bars.

Thermal etching revealed that the average grain sizes in the microstructures of D-96 GHT and D-100 GHT alumina ceramics were similar to those of their AG counterparts. Therefore, the 8.4% decrease in the average flexural strength of D-100 GHT standard alumina test bars may not have resulted from discontinuous grain growth. Despite the post-grinding heat treatment temperature of 1550°C, ~100°C lower than the original sintering temperature, the heat treatment time of 15 min may have been insufficient to enable solid-state sintering of the D-100 GHT standard alumina test bars. This suggests that re-sintering is not the mechanism at play during the post-grinding heat treatment. Instead, the liquid forming secondary phase in D-96 GHT alumina, absent in D-100 GHT alumina, may have been the distinguishing factor responsible for the changes in flexural strength observed.

Although all of the failures in D-96 GHT standard alumina test bars originated at the tensile surfaces, not all failures were within the inner span of the test fixture (Figure 4.25a). In these cases, failure origins were outside of the gage length away from the region of maximum stress. Therefore, the post-grinding heat treatment may have introduced a new type of defect in the surfaces of some of the D-96 GHT standard alumina test bars. It is postulated that the post-grinding heat treatment applied to D-96 GHT alumina, while predominantly healing at least one type of flaw associated to grinding damage or otherwise, simultaneously led to the introduction of at least one new type of flaw in the alumina surface. In this way, two competing processes resulted in relatively wider scatter in the location of failure origins, and limited the increase in the flexural strength observed to 5.2%. Such a mechanism would be consistent with the re-distribution of a finite liquid forming phase during the heat treatment (see Section 2.4.2.3).

The distribution of failure origins observed in D-100 GHT standard alumina test bars, following the heat treatment, was similar to that observed in the D-100 AG standard alumina test bars.

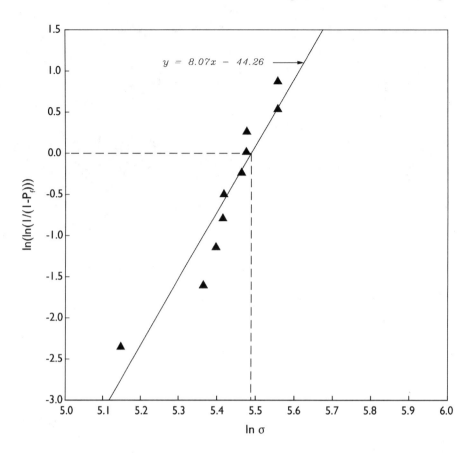

FIGURE 4.24 Weibull distribution plot of D-100 GHT standard alumina test bars.

All of the failures originated at the tensile surfaces within the inner span of the test fixture (Figure 4.25b). The 8.4% decrease in the average flexural strength of D-100 GHT standard alumina test bars did not appear to be correlated to the introduction of a new type of flaw, according to fractography analysis. Furthermore, the reduction in strength was earlier discussed not to be associated to surface roughness (see Section 4.4.1). Therefore, another mechanism may have led to this decrease in flexural strength observed.

Post-grinding heat treatment led to an improvement in the average flexural strength of D-96 AG standard alumina test bars but degraded the average flexural strength of D-100 AG standard alumina test bars (Figure 4.26). These results appeared to be statistically significant.

Post-grinding heat treatment applied to liquid phase sintered alumina ceramics may, through the formation of a liquid phase, heal grinding-induced surface defects. In addition, if the CTE of the liquid phase is higher than that of adjacent alumina grains, a clamping effect at the alumina grain boundaries may provide a strengthening effect. For solid-state sintered alumina ceramics, post-grinding heat treatment may have simply annealled any grinding-induced compressive residual stresses, un-pinning and activating surface cracks leading to a degradation in the average flexural strength (Morrell, 2015).

Fractography analysis revealed that failures were initiated at the tensile surfaces of all standard test bars. This was consistent with other studies where fractography combined with confocal scanning laser microscopy and dye penetrant inspection showed that 85% of fractures in the flexural testing of alumina originate from surface defects (Nakamura et al., 2009). Moreover, failures tend to originate from surface defects in alumina ceramics as opposed to volume defects, during flexural testing (Kirchner et al., 1970).

Alumina Ceramics

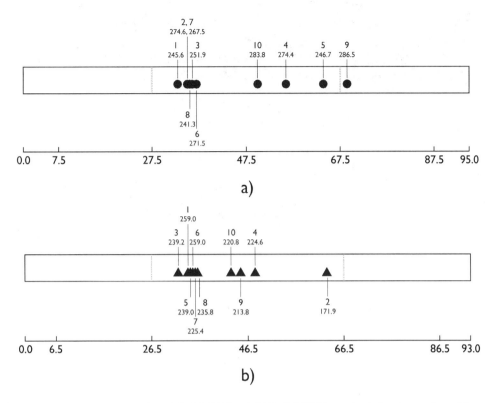

FIGURE 4.25 Failure locations in (a) D-96 GHT and (b) D-100 GHT, standard alumina test bars. The specimen numbers and failure strengths correspond to Tables 4.7 and 4.8.

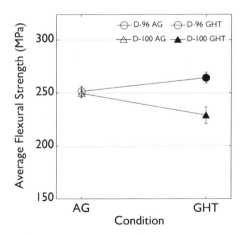

FIGURE 4.26 Average flexural strengths (MPa) of D-96 AG, D-96 GHT, D-100 AG and D-100 GHT standard alumina test bars.

Fractography analysis of the D-96 GHT standard alumina test bars showed that failures had mainly initiated from microstructural defects and voids in the alumina surface. This occurred at an increased failure load as compared with the D-96 AG, D-100 AG and D-100 GHT standard alumina test bars in which failures were not found to have initiated from microstructural defects. Figure 4.27 shows an example of the failure origin in the highest strength D-96 GHT standard alumina test bar

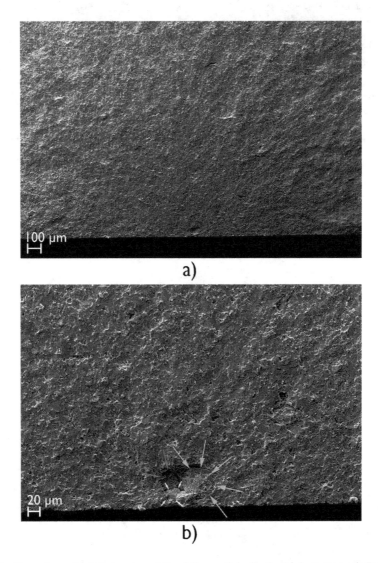

FIGURE 4.27 Secondary electron images of the failure origin in the highest strength D-96 GHT standard test bar (286 MPa) showing a clustered processing defect which comprised of a small group of subsurface agglomerates (represented by arrows) close to the tensile surface (a) lower magnification and (b) higher magnification.

(286 MPa). In this case, failure initiated from a weakly bonded sub-surface agglomerate ~20 μm beneath the tensile surface.

These results indicate that grinding-induced defects may have been responsible for the failures in both the D-96 and D-100 standard alumina test bars in AG and in GHT conditions, except for the D-96 GHT standard alumina test bars, where failures occurred at relatively higher loads but were initiated at other sites, i.e., microstructural defects or voids. These defects may have been caused by the migration of the secondary phase during the heat treatment, which may have flowed into crevices, or evaporated from the surface, creating new fissures or voids in the alumina surface.

Thermal etching had previously been observed to have altered the surface chemistry of D-96 AG alumina; it was postulated that the secondary phase may have burned off during the thermal etching procedure thereby enabling grain boundary contrast (see Section 4.3.1). The post-grinding heat treatment was performed at the same temperature as the thermal etching procedure (1550°C)

Alumina Ceramics

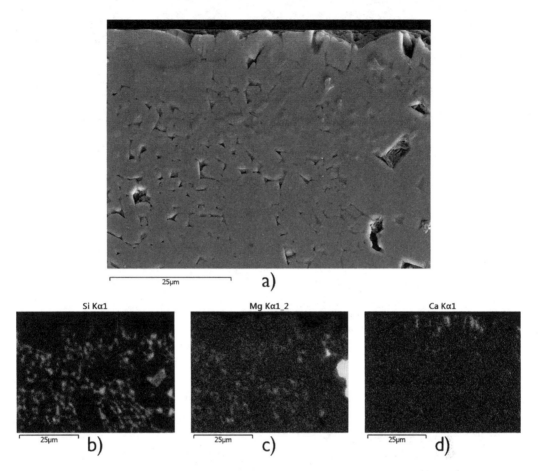

FIGURE 4.28 (a) Backscattered electron image of a chemically etched cross-section of D-96 GHT alumina, and SEM-EDX maps showing corresponding distributions of (b) Si, (c) Mg and (d) Ca.

and this may have caused a similar effect, thereby creating voids in the D-96 GHT alumina surface. The identification of grinding-induced defects in D-96 AG alumina through SEM examination, and any subsequent healing of these defects in D-96 GHT alumina as a result of the post-grinding heat treatment, was challenging due to the abrasive methods as required for ceramography.

In the SEM-EDX maps performed on the chemically etched D-96 GHT alumina specimens, Si and Mg distributions both appeared to be migrating away from the alumina surface. The relatively high counts of Ca were also unusual, appearing as if these elements were agglomerating (Figure 4.28).

4.6 CONCLUSIONS

The properties of liquid-phase sintered D-96 and solid-state sintered D-100 alumina ceramics have been studied. The effect of post-grinding heat treatment on the surface condition of D-96 AG and D-100 AG alumina ceramics was investigated. The following conclusions and observations have been derived:

1. The D-96 AG and D-100 AG standard alumina test bars were both prepared and ground according to ASTM C1161-13. D-96 AG alumina was composed of 96.0 wt.% Al_2O_3 alumina with 3.2 wt.% SiO_2 silica as the main secondary phase whereas D-100 AG alumina was composed of 99.7 wt.% Al_2O_3 alumina.

2. D-96 AG alumina consisted of an Al-Si-Mg-rich intergranular secondary phase and comprised relatively lower levels of porosity as compared with D-100 AG alumina.
3. Both grades of alumina consisted of bi-modal grain size distributions. The average grain sizes of 1.5 and 6.0 µm in D-96 AG alumina were relatively smaller than the average grain sizes of 2.0 and 9.0 µm in D-100 AG alumina.
4. The average Ra values were found to be higher for D-96 AG standard alumina test bars than for D-100 AG standard alumina test bars in both the *L*- and *T*- directions by 17% and 6%, respectively.
5. The average flexural strengths of sets of 10 standard alumina test bars were found to be 252 and 250 MPa for D-96 AG and D-100 AG standard alumina test bars, respectively.
6. The post-grinding heat treatment performed at 1550°C for 1 h resulted in a net change of just 0.2% in the average Ra values of both D-96 and D-100 standard and short alumina test bars.
7. The post-grinding heat treatment led to a 5.2% increase in the average flexural strength of the D-96 AG standard alumina test bars and an 8.4% decrease in that of the D-100 AG standard alumina test bars. The average flexural strengths of sets of 10 standard alumina test bars were found to be 265 and 228 MPa for D-96 GHT and D-100 GHT standard alumina test bars, respectively.
8. Post-grinding heat treatment was used to recover grinding damage in D-96 GHT alumina; the re-distribution of the liquid phase to areas comprising grinding-induced damage being the likely mechanism. The improvement in flexural strength was unrelated to surface roughness. However, the secondary phase elements including both Si and Mg appeared to migrate away from the alumina surface creating fissures which subsequently acted as failure initiation sites.
9. Post-grinding heat treatment may have annealed any grinding-induced compressive residual stresses in D-100 GHT alumina, un-pinning surface cracks and degrading the average flexural strength.
10. The D-96 AG and D-100 AG standard alumina test bars exhibited similar flexural strengths despite differences in both the compositions and microstructures. The flexural strength of D-96 alumina was favourably responsive to post-grinding heat treatment, more so than that of D-100 alumina, which appeared to be somewhat adversely affected.

5 Microstructural Evolution

5.1 AS-RECEIVED TICUSIL® BRAZE FOILS

The TICUSIL®, braze alloy, with nominal composition of 68.8Ag-26.7Cu-4.5Ti wt.% is manufactured by Wesgo Metals, Hayward, CA, USA (Table 3.3). In studies relating to the AMB of alumina, TICUSIL® has been one of the most commonly used Ag-Cu-Ti braze alloys – in the form of 50- to 200-μm-thick braze foils (see Section 2.2.2.1).

The cross-sectional microstructures of the as-received 50-, 100-, 150- and 250-μm-thick TICUSIL® braze foils used in this study are shown in Figures 5.1, 5.2, 5.3 and 5.4, respectively. In all of these braze foils, three phases were characterised using SEM-EDX: Ti (Figure 5.1b, A), Ag-Cu eutectic (Figure 5.1b, B) and Cu_4Ti_3 (Figure 5.1b, C). The average chemical compositions of these phases, characterised using EPMA, are shown in Table 5.1.

In several regions of the 50- and 100-μm-thick braze foil microstructures, Ti was found to be somewhat depleted and the compositions of these regions varied from that of TICUSIL®. A higher degree of uniformity in the distribution of Ti was observed in the microstructures of the 150- and 250-μm-thick braze foils. The use of Ti ribbons that are less than ~10 μm thick as required in the manufacture of the 50- and 100-μm-thick TICUSIL® braze foils may have caused the non-uniformity in the Ti distribution that was observed, particularly during roll-forming (Yu et al., 2014). TICUSIL®

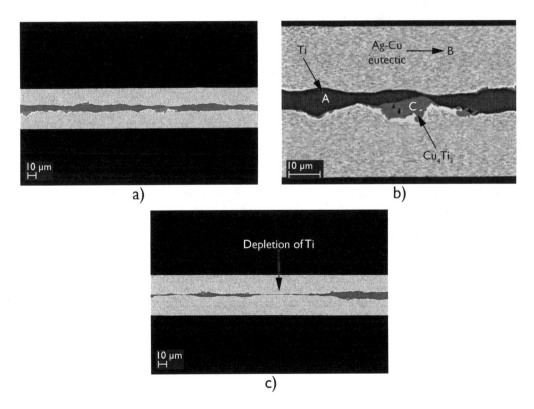

FIGURE 5.1 Backscattered electron images showing the cross-sectional microstructure of an as-received 50-μm-thick TICUSIL® braze foil at (a) typical regions, (b) higher magnification showing formation of Cu_4Ti_3 and (c) regions where Ti was observed to be depleted.

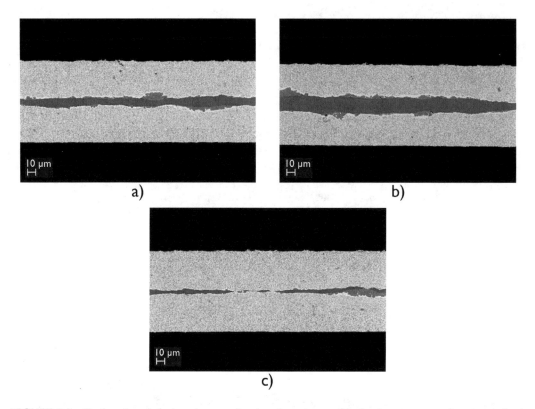

FIGURE 5.2 Backscattered electron images showing the cross-sectional microstructure of an as-received 100-µm-thick TICUSIL® braze foil at (a) typical regions, (b) regions where higher Ti concentrations were observed and (c) regions where Ti was observed to be depleted.

braze foil is manufactured by a cladding process, whereby a Ti ribbon is sandwiched between Ag-Cu eutectic layers before being rapidly solidified. The non-uniformity of the Ti structure at the centre of the 50- and 100-µm-thick braze foils is unlikely to affect the brazing of alumina since this is only its starting position during the brazing process. Furthermore, the average compositions of the braze foils are consistent with that of TICUSIL® when larger and more typical dimensions of the braze foils, as those applied during brazing are characterised.

Considering a fixed alumina surface area, an increase in the TICUSIL® braze foil thickness provides relatively more Ti per unit area of the alumina surface, hereby defined as the relative Ti concentration factor. For example, while the TICUSIL® composition may remain constant with an increase in the braze foil thickness from 50 to 100 µm, the total amount of Ti that is available to diffuse to the alumina surface increases twofold (Figure 5.5).

Zhu and Chung (1994) suggested that the relative Ti concentration factor, which increased with the thickness of braze applied, can result in excessive reactions at the alumina/Ag-Cu-Ti joint interface. The addition of carbon fibres made to the Ag-Cu-Ti braze alloy, which reacted with Ti in the braze interlayer, was used to control the extent of any excessive reactions at the joint interface.

5.1.1 Cu$_4$Ti$_3$ in TICUSIL® Braze Foil

The cross-sectional microstructures of all of the TICUSIL® braze foils comprised the Cu$_4$Ti$_3$ intermetallic phase in regions tied to the Ti ribbons (Figure 5.1b, C). The formation of this phase is consistent with the excellent chemical affinity between Ti and Cu (see Section 2.2.1); chemical

Microstructural Evolution 111

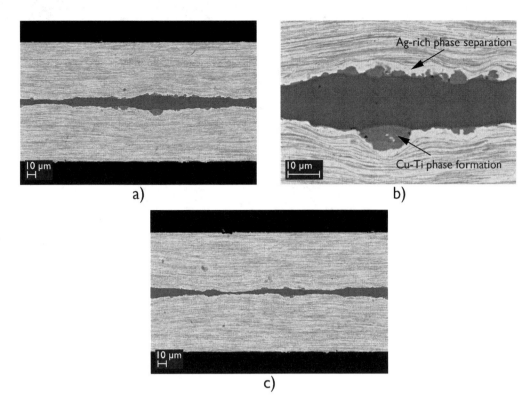

FIGURE 5.3 Backscattered electron images showing the cross-sectional microstructure of an as-received 150-μm-thick TICUSIL® braze foil at (a) typical regions, (b) higher magnification showing formation of Cu$_4$Ti$_3$ and Ag-rich phase separation and (c) regions where lower Ti concentrations were observed.

reactions between Ti and Cu are thermodynamically more favourable than those between Ti and Ag due to a more negative heat of mixing (Hirnyj and Indacochea, 2008).

In the Ag-Cu binary system, the heat of mixing is endothermic indicating that Ag and Cu atoms repel each other. Despite being practically immiscible in the solid state at room temperature, Ag and Cu form a single liquid phase due to a large entropic contribution to the free energy of mixing at higher temperatures – above the eutectic temperature of 780°C (72Ag-28Cu wt.%) (Figure 5.6).

The molten behaviour of the Ag-Cu-Ti ternary alloy can be determined from the heat of mixing in the Ag-Ti and Cu-Ti binary systems; both of these systems are compound forming.

The Ag-Ti binary system is characterised by the formation of two intermetallic compounds – AgTi and AgTi$_2$ (Figure 5.7a), and the Cu-Ti binary system is characterised by that of the six intermetallic compounds – CuTi$_2$, CuTi, Cu$_4$Ti$_3$, Cu$_3$Ti$_2$, Cu$_2$Ti and Cu$_4$Ti (Figure 5.7b).

The Ag-Cu-Ti ternary system is characterised by a miscibility gap in the liquid phase and the numerous intermetallics of the Cu-Ti binary system. There are no ternary compounds reported to form in the Ag-Cu-Ti system. The miscibility gap was first described by Eremenko et al. (1969) and later re-evaluated by Chang et al. (1977), confirmed experimentally in Paulasto et al. (1995). The liquidus boundary of the miscibility gap in the Ag-Cu-Ti ternary phase diagram is shown in Figure 5.8.

The Ag-Cu eutectic braze alloy becomes saturated with the addition of ~2 wt.% Ti (Nicholas, 1990). This (72Ag-28Cu)2Ti wt.% composition lies on the boundary of the miscibility gap. With a further increase in the Ti concentration, an inhomogeneous two-phase melt (L) is formed consisting

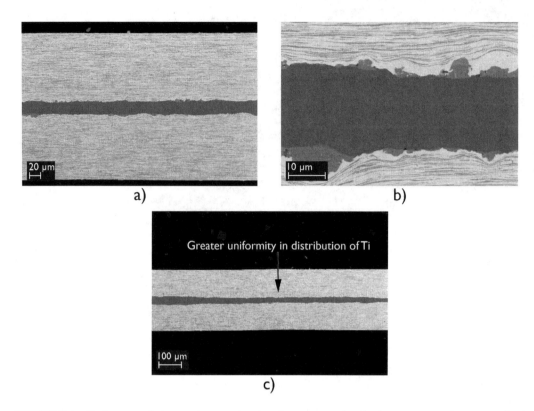

FIGURE 5.4 Backscattered electron images showing the cross-sectional microstructure of an as-received 150-μm-thick TICUSIL® braze foil at (a) typical regions, (b) higher magnification showing formation of Cu_4Ti_3 and Ag-rich phase separation and (c) lower magnification showing greater uniformity in the Ti concentration along the length of the braze foil.

TABLE 5.1
Average Chemical Compositions (wt.%) of Phases Observed in 50-, 100-, 150- and 250-μm-Thick TICUSIL® Braze Foils

Phases	Ag	Cu	Ti
A – Ti	0.2	0.3	99.5
B – Ag-Cu eutectic	71.7	28.3	–
C – Cu_4Ti_3	3.5	36.2	60.3

Labels A to C Refer to Figure 5.1b.
Analysis performed using EPMA and average values based on ten measurements.

of an Ag-rich liquid (L_1 = Ag-27Cu-2Ti wt.%) and a Ti-rich liquid (L_2 = Ag-66Cu-22Ti wt.%), where $L = L_1 + L_2$. The Ti-rich liquid, with 22 wt.% Ti, is reported to be the phase that significantly improves the wettability of alumina (Jacobson and Humpston, 2005).

The TICUSIL® composition of 68.8Ag-26.7Cu-4.5Ti wt.%, lies within the miscibility gap of the Ag-Cu-Ti system. At 900°C, the resulting inhomogeneous two-phase melt (*L*) consists of an Ag-rich phase (L_1 = Ag-23Cu-2Ti wt.%) and a Ti-rich phase (L_2 = Ag-63Cu-27Ti wt.%), where $L = L_1 + L_2$. The lever rule can be used to calculate the relative proportions of each liquid; the liquid (*L*) is segregated into 91% L_1 and 9% L_2.

Microstructural Evolution

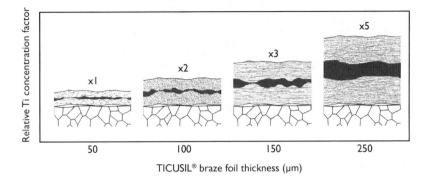

FIGURE 5.5 The effect of TICUSIL® braze foil thickness on the relative Ti concentration factor; the amount of Ti available to diffuse to a fixed alumina surface area.

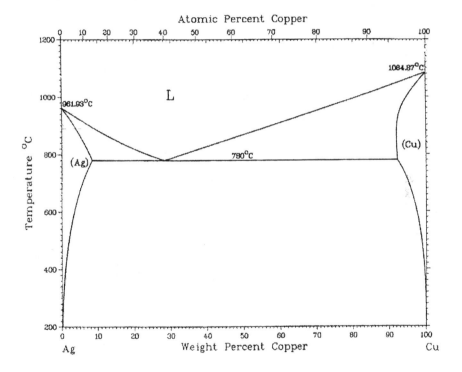

FIGURE 5.6 Equilibrium Ag-Cu binary phase diagram (Murray, 1984).

The chemical interactions in TICUSIL® braze foil are limited during its manufacture to those between Ag, Cu and Ti, unlike in the AMB process where Ti diffuses to and reduces the alumina surface. At the regions where the Ti ribbon is in contact with the Ag-Cu eutectic layers, the two-phase melt (L) can form, leading to the precipitation of Cu_4Ti_3 (Figure 5.1b, C). This process can be described by considering the 60Ag-Cu-Ti at.% section ($X - X'$) of the Ag-Cu-Ti ternary phase diagram which is perpendicular to the segregation path on which TICUSIL® lies (Figure 5.9).

The normal projection of the 60Ag-Cu-Ti at.% section ($X - X'$) can be used to study the cooling path of TICUSIL® (Figure 5.10). The liquidus temperature of TICUSIL® is 900°C, and this corresponds to the composition 34Cu-6Ti at.% on the 60Ag-Cu-Ti at.% section.

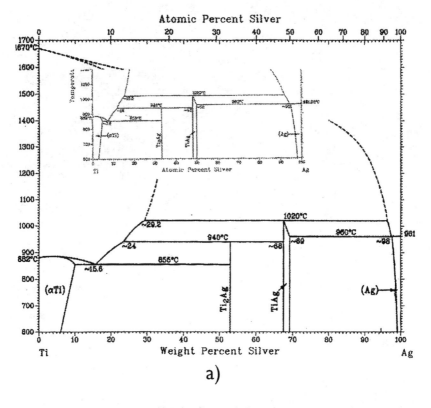

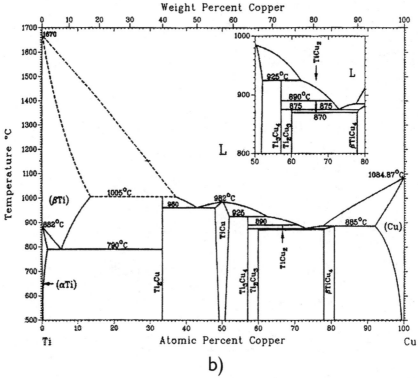

FIGURE 5.7 Equilibrium binary phase diagrams for (a) Ag-Ti (Murray and Bhansali, 1990) and (b) Cu-Ti (Murray, 1990).

Microstructural Evolution 115

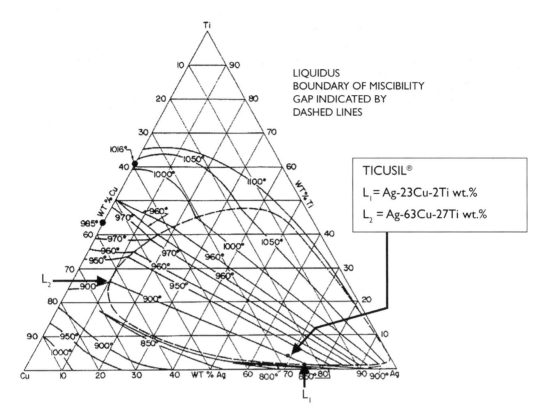

FIGURE 5.8 Ag-Cu-Ti ternary phase diagram (wt.%) showing projections of the liquidus surface (adapted from Conquest, 2003). The composition of TICUSIL® (68.8Ag-26.7Cu-4.5Ti wt.%) indicated by an arrow, lies within the miscibility gap (dotted line) and on the segregation path at the liquidus temperature of 900°C. The ends of the segregation path provide the compositions of an Ag-rich liquid (L_1 = Ag-27Cu-2Ti wt.%) and a Ti-rich liquid (L_2 = Ag-66Cu-22Ti wt.%), which form due to liquid immiscibility.

At 900°C, TICUSIL® exists as 91% L_1, an Ag-rich phase with composition Ag-23Cu-2Ti wt.% and as 9% L_2, a Ti-rich phase with the composition Ag-63Cu-27Ti wt.%. The cooling path enters the $(L_1 + L_2 + Cu_4Ti_3)$ domain with decreasing the temperature. This shows that Cu_4Ti_3 is the first intermetallic compound to form from the Ti-rich liquid L_2. The excess Ag and Cu from L_2 can form a homogeneous solution with L_1 to form a new liquid (L) which is in equilibrium, balanced by the precipitation of Cu_4Ti_3 in the $(L + Cu_4Ti_3)$ domain. At temperatures below ~858°C, the cooling path enters the $(L + Cu_4Ti_3 + Ag)$ domain where an Ag-rich solid precipitates. This Ag-rich phase was observed as a 2-μm-thick layer, adjacent to the Cu_4Ti_3 phases in all of the TICUSIL® braze foils (Figure 5.3b).

The tendency for Cu_4Ti_3 phase formation in TICUSIL® braze foil is indicative of the likely phase formations that could occur when Ti is retained in the braze interlayer of alumina/Ag-Cu-Ti brazed joints. In AMB, Ti diffuses to the alumina surface as it exhibits a relatively higher affinity to its oxygen than with either Ag or Cu in the braze alloy. Typically, while the reaction layer comprises Ti-rich phases, the braze interlayer consists of Ag-rich and Cu-rich phases in a eutectic-like distribution. The formation of stable Ti-O and Ti-Cu-O reaction layers at the joint interface creates diffusion barriers between the remaining Ti in the braze alloy and the alumina surface. Therefore, excess Ti in the braze alloy would be expected to form the Cu_4Ti_3 and Ag-rich phases in the braze interlayer, due to the chemical interactions between Ag, Cu and Ti, as observed in the as-received TICUSIL® braze foil.

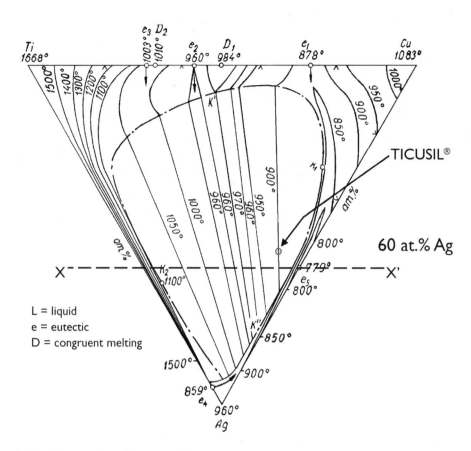

FIGURE 5.9 Ternary phase diagram (at.%) showing the projection of the liquidus surface in the Ag-Cu-Ti system. The 60Ag-Cu-Ti at.% section ($X - X'$) is shown (adapted from Hirnyj and Indacochea, 2008).

5.2 MICROSTRUCTURES OF BRAZED JOINTS IN AS-GROUND CONDITION

5.2.1 50-μm-THICK TICUSIL® BRAZE PREFORMS

The microstructures of D-96 AG and D-100 AG brazed joints made using 50-μm-thick TICUSIL® braze preforms were both found to be similar to those commonly reported in the literature (Ali et al., 2015; Lin et al., 2014). In both sets of these joints, three phases were characterised using SEM-EDX: an Ag-rich phase, a Cu-rich phase and a reaction layer phase. The average chemical compositions of these phases are listed in Table 5.2.

The D-96 AG brazed joints made using 50-μm-thick TICUSIL® braze preforms had an average brazed joint thickness of 26 μm (Figure 5.11) and an average reaction layer thickness of 1.7 μm (Figure 5.12a). The braze interlayer consisted of an Ag-rich phase with composition 96.5Ag-3.5Cu wt.% (Figure 5.11, D), and a Cu-rich phase with composition 93.9Cu-5.6Ag-0.5Ti wt.% (Figure 5.11, E). The reaction layer phase with composition 49.7Cu-44.8Ti-5.5Al wt.% was characterised as $Ti_3(Cu + Al)_3O$ (Figures 5.11 and 5.12, F).

Similarly, the D-100 AG brazed joints made using 50-μm-thick TICUSIL® braze preforms had an average brazed joint thickness of 21 μm and an average reaction layer thickness of 1.6 μm (Figure 5.13a). The braze interlayer consisted of an Ag-rich phase and a Cu-rich phase with similar compositions to those characterised in the D-96 AG brazed joints. The reaction layer phase with composition 46.4Cu-47.9Ti-6.0Al wt.% was also characterised as $Ti_3(Cu+Al)_3O$ (Figure 5.13a, G).

Microstructural Evolution

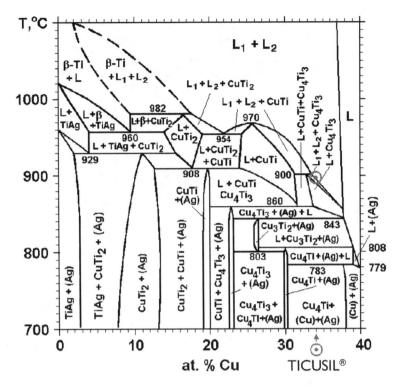

FIGURE 5.10 Vertical 60Ag-Cu-Ti at.% section of the Ag-Cu-Ti ternary phase diagram, corresponding to $(X - X')$ on Figure 5.9 (adapted from Hirnyj and Indacochea, 2008).

TABLE 5.2
Average Chemical Compositions (wt.%) of Phases Observed in D-96 AG and D-100 AG Brazed Joints Made Using 50-μm-Thick TICUSIL® Braze Preforms

Phases	Ag	Cu	Ti	Al
D – Ag-rich	96.5	3.5	–	–
E – Cu-rich	5.6	93.9	0.5	–
F – Ti$_3$Cu$_3$O	–	49.7	44.8	5.5
G – Ti$_3$Cu$_3$O	–	46.4	47.9	6.0

Labels D to G Correspond to Figures 5.11 to 5.13.

For both sets of brazed joints made using 50-μm-thick TICUSIL® braze preforms, SEM-EDX mapping was used to confirm that all of the Ti in the braze alloy had diffused to the joint interfaces (Figures 5.12 and 5.14). This led to the formation of Ti$_3$(Cu + Al)$_3$O reaction layers which is consistent with literature studies in which Ag-Cu-Ti braze alloys with intermediate Ti concentrations (1.75 < wt.% Ti < 4.5) have been selected (see Table 2.6 in Section 2.2.2.3). The presence of Al in the Ti$_3$(Cu + Al)$_3$O reaction layer is likely to have been due to the reduction of the alumina surface by Ti and enables the chemical mass balance to be maintained. According to the majority of joining mechanisms proposed, the liberation of Al from the alumina surface is associated with the formation of a TiO reaction product (see Section 2.1.7). In this study,

Alumina/Ag-Cu-Ti Active Metal Brazing—Review

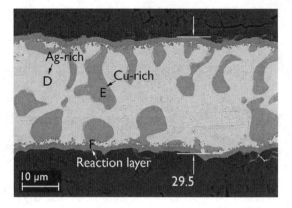

FIGURE 5.11 Backscattered electron image showing the microstructure of a typical D-96 AG brazed joint made using a 50-μm-thick TICUSIL® braze preform.

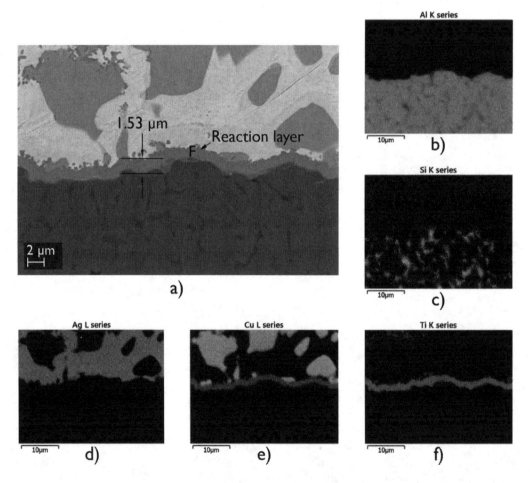

FIGURE 5.12 (a) Backscattered electron image of the joint interface in a typical D-96 AG brazed joint made using a 50-μm-thick TICUSIL® braze preform and corresponding SEM-EDX maps showing distributions of (b) Al, (c) Si, (d) Ag, (e) Cu and (f) Ti.

Microstructural Evolution

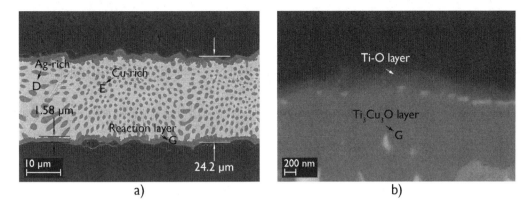

FIGURE 5.13 Backscattered electron images showing (a) the microstructure of a typical D-100 AG brazed joint made using a 50-μm-thick TICUSIL® braze preform and (b) the joint interface at high magnification showing the Ti-O and Ti_3Cu_3O layers.

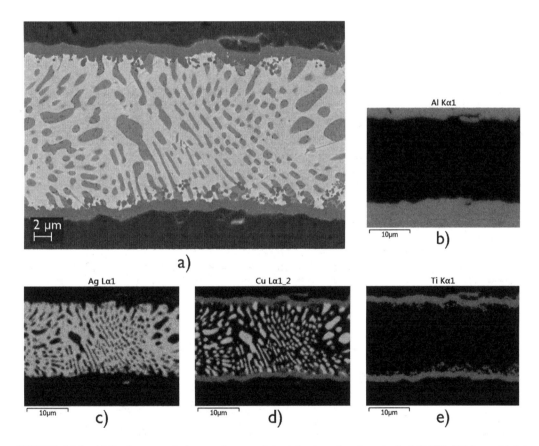

FIGURE 5.14 (a) Backscattered electron image of the joint interface in a typical D-100 AG brazed joint made using a 50-μm-thick TICUSIL® braze preform and corresponding SEM-EDX maps showing distributions of (b) Al, (c) Ag, (d) Cu and (e) Ti.

the TiO layer could not be adequately characterised, however, using SEM-EDX techniques (Figure 5.13b) - see Section 5.3, where the TiO layer is characterised using TEM techniques.

5.2.2 100-μm-Thick TICUSIL® Braze Preforms

The microstructures of D-96 AG and D-100 AG brazed joints made using 100-μm-thick TICUSIL® braze preforms were both found to be dissimilar to those commonly reported in the literature. In both sets of these joints, five phases were characterised using SEM-EDX as: an Ag-rich phase, a Cu-rich phase, two Cu-Ti phases and a reaction layer phase. The average chemical compositions of these phases are listed in Tables 5.3 and 5.4 for the D-96 AG and D-100 AG brazed joints, respectively.

D-96 AG brazed joints made using 100-μm-thick TICUSIL® braze preforms had an average brazed joint thickness of 39 μm and an average reaction layer thickness of 2.3 μm (Figure 5.15). The reaction layer phase, with a similar composition to that of the reaction layer phase observed in the brazed joints made using 50-μm-thick TICUSIL® braze preforms, was characterised as $Ti_3(Cu + Al)_3O$.

The average chemical compositions of the phases observed in the microstructures of D-96 AG brazed joints made using 100-μm-thick TICUSIL® braze preforms are shown in Table 5.3. The braze interlayer consisted of an Ag-rich phase with composition 94.4Ag-5.6Cu wt.% (Figure 5.15a, H), a Cu-rich phase with composition 93.3Cu-6.1Ag-0.6Ti wt.% (Figure 5.15a, I), a Cu_4Ti_3 phase with composition 62.6Cu-33.4Ti-4.0Ag wt.% (Figure 5.15a, J) and a Cu_4Ti phase with composition 81.9Cu-15.6Ti wt.% (Figure 5.15a, K).

The distributions of the Ag-rich and Cu-rich phases, which had similar compositions to those of the Ag-rich and Cu-rich phases observed in joints made using 50-μm-thick TICUSIL® braze preforms, no longer represented a eutectic-like structure. The volume fraction of the Ag-phase was significantly increased, and the Cu-rich phase was distributed close to the reaction layer. The Cu-Ti phases, therefore, were the predominant phases in the microstructures of the D-96 AG brazed joints made using 100-μm-thick TICUSIL® braze preforms.

The distribution of the Cu-Ti phases appeared to represent a micro-segregated structure. The core phase, characterised as Cu_4Ti_3 (Figure 5.15a, J), was always surrounded by a shell phase characterised as Cu_4Ti (Figure 5.15b, K).

Transition regions (Figure 5.15b) were observed between the Cu-Ti phases and Ag-rich globules (Figure 5.15c) at the edges of the joints. These transition regions consisted of Ag-rich and Cu-rich phases without any Cu-Ti phases, and their distributions represented the eutectic-like structure, similar to that observed throughout the joints made using the 50-μm-thick TICUSIL® braze preforms.

The reaction layer in D-96 AG brazed joints made using 100-μm-thick TICUSIL® braze preforms was not found to be deficient in Ti despite a significant amount of Ti being retained in the braze interlayer (Figure 5.16).

TABLE 5.3
Average Chemical Compositions (wt.%) of Phases Observed in D-96 AG Brazed Joints Made Using 100-μm-Thick TICUSIL® Braze Preforms

Phases	Ag	Cu	Ti
H – Ag-rich	94.4	5.6	–
I – Cu-rich	6.1	93.3	0.6
J – Cu_4Ti_3	4.0	62.6	33.4
K – Cu_4Ti	2.5	81.9	15.6

Labels H to K Refer to Figure 5.15.

Microstructural Evolution

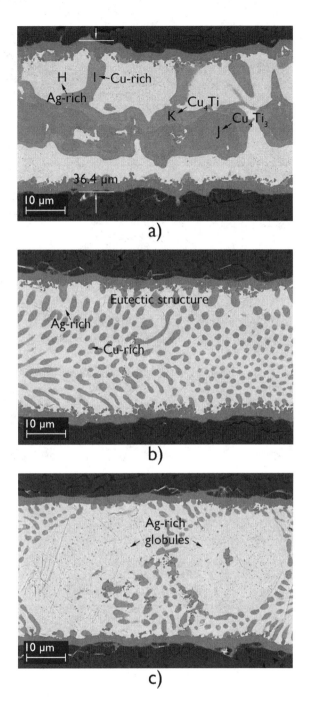

FIGURE 5.15 Backscattered electron images of typical regions in D-96 AG brazed joints made using 100-μm-thick TICUSIL® braze preforms (a) central regions showing Cu-Ti phases, (b) transition regions showing Ag-Cu eutectic and (c) joint edges showing Ag-rich globules.

D-100 AG brazed joints made using 100-μm-thick TICUSIL® preforms had an average brazed joint thickness of 39 μm and an average reaction layer thickness of 2.2 μm (Figure 5.17). The reaction layer phase, with a similar composition to that of the reaction layer phases observed in joints made using 50-μm-thick TICUSIL® braze preforms and in D-96 AG brazed joints made using 100-μm-thick TICUSIL® braze preforms, was characterised as $Ti_3(Cu + Al)_3O$.

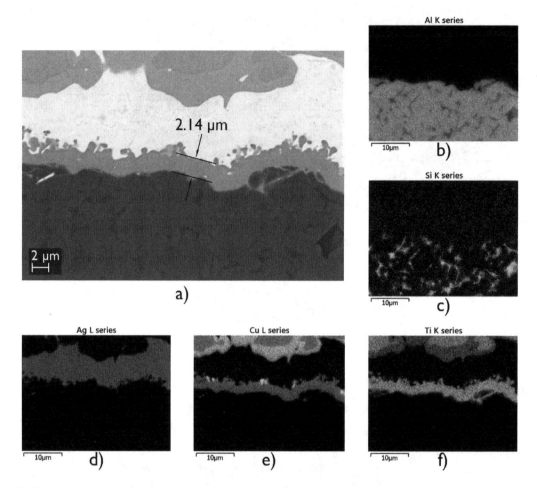

FIGURE 5.16 (a) Backscattered electron image of the joint interface in a typical D-96 AG brazed joint made using a 100-μm-thick TICUSIL® braze preform and corresponding SEM-EDX maps showing distributions of (b) Al, (c) Si, (d) Ag, (e) Cu and (f) Ti.

The average chemical compositions of the phases observed in the microstructures of D-100 AG brazed joints made using 100-μm-thick TICUSIL® braze preforms are shown in Table 5.4. The braze interlayer consisted of a Cu_4Ti_3 phase with composition 62.4Cu-33.8Ti-3.8Ag wt.% (Figure 5.17a, L), a Cu_4Ti phase with composition 82.4Cu-14.9Ti-2.7 wt.% (Figure 5.17a, M), a Cu-rich phase with composition 92.6Ag-5.4Cu-2.0 wt.% (Figure 5.17a, N) and an Ag-rich phase with composition 94.7Ag-5.3Cu wt.% (Figure 5.17a, O). The transition regions towards the edges of the joints (Figure 5.17b) were again observed comprising the Cu-Ti phases and the Ag-rich globules (Figure 5.17c).

The microstructures of D-100 AG brazed joints made using 100-μm-thick TICUSIL® braze preforms were identical to those of D-96 AG brazed joints made using 100-μm-thick TICUSIL® braze preforms. The microstructural evolution that occurred as a result of an increase in the TICUSIL® braze preform thickness from 50 to 100 μm, was found to be independent of alumina purity and was likely to be associated to the incomplete dissolution of Ti in the braze alloy under the brazing conditions used.

Cu-Ti phase formation in the braze interlayer of D-96 AG and D-100 AG brazed joints made using 100-μm-thick TICUSIL® braze preforms may have been influenced by the relative Ti concentration factor, i.e. excess Ti relative to the faying alumina surfaces, coupled with an insufficient brazing time of 10 min at the sub-liquidus brazing temperature of 850°C.

Microstructural Evolution

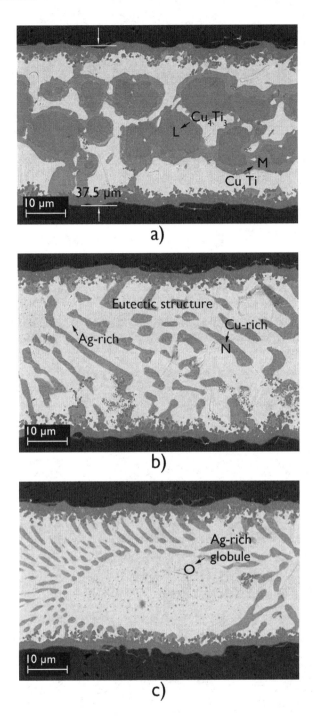

FIGURE 5.17 Backscattered electron images of typical regions in D-100 AG brazed joints made using 100-μm-thick TICUSIL® braze preforms (a) central regions showing Cu-Ti phases, (b) transition regions showing Ag-Cu eutectic and (c) joint edges showing Ag-rich globules.

The brazing conditions were adequate for the formation of ~1.7-μm-thick reaction layers in both D-96 AG and D-100 AG brazed joints made using 50-μm-thick TICUSIL® preforms. In these joints, Ti was observed to have completely diffused to the joint interfaces leading to an Ag-Cu eutectic-like braze interlayer (Figure 5.14). Therefore, the microstructural evolution that occurred as a result of

TABLE 5.4
Average Chemical Compositions (wt.%) of Phases Observed in D-100 AG Brazed Joints Made Using 100-μm-Thick TICUSIL® Braze Preforms

Phases	Ag	Cu	Ti
L – Cu$_4$Ti$_3$	3.8	62.4	33.8
M – Cu$_4$Ti	2.7	82.4	14.9
N – Cu-rich	5.4	92.6	2.0
O – Ag-rich	94.7	5.3	–

Labels L to O Refer to Figure 5.17.

an increase in the TICUSIL® braze preform thickness from 50 to 100 μm may not have been solely due to insufficient brazing conditions.

Formed under the same brazing conditions, D-96 AG and D-100 AG brazed joints made using 100-μm-thick TICUSIL® braze preforms were comprised of reaction layers that were ~2.3 μm thick. Despite an increase in the reaction layer thickness, a significant amount of Ti was retained in the braze interlayer which led to the formation of the intermetallic Cu$_4$Ti$_3$ and Cu$_4$Ti phases and an Ag-rich phase (Figure 5.18). Excess Ti in the braze interlayer as a result of an increase in

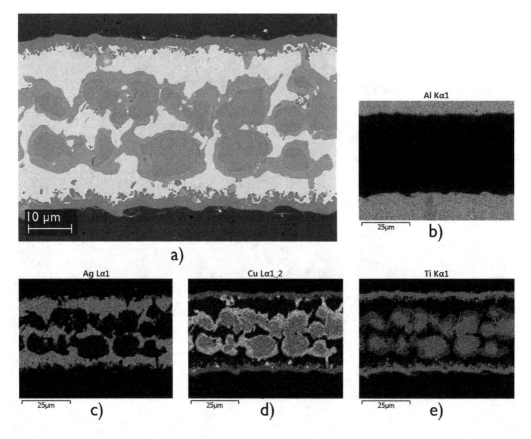

FIGURE 5.18 (a) Backscattered electron image of the joint interface in a typical D-100 AG brazed joint made using a 100-μm-thick TICUSIL® braze preform and corresponding SEM-EDX maps showing distributions of (b) Al, (c) Ag, (d) Cu and (e) Ti.

Microstructural Evolution

the TICUSIL® braze preform thickness from 50 to 100 μm led to both an increase in the reaction layer thickness and a microstructural evolution in the braze interlayer. Therefore, the microstructural evolution that occurred as a result of an increase in the TICUSIL® braze preform thickness from 50 to 100 μm, was predominantly due to excess amounts of Ti relative to the faying alumina surfaces.

According to the literature, an increase in the brazing time may have reduced the volume fraction of Cu-Ti phases in the braze interlayer by promoting further growth of the reaction layer. However, this may have led to excessively thick reaction layers which can be detrimental to joint strength (see Section 2.3.1.3). Joint performance is discussed in Chapter 6.

5.2.2.1 Cu-Ti Phase Formation

The micro-segregated Cu-Ti phases observed in joints made using 100-μm-thick TICUSIL® braze preforms consisted of a Cu_4Ti_3 core surrounded by a Cu_4Ti shell. Kozlova et al. (2010) also observed this phase, in the microstructure of alumina-to-copper brazed joints made using an Ag-Cu eutectic braze alloy with 4.5 wt.% Ti (Figure 5.19). It was suggested that these phases, due to their locations and shapes, had precipitated during brazing rather than during cooling, as in the latter, they would otherwise have been expected to grow as thin elongated dendrites. Based on this, the isolated Cu_4Ti_3 and Cu_4Ti phases observed in this study may have formed at the brazing temperature, in a similar way to that of the $Ti_3(Cu + Al)_3O$ layer (Voytovych et al. 2004; see Section 2.3.1.2).

Andrieux et al. (2009) observed the formation of a 3-μm-thick reaction layer at the joint interface, following a brazing experiment whereby the 72Ag-28Cu wt.% eutectic braze alloy was melted onto a Ti plate at 790°C for 2 min. In order of increasing distance towards the Ti plate, several Cu-Ti phases; Cu_4Ti, Cu_4Ti_3, $CuTi$, $CuTi_2$ and α-TiO were observed. According to phase analysis performed using the thermodynamic software Thermo-Calc, the Ti activity level was found to increase while the Cu activity level was found to decrease, in a direction towards the Ti plate. With Ti and Cu atoms diffusing in opposite directions, the resulting phase equilibria established at the joint interface resulted in the formation and ordering of these foregoing Cu-Ti phases.

The first Cu-Ti phase to form at the Ag-Cu/Ti interface was Cu_4Ti as this requires the lowest Ti activity in the Ag-Cu alloy (Figure 5.20). This was confirmed in earlier work by Andrieux et al., 2008, which showed using differential thermal analysis that Cu_4Ti was the first phase to form when an Ag-Cu eutectic braze alloy is melted onto a solid Ti substrate. It should be noted that the brazing temperature of 790°C in Andrieux et al. (2009) is only 10°C higher than the 780°C melting temperature of the eutectic Ag-Cu braze alloy. In this study, the formation of Cu_4Ti may have occurred in a similar way; therefore, since at the sub-liquidus brazing

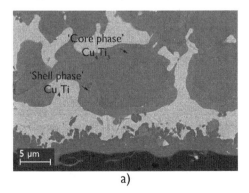
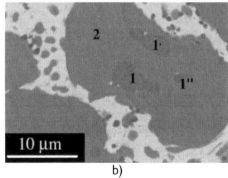

FIGURE 5.19 Cu-Ti structure consisting of a core phase (1-Cu_4Ti_3) and shell phase (2-Cu_4Ti) observed in (a) the braze interlayer of typical D-96 AG and D-100 AG brazed joints made using 100-μm-thick TICUSIL® braze preforms and (b) similar phases observed by Kozlova et al. (2010) in the braze interlayer of alumina-to-copper brazed joints made using an Ag-Cu-4.5 wt.% Ti braze alloy.

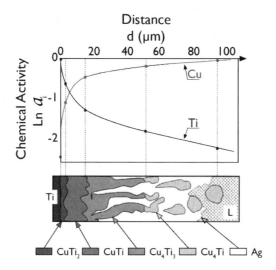

FIGURE 5.20 The formation of Cu-Ti phases at the interface of a pure Ti plate and a 72Ag-28Cu wt.% eutectic braze alloy, following brazing at 790°C for 2 min (Andrieux et al., 2009).

temperature of 850°C, excess solid Ti was still available in the braze interlayer of joints made using 100-μm-thick TICUSIL® braze preforms, following the formation of stable reaction layers at the joint interfaces.

Andrieux et al. (2009) proposed that two competing processes occur through which Cu_4Ti forms at the Ag-Cu/Ti interface. In this study, these processes are applicable once stable Ti-rich reaction layers are formed at the alumina/Ag-Cu-Ti joint interfaces. At this stage, the excess solid Ti phases in the braze interlayer thereby form multiple Ag-Cu/Ti interfaces with the surrounding Ag-Cu braze alloy. The first process involves the precipitation of Cu-Ti phases through solid-state diffusion mechanisms, along the path shown in the isothermal section of the Ag-Cu-Ti ternary phase diagram at the brazing temperature of 850°C (Figure 5.21).

The solid-state diffusion mechanisms are limited due to a second competing process whereby the excess solid Ti phases in the braze interlayer partially dissolves in the surrounding Ag-Cu eutectic braze alloy. This second process is limited since the dissolution of the solid Ti phases in the surrounding Ag-Cu eutectic alloy leads to a local saturation; i.e., the Ag-Cu eutectic braze alloy becomes saturated with a Ti addition of ~2 wt.% Ti (Nicholas, 1990). Therefore, the excess solid Ti phases react with Cu in the braze alloy to form of the Cu_4Ti phases. Furthermore, Cu-Ti phase formation causes the local depletion of Cu in the surrounding Ag-Cu eutectic braze alloy which leads to the separation of an Ag-rich phase.

While the dissolution of the solid Ti phases in the surrounding Ag-Cu eutectic alloy is limited, the solid-state diffusion process continues through the precipitating Cu_4Ti layers. However, the growth rate of the Cu_4Ti layer, which acts as a diffusion barrier to Ti, begins to decrease. This may be used to explain why the morphology of the Cu_4Ti phases was observed to be globular (Kozlova et al., 2010).

The subsequent long ranged cross-diffusion of Cu and Ti leads to the conversion of the centre of the Cu_4Ti phase into Cu_4Ti_3 (Andrieux et al., 2009). Therefore, while Cu_4Ti is the first phase to form, sufficient supply of Ti in the braze alloy leads to the formation of the Cu_4Ti_3 phases. These processes can be used to describe the formation of the Cu_4Ti, Cu_4Ti_3 and Ag-rich phases which were observed in the braze interlayer of joints made using 100-μm-thick TICUSIL® braze preforms, due to the presence of excess solid Ti and at the brazing temperature of 850°C.

Microstructural Evolution

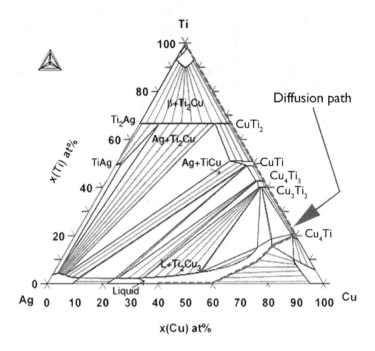

FIGURE 5.21 Isothermal section of the Ag-Cu-Ti ternary phase diagram at 850°C showing a diffusion path corresponding the 72Ag-28Cu wt.% (60Ag-40Cu at.%) alloy as it becomes saturated with Ti (adapted from Dezellus et al., 2011).

The excellent chemical affinity that Ti has towards Cu may have increased the tendency for Cu-Ti phase formation. As the Ag concentration in an Ag-Cu-Ti braze alloy increases, the Ti activity level also increases. On the contrary, as the Cu concentration in an Ag-Cu-Ti braze alloy increases, the Ti activity level decreases. The solubility of Ti in Ag is significantly lower than that of Ti in Cu. For example, at 1,150°C the solubility of Ti in Ag is ~3.7 wt.%, whereas the solubility of Ti in Cu is ~67 wt.% (Hansen and Anderko, 1958).

Kozlova et al. (2010) found that the minimum molar fraction of Ti (x_{Ti}^*) required for the formation of Cu_4Ti and Cu_4Ti_3 phases in an Ag-Cu-Ti melt significantly decreases with increasing concentrations of Ag and decreasing concentrations of Cu (Table 5.5). Therefore, the relatively higher 68.8 wt.% Ag concentration in TICUSIL®, as compared to the 63.0 wt.% Ag in Cusil ABA®, may have further promoted the formation of these Cu-Ti phases.

TABLE 5.5
Calculated Values of Minimum Mole Fraction of Ti (x_{Ti}^*) for Intermetallic Compound Formation from the Ag-Cu-Ti Melt

			x_{Ti}^*	
x_{Ag}	x_{Cu}	T_L (°C)	Cu_4Ti	Cu_4Ti_3
0	0.57	925	–	0.43
0	0.80	880	0.20	–
0.42	0.58	850	0.072	0.146
0.50	0.50	820	0.050	0.106
0.60	0.40	780	0.027	0.065

As the TICUSIL® braze preform thickness was increased from 50 to 100 μm, the thickness of the Ti$_3$(Cu + Al)$_3$O reaction layer was also observed to increase, by 35% from 1.7 to 2.3 μm. With the twofold increase in the relative Ti concentration factor, the brazing conditions were adequate to enable a further amount of Ti to diffuse to the joint interface; however, the complete diffusion of Ti in this manner was inhibited. While the brazing conditions enabled the formation of adequately thick reaction layers at the joint interfaces, the amount of Ti which remained in the braze alloy was excessive. The interaction between the excess solid Ti, with a somewhat reduced tendency to diffuse to the alumina surface, with the surrounding Ag-Cu braze alloy may have thermodynamically favoured the Cu-Ti phase formation.

The partial dissolution of the Ti phases in the surrounding Ag-Cu braze alloy led to formation of the Cu$_4$Ti phases. This occurred when the surrounding Ag-Cu braze alloy became locally saturated in Ti as this is when Cu-Ti phase formation is thermodynamically favourable in the Ag-Cu-Ti ternary alloy system. Therefore, due to the excellent chemical affinity between Ti and Cu, solid-state diffusion mechanisms enabled the formation of Cu$_4$Ti$_3$ phases at the centre of the once-Ti-phases, with Cu$_4$Ti at the outskirts. These Cu-Ti structures are hereby referred to as isolated Cu-Ti phases, due to their rounded and discontinuous nature in the braze interlayer. The consumption of Cu from the surrounding Ag-Cu braze alloy in the formation of these isolated Cu-Ti phases was observed to have led to the local separation of an Ag-rich phase, which flowed towards the joint edges in the form of Ag-rich globules.

5.2.2.2 Ag-Rich Braze Outflow

In the literature, the wetting and spreading behaviour of TICUSIL® on alumina has been found to be led by a Ti-depleted Ag-rich phase when Ti is retained in the braze interlayer (Figure 2.10). In this study, the localised separation of an Ag-rich phase resulted from the consumption of Cu in the formation of Cu-Ti phases in the braze interlayer, in joints made using and 100-μm-thick TICUSIL® braze preforms. This Ag-rich phase flowed towards the joint edges in the form of globules which indicated that this Ag-rich phase may have been relatively fluid at the brazing temperature.

Braze outflow was observed in all of the D-96 and D-100 brazed joints made using 50- to 150-μm-thick TICUSIL® braze preforms. Appendix 2 shows the macro images of each side of each brazed joint. In these images, captured to study the alignment of the brazed butt-joints, the braze outflow can be clearly seen.

The average brazed joint thickness of joints made using 100-μm-thick TICUSIL® braze preforms was ~39 μm. This thickness was observed to be uniform across the entire bond line. Therefore, assuming that the 7 mm × 5 mm × 0.1 mm braze preform had completely filled the joint gap between the two 8 mm × 6 mm faying alumina surfaces, the volume of braze outflow was estimated to be ~46 vol.%.

In the brazed joints made using 50-μm-thick TICUSIL® preforms, the microstructure of the braze outflow featured the Ag-rich and Cu-rich phases consistent with those phases observed throughout the joint microstructure (Figure 5.22a). However, in the brazed joints made using 100-μm-thick TICUSIL® braze preforms, the microstructure of the braze outflow was Ag-rich, in the form of Ag-rich globules, dissimilar to that observed in the central region of the brazed joints (Figures 5.22b and 5.23).

The Ag-rich braze outflow observed in the joints made using 100-μm-thick TICUSIL® braze preforms may have led to a disproportionate increase in the Cu concentration in the braze alloy. In the literature, it is known that an increase in the Cu concentration of an Ag-Cu-Ti braze alloy can adversely affect the Ti activity level; hence, Ti can become retained in the braze interlayer (Mandal et al., 2004a; Pak et al., 1990). Therefore, the Ag-rich braze outflow observed may have further promoted the formation of the isolated Cu-Ti phases, which featured the Cu$_4$Ti and Cu$_4$Ti$_3$ phases.

Microstructural Evolution

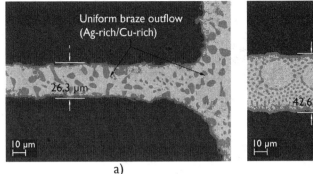
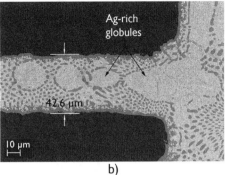

FIGURE 5.22 Backscattered electron images showing typical braze outflow which occurred in both, D-96 AG and D-100 AG brazed joints made using (a) 50-µm- and (b) 100-µm-thick TICUSIL® braze preforms.

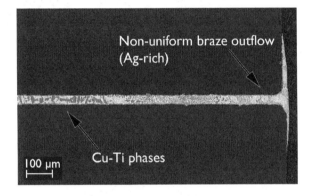

FIGURE 5.23 Backscattered electron image showing typical braze outflow which occurred in both, D-96 AG and D-100 AG brazed joints made using 100-µm-thick TICUSIL® braze preforms (low magnification).

5.2.3 150-µm-Thick TICUSIL® Braze Preforms

D-96 AG brazed joints made using 150-µm-thick TICUSIL® preforms had an average brazed joint thickness of 65 µm (Figure 5.24a) and an average reaction layer thickness of 2.8 µm (Figure 5.24b). The reaction layer phase was characterised as $Ti_3(Cu + Al)_3O$; however, the TiO layer on the alumina side of the joint interface was somewhat more apparent (Figure 5.24c). Similarly, D-100 AG brazed joints made using 100-µm-thick TICUSIL® preforms had an average brazed joint thickness of 68 µm (Figure 5.24d) and an average reaction layer thickness of 3.0 µm (Figure 5.24e).

The microstructures of D-96 AG and D-100 AG brazed joints made using 150-µm-thick TICUSIL® braze preforms consisted of the typical Ag-rich (Figure 5.24a, Q) and Cu-rich phases (Figure 5.24a, R). At several positions along the braze interlayer in both sets of joints, Cu-Ti structures with a significantly higher degree of micro-segregation than that of the isolated Cu-Ti phases, in joints made using 100-µm-thick TICUSIL® braze preforms, were observed. These structures, hereby referred to as the multi-layered Cu-Ti structures, consisted of a Ti core that was surrounded by four Cu-Ti shell phases. In each of the multi-layered Cu-Ti structures, the Ti concentration was found to decrease in an outwardly direction, away from the Ti core and towards the joint interfaces. In order of increasing Ti concentration, these phases were characterised using SEM-EDX as follows: Cu_4Ti (Figure 5.25, W), Cu_4Ti_3 (Figure 5.25, V), CuTi (Figure 5.25, U), $CuTi_2$ (Figure 5.25, T) and Ti (Figure 5.25, S). The average chemical compositions of these phases are listed in Table 5.6.

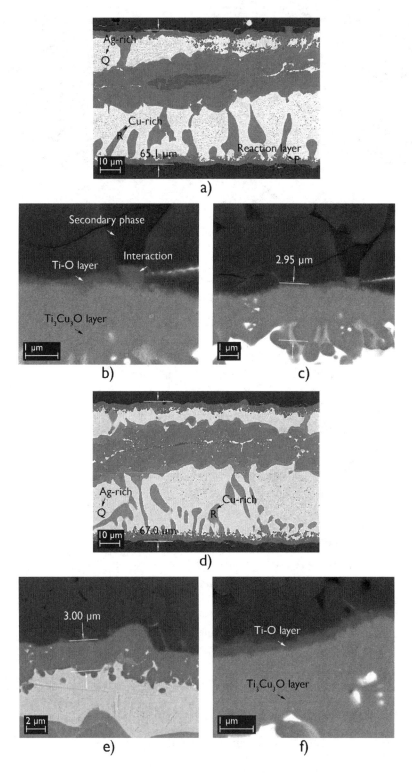

FIGURE 5.24 Backscattered electron images showing (a) the microstructure of a typical D-96 AG brazed joint made using a 150-μm-thick TICUSIL® braze preform and the joint interface at (b) high magnification and (c) higher magnification, (d) the microstructure of a typical D-100 AG brazed joint made using a 150-μm-thick TICUSIL® braze preform and the joint interface at (e) high magnification and (f) higher magnification.

Microstructural Evolution

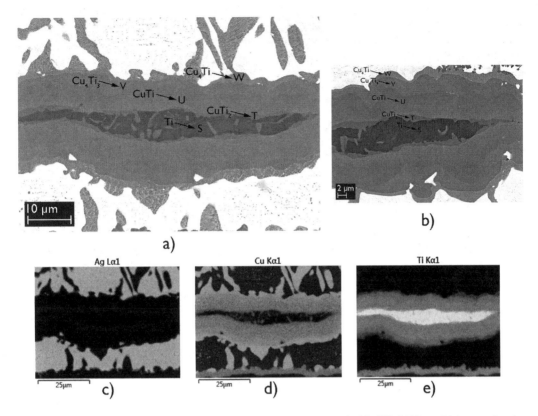

FIGURE 5.25 Backscattered electron images of typical regions in typical D-100 AG brazed joints made using 150-µm-thick TICUSIL® braze preforms (a) central regions showing Cu-Ti phases, (b) higher magnification and contrast, and corresponding SEM-EDX maps showing distributions of (c) Ag, (d) Cu and (e) Ti.

TABLE 5.6
Average Chemical Compositions (wt.%) of Phases Observed in D-96 AG and D-100 AG Brazed Joints Made Using 150-µm-Thick TICUSIL® Braze Preforms

Phases	Ag	Cu	Ti	Al
P – Ti$_3$(Cu + Al)$_3$O	–	52.6	42.6	4.8
Q – Ag-rich	94.8	5.2	–	–
R – Cu-rich	6.4	91.8	1.8	–
S – Ti	2.4	1.5	96.1	–
T – CuTi$_2$	5.5	39.5	55.0	–
U – CuTi	5.7	55.4	38.9	–
V – Cu$_4$Ti$_3$	3.1	64.0	32.9	–
W – Cu$_4$Ti	3.8	83.6	12.6	–

Labels P to W Refer to Figures 5.24 and 5.25.

The graded ordering of the Cu-Ti phases in the multi-layered Cu-Ti structure was similar to that observed by Andrieux et al. (2009) (Figure 5.20). Therefore the interaction between the excess solid Ti and the surrounding Ag-Cu eutectic in the braze alloy produced a similar interface to that observed when an Ag-Cu eutectic braze alloy is melted onto a Ti substrate. The presence of Ti at the core of the multi-layered Cu-Ti structure is likely to have been that of the Ti ribbon, in TICUSIL® braze foil.

5.2.3.1 Multi-Layered Cu-Ti Structure

The multi-layered Cu-Ti structure observed in joints made using 150-μm-thick TICUSIL® braze preforms consisted of a Ti core surrounded by layers of $CuTi_2$, CuTi, Cu_4Ti_3, and Cu_4Ti. The processes involved in the formation of this phase may have been similar to that which led to the formation of the isolated Cu-Ti phases, observed in joints made using 100-μm-thick TICUSIL® braze preforms (see Section 5.2.2.1).

Following the formation of stable reaction layers at the joint interfaces, the partial dissolution of the excess solid Ti in the surrounding Ag-Cu braze alloy leads to the formation of a first Cu_4Ti layer. This first Cu_4Ti layer acts as a diffusion pathway for Ti and results when the surrounding Ag-Cu braze alloy becomes locally saturated in Ti; Cu-Ti phase formation is thermodynamically favourable in the Ag-Cu-Ti ternary alloy system. Surrounding the Ti core, other Cu-Ti layers characterised as $CuTi_2$, CuTi and Cu_4Ti_3 can form via solid-state diffusion processes along the path shown in the isothermal section of the Ag-Cu-Ti ternary phase diagram at the brazing temperature of 850°C (Figure 5.21). While Cu_3Ti_2 would also have been expected to form in this way, it was not observed. However, Cu_3Ti_2 phase formation has been shown to occur at considerably slower rates than that of the Cu_4Ti and Cu_4Ti_3 phases (Andrieux et al., 2008). During the solid-state diffusion process, the dissolution of Ti in the surrounding Ag-Cu eutectic alloy occurs simultaneously. While this limits the thicknesses of the $CuTi_2$ and CuTi layers, it can lead to a second Cu_4Ti layer which surrounds the entire multi-layered Cu-Ti structure.

The formation of both the first and second Cu_4Ti layers occurs rapidly resulting in the local depletion of Cu, hence the local separation of an Ag-rich phase. With the formation of a relatively thick Cu_4Ti layer embedded in an Ag-rich matrix, further precipitation of the Cu_4Ti phase would require the solid-state diffusion of Ti through the $CuTi_2$, CuTi and Cu_4Ti_3 layers, which may have increased their relative thicknesses. Therefore, the rate of Cu_4Ti phase formation and the local Ag-rich phase separation gradually decreases; however, the first Cu_4Ti layer may convert into Cu_4Ti_3 with long ranged cross-diffusion of Cu and Ti.

The graded ordering of the Cu-Ti phases in the multi-layered Cu-Ti structure is manifested from the opposing solid-state diffusion directions of Ti and Cu. The dissolution of Ti in the surrounding Ag-Cu braze alloy results in the formation of a Cu_4Ti layer which surrounds the entire multi-layered Cu-Ti structure. These processes are illustrated in Figure 5.26 which shows how the multi-layered Cu-Ti structure, observed in the braze interlayer of joints made using 150- and 250-μm-thick TICUSIL® braze preforms, may have formed.

These results show that an increase in the TICUSIL® braze preform thickness can induce an increase in the a relative Ti concentration factor such that excess solid Ti in the braze interlayer can lead to internal Ag-Cu/Ti interfaces.

The formation of such a multi-layered Cu-Ti structure in the braze interlayer of alumina-to-alumina brazed joints made using Ag-Cu-Ti braze alloys could not be found in the literature. Typically, Ti retained in the braze interlayer in the alumina/Ag-Cu-Ti system may be viewed as an incomplete brazing process where the role of Ti in diffusing to the joint interfaces is not fully accomplished.

The effect of Cu-Ti phases in the braze interlayer on the strength of alumina-to-alumina brazed joints could not be found in the literature. However, the presence of relatively harder phases in the braze interlayer was postulated to strengthen the joints in a similar way to the strengthening mechanisms that are provided by composite braze alloys (Zhu and Chung, 1994). Furthermore, the separation of a relatively ductile Ag-rich phase, as a matrix phase in the braze interlayer and at the joint edges may also benefit the joint strength (see Chapter 6). These phenomena would undoubtedly make the stress distribution in the braze interlayer relatively complex as compared to that in a typical alumina/Ag-Cu-Ti/alumina brazed joint (Figure 2.4).

In the brazed joints made using 150-μm-thick TICUSIL® preforms, transition regions (Figure 5.27a) were again observed between the Cu-Ti phases in the central region of the braze

Microstructural Evolution

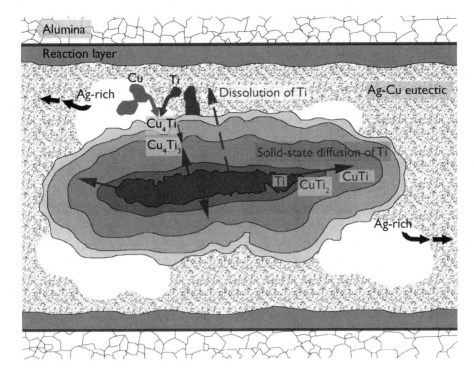

FIGURE 5.26 Schematic showing the likely mechanisms which led to the formation of a multi-layered Cu-Ti structure in joints made using 150-μm-thick TICUSIL® braze preforms.

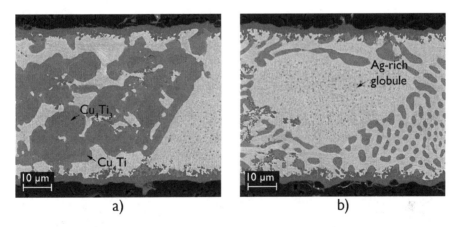

FIGURE 5.27 Backscattered electron images showing typical regions in both D-96 AG and D-100 AG brazed joints made using 150-μm-thick TICUSIL® braze preforms (a) transition region showing Cu-Ti phases and (b) joint edges showing Ag-rich globules.

interlayer and the Ag-rich globules at the joint edges (Figure 5.27b). These transition regions consisted of the isolated Cu_4Ti_3 and Cu_4Ti phases which were previously observed in the central regions of the microstructures of joints made using 100-μm-thick TICUSIL® braze preforms.

The estimated volume of braze outflow in joints made using 150-μm-thick TICUSIL® braze preforms had decreased from 46 to 42 vol.% as compared to joints made using 100-μm-thick TICUSIL® braze preforms. However, the microstructure of the braze outflow was again found to consist of the Ag-rich globules that were surrounded by an Ag-Cu eutectic-like structure.

The phases of the multi-layered Cu-Ti structure in joints made using 150-μm-thick TICUSIL® braze preforms had relatively higher Ti to Cu ratios as compared to that of the isolated Cu_4Ti_3 and Cu_4Ti phases in joints made using 100-μm-thick TICUSIL® braze preforms. Therefore, with an overall relatively lower Cu consumption in joints made using 150-μm-thick TICUSIL® braze preforms, a reduction in the Ag-rich phase separation and subsequent Ag-rich braze outflow was observed.

5.2.4 250-μm-Thick TICUSIL® Braze Preforms

D-96 AG brazed joints made using 250-μm-thick TICUSIL® braze preforms had an average brazed joint thickness of 182 μm (Figure 5.28a) and an average reaction layer thickness of 3.2 μm (Figure 5.28b). The reaction layer phase, with a similar composition to that of the reaction layer phases observed in joints made using 50-, 100- and 150-μm-thick TICUSIL® braze preforms, was characterised as $Ti_3(Cu + Al)_3O$.

The braze interlayer consisted of the same seven phases as observed in joints made using 150-μm-thick TICUSIL® braze preforms: an Ag-rich phase, a Cu-rich phase, and the multi-layered Cu-Ti structure comprising a Ti core surrounded by $CuTi_2$, $CuTi$, Cu_4Ti_3 and Cu_4Ti layers. In joints made using 150-μm-thick TICUSIL® braze preforms, the multi-layered Cu-Ti structure was discontinuous, whereas in joints made using 250-μm-thick TICUSIL® braze preforms, it was observed to be a continuous structure across the entire bond line (Figure 5.29).

D-100 AG brazed joints made using 250-μm-thick TICUSIL® braze preforms had an average brazed joint thickness of 191 μm (Figure 5.30a) and an average reaction layer thickness of 3.2 μm (Figure 5.30b). The phases observed in the microstructures of these joints were similar to those of the D-96 AG joints made using 250-μm-thick TICUSIL® braze preforms.

The microstructures of both the D-96 AG and D-100 AG brazed joints made using 250-μm-thick TICUSIL® braze preforms somewhat resembled that of the as-received 250-μm-thick TICUSIL® braze foil (see Section 5.1). This result is indicative of the partial dissolution of the Ti ribbon at the centre of the TICUSIL® braze foil during brazing. While a fraction of the Ti, from the Ti ribbon, had diffused to the joint interfaces in the formation of 3.2 μm thick reaction layers, the rest of the Ti ribbon remained as solid excess Ti that interacted with the surrounding Ag-Cu braze alloy to form the multi-layered Cu-Ti structure (see Section 5.3.3.1).

Due to the relatively thick braze interlayer in joints made using 250-μm-thick TICUSIL® braze performs, the flow of the separated Ag-rich phase was localised, and contained within the joint. The Ag-rich phase which separated as a consequence of the Cu-Ti phase formation appeared as globules surrounded by an Ag-Cu eutectic-like structure radiating away from the continuous multi-layered Cu-Ti structure.

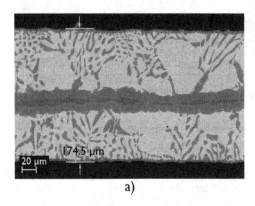 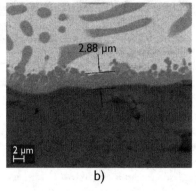

FIGURE 5.28 Backscattered electron images showing (a) the microstructure of a typical D-96 AG brazed joint made using a 250-μm-thick TICUSIL® braze preform and (b) the joint interface at high magnification.

Microstructural Evolution 135

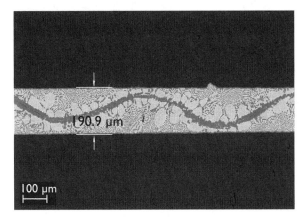

FIGURE 5.29 Backscattered electron image showing the microstructure of a typical D-96 AG brazed joint made using a 250-μm-thick TICUSIL® braze preform (lower magnification).

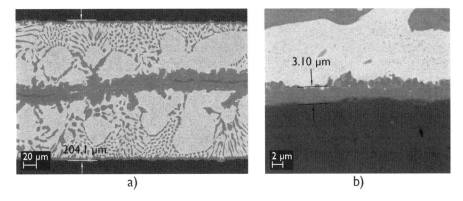

FIGURE 5.30 Backscattered electron images showing (a) the microstructure of a typical D-100 AG brazed joint made using a 250-μm-thick TICUSIL® braze preform and (b) the joint interface at high magnification.

The reduction in braze outflow observed in joints made using 150-μm-thick TICUSIL® braze preforms as compared to that in joints made using 100-μm-thick TICUSIL® braze preforms was postulated to be due to a reduction the volume fraction of the isolated Cu_4Ti_3 and Cu_4Ti phases which formed (Figure 5.31a). The continuous multi-layered Cu-Ti structure, in joints made using 250-μm-thick TICUSIL® braze preforms, may have consumed a relatively lower concentration of Cu from the braze alloy. In these joints, the average brazed joint thickness of ~182 μm enabled the separated Ag-rich phase to be accommodated within the joint, and this completely prevented any braze outflow (Figure 5.31b). Instead, the joint edges comprised the typical Ag-rich and Cu-rich phases in a eutectic-like distribution.

The properties of D-96 AG and D-100 AG brazed joints made using 50- to 250-μm-thick TICUSIL® braze preforms are summarised in Table 5.7. The microstructural evolution that occurred as the TICUSIL® braze preform thickness was increased from 50 to 250 μm is illustrated in Figure 5.32.

The microstructural evolution can be summarised as follows: An increase in the TICUSIL® braze preform thickness from 50 to 250 μm led to an increase in the average brazed joint thickness; from 26 to 182 μm in D-96 AG brazed joints and from 21 to 191 μm in D-100 AG brazed joints. A corresponding increase in the average reaction layer thickness was also observed; from 1.7 to 3.2 μm in D-96 AG brazed joints and from 1.6 to 3.2 μm in D-100 AG brazed joints.

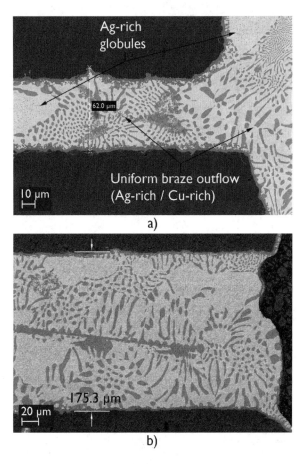

FIGURE 5.31 Backscattered electron images showing typical braze outflow which occurred in D-96 AG and D-100 AG brazed joints made using (a) 150-μm- and (b) 250-μm-thick TICUSIL® braze preforms.

The microstructure of joints made using 50-μm-thick TICUSIL® braze preforms consisted of an Ag-Cu eutectic-like structure and Ti was observed to have completely diffused to the joint interfaces. This led to the formation of $Ti_3(Cu + Al)_3O$ reaction layers.

An increase in the TICUSIL® braze preform thickness from 50 to 100 μm doubled the amount of Ti available to diffuse to the joint interfaces. While the thickness of the $Ti_3(Cu + Al)_3O$ reaction layers increased by 35% from 1.7 to 2.3 μm, excess Ti was retained in the braze interlayer. This led to the formation of isolated Cu_4Ti_3 and Cu_4Ti phases and the separation of an Ag-rich phase which flowed towards the joint edges. Ag-rich braze outflow accounted for ~46% of the starting braze preform volume.

The relative Ti concentration factor in joints made using 150-μm-thick TICUSIL® braze preforms increased by a factor of three as compared to that in 50-μm-thick TICUSIL® braze preforms. This led to an increase in the thickness of the $Ti_3(Cu + Al)_3O$ reaction layer from 2.3 to 2.8 μm as compared to joints made using 100-μm-thick TICUSIL® braze preforms. The excess Ti in the braze interlayer led to the formation of a multi-layered Cu-Ti structure composed of Ti, $CuTi_2$, $CuTi$, Cu_4Ti_3 and Cu_4Ti phases which was observed to be discontinuous across the bond line. The isolated Cu_4Ti_3 and Cu_4Ti phases, observed earlier in joints made using 100-μm-thick TICUSIL® braze preforms, were located in regions between the multi-layered Cu-Ti structures. Due to an overall relatively lower consumption of Cu in the formation of these Cu-Ti phases, the Ag-rich braze outflow reduced to ~40 vol.% of the starting braze preform volume.

TABLE 5.7
Properties of D-96 AG and D-100 AG Brazed Joints Made Using 50- to 250-µm-Thick TICUSIL® Braze Preforms

Sample Set	Alumina Purity, (wt.% Al_2O_3)	Surface Condition	Braze Preform Thickness (µm)	Specimens Characterised	Reaction Layer Thickness (µm)	Average Brazed Joint Thickness (µm)	Estimated Volume of Braze Outflow (%)
1	D-96	AG	50	1	1.7 ± 0.1	26 ± 1	29
2	D-96	AG	100	3	2.3 ± 0.1	39 ± 1	46
3	D-96	AG	150	1	2.8 ± 0.1	65 ± 2	40
4	D-96	AG	250	1	3.2 ± 0.2	182 ± 3	0
9	D-100	AG	50	1	1.6 ± 0.1	21 ± 1	42
10	D-100	AG	100	2	2.2 ± 0.1	39 ± 1	46
11	D-100	AG	150	1	3.0 ± 0.1	68 ± 2	40
12	D-100	AG	250	1	3.2 ± 0.2	191 ± 6	0

The relative Ti concentration factor in joints made using 250-µm-thick TICUSIL® braze preforms increased by a factor of four as compared to that in 50-µm-thick TICUSIL® braze preforms. This led to an increase in the thickness of the $Ti_3(Cu + Al)_3O$ reaction layer from 2.8 to 3.2 µm as compared to joints made using 150-µm-thick TICUSIL® braze preforms. The multi-layered Cu-Ti structure which consisted of Ti, $CuTi_2$, CuTi, Cu_4Ti_3 and Cu_4Ti phases was observed to be continuous across the bond line. The isolated Cu_4Ti_3 and Cu_4Ti phases, observed earlier in joints made using 100- and 150-µm-thick TICUSIL® braze preforms, were no longer observed. The continuous multi-layered Cu-Ti structure somewhat resembled the Ti ribbon in the as-received 250-µm-thick TICUSIL® braze foil. The Ag-rich phase separation was contained within the ~182-µm-thick brazed joint, and this completely prevented any braze outflow.

5.3 TRANSMISSION ELECTRON MICROSCOPY

Transmission Electron Microscopy (TEM) techniques were used to characterise the nm-thick Ti-O layer which formed on the alumina side of the joint interfaces in both D-96 AG and D-100 AG brazed joints. The other phases observed to have formed in the microstructures of the D-96 AG and D-100 AG brazed joints made using 50- and 100-µm-thick TICUSIL® braze preforms were also characterised using TEM techniques.

The average composition of alumina characterised using STEM-EDX was 51.7Al-48.3O wt.%. This was consistent with theoretical calculations which show that the chemical composition of alumina is 51.9Al-47.1O wt.%. Selected area diffraction patterns acquired from alumina crystals showed its rhombohedral structure. The unit cell of an alumina crystal structure is of the R-3c (167) space group with lattice parameters $a = b = 4.76$ Å and $c = 12.9$ Å (Finger and Hazen, 1977). These results were identical for all alumina crystals characterised in both the D-96 AG (Figure 5.33) and D-100 AG brazed joints.

5.3.1 50-µm-Thick TICUSIL® Braze Preforms

In D-96 AG brazed joints made using 50-µm-thick TICUSIL® braze preforms, the braze side of the joint interface consisted of a polycrystalline $Ti_3(Cu + Al)_3O$ layer with particle sizes ranging from

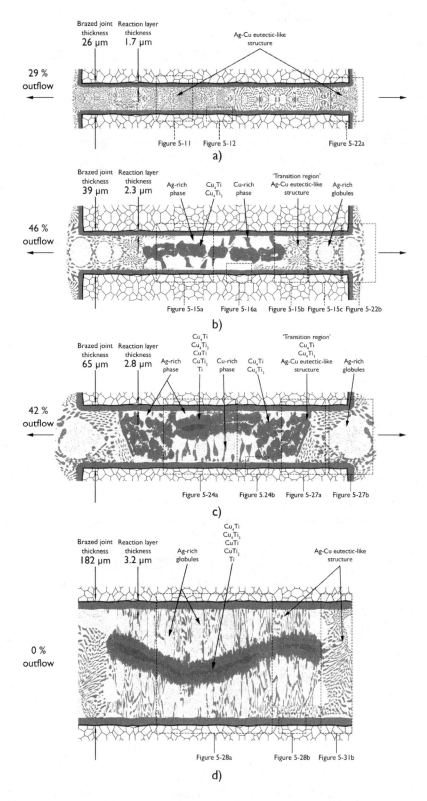

FIGURE 5.32 Schematics of the microstructures of D-96 AG and D-100 AG brazed joints made using (a) 50-μm- (b) 100-μm- (c) 150-μm- and (d) 250-μm-thick TICUSIL® braze preforms.

Microstructural Evolution 139

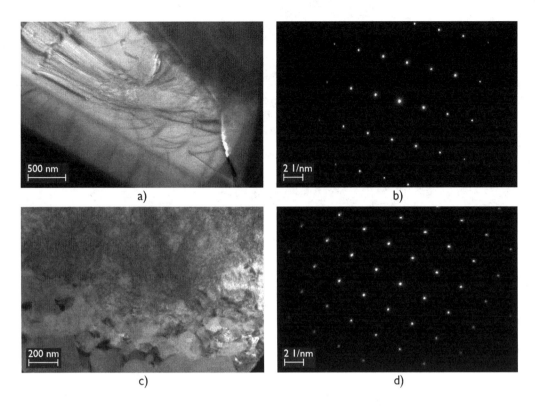

FIGURE 5.33 (a) High-resolution TEM image of a selected alumina crystal near the joint interface of a D-96 AG brazed joint, (b) the corresponding selected area diffraction pattern, (c) high-resolution TEM image of a selected alumina crystal near the joint interface of a D-100 AG brazed joint and (d) the corresponding selected area diffraction pattern.

0.2 to 1.3 μm (Figure 5.34a). STEM-EDX was used to characterise the composition of this layer as 42.4Ti-45.1Cu-5.8Al-0.4Ag-6.3O wt.% (Figure 5.34b). This somewhat validated the SEM-EDX results whereby the reaction layer was characterised with the composition of 46.4Cu-47.9Ti-6.0 Al wt.%.

The crystal structure of the particles in the $Ti_3(Cu+Al)_3O$ layer were identified using electron diffraction as that of Ti_3Cu_3O, with lattice parameters $a = b = c = 11.24$ Å (cubic, Fd-3m (227) space group). A high magnification bright field TEM image of a selected Ti_3Cu_3O particle, and the corresponding selected area diffraction pattern, are shown in Figures 5.35a and 5.35b, respectively. The diffraction pattern agreed with the diamond-cubic structure of Ti_3Cu_3O and was similar to other diffraction patterns of this phase found in the literature (Ali et al., 2015; Lin et al., 2014; Stephens et al., 2003).

The alumina side of the joint interface in D-96 AG brazed joints made using 50-μm-thick TICUSIL® braze preforms consisted of a polycrystalline TiO layer with particle sizes ranging from 50 to 200 nm (Figure 5.36a). Within this TiO layer, isolated Ti_2O particles were also observed (Figure 5.36b).

The average thickness of the TiO layer was ~177 nm, characterised using STEM-EDX to be composed of 76.1Ti-22.8O-1.1Al wt.%. Similarly, the average composition of the isolated Ti_2O particles was 83.6Ti-2.4Al-2.5Cu-11.5O wt.% (Figure 5.37). The presence of Mo in the STEM-EDX spectra acquired was due to the Mo grid on which the TEM samples were mounted.

Electron diffraction was used to confirm the hexagonal Ti_2O phase of the P6/mmm (191) space group, for which the lattice parameters $a = b = 5.06$ Å and $c = 2.92$ Å were reported in the literature (Lin et al., 2014). The d-spacings in the selected diffraction pattern for Ti_2O matched those of both simulated and experimental diffraction patterns reported, with a zone axis of [310] (Figure 5.38).

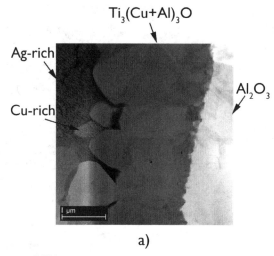
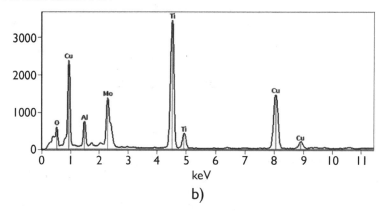

FIGURE 5.34 (a) bright field STEM image of a typical region at the joint interface in a D-96 AG brazed joint made using a 50-μm-thick TICUSIL® braze preform and (b) STEM-EDX spectrum acquired from the corresponding $Ti_3(Cu+Al)_3O$ layer.

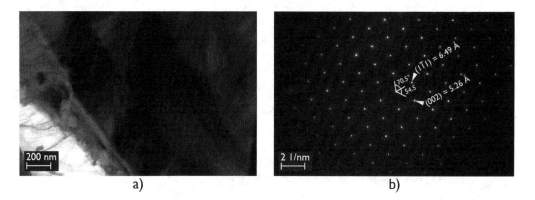

FIGURE 5.35 (a) high magnification bright field TEM image of a selected Ti_3Cu_3O particle and (b) the corresponding selected area diffraction pattern acquired, with zone axes of [110].

Microstructural Evolution 141

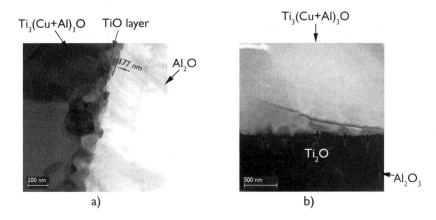

FIGURE 5.36 (a) bright field STEM image of a typical region at joint interface in a D-96 AG brazed joint made using a 50-μm-thick TICUSIL® braze preform showing the TiO layer and (b) HAADF image showing isolated Ti$_2$O particles in the TiO layer.

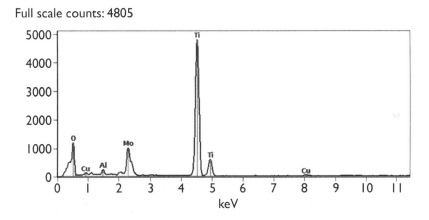

FIGURE 5.37 STEM-EDX spectrum acquired from the isolated Ti$_2$O particles.

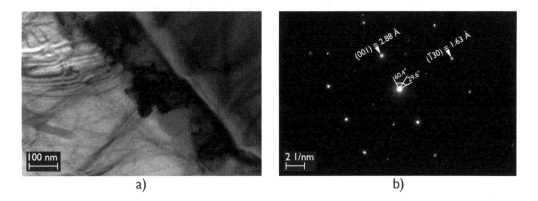

FIGURE 5.38 (a) high magnification bright field TEM image of a selected Ti$_2$O particle and (b) the corresponding selected area diffraction pattern acquired, with zone axes of [310].

STEM-EDX maps of the joint interface in D-96 AG brazed joints made using 50-μm-thick TICUSIL® braze preforms showed that the thickness of the TiO layer was non-uniform in these joints (Figure 5.39b).

Electron diffraction was used to confirm that the TiO layer was composed of face-centred cubic γ-TiO particles of the Fm-3m (225) space group with the lattice parameters $a = b = c = 4.21$ Å. A high magnification bright field TEM image of a selected γ-TiO particle, and the corresponding selected area diffraction pattern, are shown in Figures 5.40a and 5.40b, respectively. The diffraction pattern acquired resembled the NaCl-type structure and was consistent with that of other diffraction patterns of this phase found in the literature (Ali et al., 2015; Stephens et al., 2003).

In D-100 AG brazed joints made using 50-μm-thick TICUSIL® braze preforms, the braze side of the joint interface consisted of a polycrystalline $Ti_3(Cu + Al)_3O$ layer with particle sizes ranging from 0.2 to 1.4 μm (Figure 5.41). STEM-EDX was used to characterise the composition of this layer as 44.6Ti-43.5Cu-6.4Al-5.5O wt.%.

The crystal structures of the particles in the reaction layer, on the braze side of the joint interface, were identified using electron diffraction techniques as that of Ti_3Cu_3O (cubic, Fd-3m (227) space group). A high magnification bright field TEM image of a selected Ti_3Cu_3O particle and the corresponding selected area diffraction pattern are shown in Figures 5.42a and 5.42b, respectively.

In D-100 AG brazed joints made using 50-μm-thick TICUSIL® braze preforms, the alumina side of the joint interface consisted of a TiO layer composed of γ-TiO, with isolated Ti_2O particles. The thickness of this TiO layer was again non-uniform; however, an average thickness of 155 nm was measured.

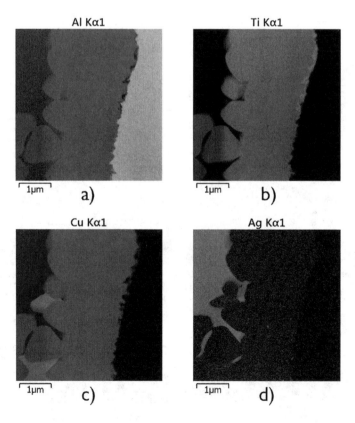

FIGURE 5.39 STEM-EDX map, corresponding to the selected area shown in Figure 5.36a of the joint interface in a D-96 AG brazed joint made using a 50-μm-thick TICUSIL® braze preform, showing the distributions of (a) Al, (b) Ti, (c) Cu and (d) Ag.

Microstructural Evolution 143

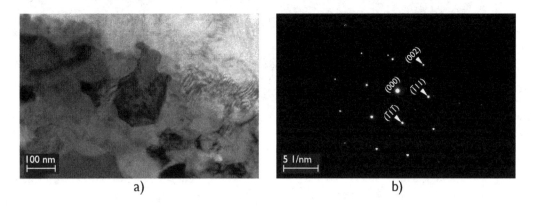

FIGURE 5.40 (a) High magnification bright field TEM image of a selected γ-TiO particle and (b) The corresponding selected area diffraction pattern acquired, with zone axes of [110].

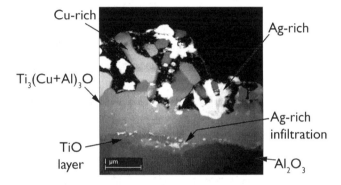

FIGURE 5.41 HAADF image of a typical region at the joint interface in a D-100 AG brazed joint made using a 50-μm-thick TICUSIL® braze preform. The thickness of this TEM specimen was significantly reduced during sample preparation which led to failure in the Ag-rich regions of the braze interlayer.

FIGURE 5.42 (a) High magnification bright field TEM image of a selected Ti$_3$Cu$_3$O particle and (b) The corresponding selected area diffraction pattern acquired, with zone axes of [110].

The volume fractions of the Ti$_2$O particles in the TiO layers of D-96 AG and D-100 AG brazed joints made using 50-μm-thick TICUSIL® braze preforms were similar. However, in D-100 AG brazed joints, some Ag-rich braze infiltration was observed between the TiO layer and the Ti$_3$(Cu + Al)$_3$O layer (Figure 5.41). This Ag-rich braze infiltration was not observed in

D-96 AG brazed joints (based on both SEM and TEM analysis); however, isolated Ag-rich particles were instead observed in the Ti$_3$Cu$_3$O layer. These differences in the joint interfaces of the D-96 AG and D-100 AG brazed joints made using 50-μm-thick TICUSIL® braze preforms could not be confidently attributed to alumina purity, due to the limited number of TEM specimens analysed.

The joint interfaces of both D-96 AG and D-100 AG brazed joints made using 50-μm-thick TICUSIL® braze preforms appeared to be similar. Both sets of joints consisted of a γ-TiO layer with isolated Ti$_2$O particles on the alumina side of the joint interface (layer I), and a Ti$_3$(Cu + Al)$_3$O layer on the braze side of the joint interface (layer II). These results were similar to those reported in the literature (Ali et al., 2016; Ichimori et al., 1999; Stephens et al., 2003).

The braze interlayer in both D-96 AG and D-100 AG brazed joints made using 50-μm-thick TICUSIL® braze preforms consisted of an Ag-Cu eutectic-like structure. The Ag-rich and Cu-rich phases were characterised using STEM-EDX (Figure 5.43). In D-96 AG brazed joints, the composition of the Ag-rich phase was 96.2Ag-3.7Cu-0.1Ti wt.% (Figure 5.43b) while that of the Cu-rich phase was 92.6Cu-6.2Ag-1.2Ti wt.% (Figure 5.43c). In D-100 AG brazed joints, the composition of the Ag-rich phase was 95.6Ag-4.4Cu wt.% while that of the Cu-rich phase was 93.3Cu-5.5Ag-1.2Ti wt.%. Therefore, the Ag-rich and Cu-rich phases which formed in the braze interlayer of these D-96 AG and D-100 AG brazed joints were similar. Their compositions were also similar to those of these phases characterised earlier in Section 5.2 using SEM-EDX, which were 95.6Ag-3.4Cu wt.% for the Ag-rich phase and 93.9Cu-5.6Ag-0.5Ti wt.% for the Cu-rich phase.

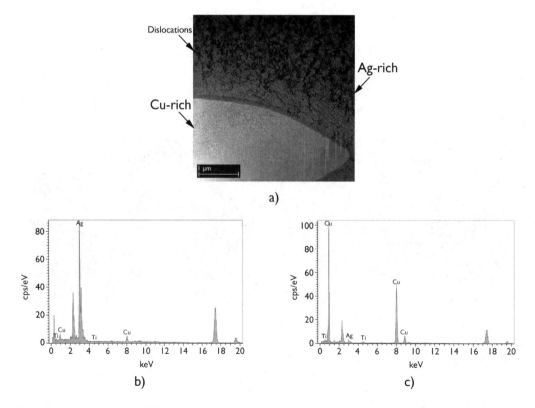

FIGURE 5.43 Bright field STEM image of a typical region in the braze interlayer of a D-96 AG brazed joint made using a 50-μm-thick TICUSIL® braze preform and (a) STEM-EDX spectrums acquired from the corresponding (b) Ag-rich and (c) Cu-rich phases.

Microstructural Evolution

A relatively higher dislocation density was observed in the Ag-rich phase as compared with the Cu-rich phase. These dislocations may have indicated the extent of plastic deformation in the braze interlayer in the accommodation of thermally induced residual stresses. It follows that the Ag-rich phase is relatively more ductile as compared to the Cu-rich phase (see Chapter 6).

5.3.2 100-μm-THICK TICUSIL® BRAZE PREFORMS

The braze side of the joint interface in both the D-96 AG and D-100 AG brazed joints made using 100-μm-thick TICUSIL® braze preforms consisted of a polycrystalline $Ti_3(Cu + Al)_3O$ layer. The composition of this layer measured using STEM-EDX was 42.2Ti-45.2Cu-6.1Al-6.5O wt.% in the D-96 AG brazed joints and 45.9Ti-42.7Cu-7.6Al-3.8O wt.%, in the D-100 AG brazed joints. In both sets of joints, electron diffraction was used to confirm the cubic Ti_3Cu_3O phase of the Fd-3m (227) space group (Figure 5.44). This confirmed that joints made using both 50- and 100-μm-thick TICUSIL® braze preforms consisted of Ti_3Cu_3O on the braze side of the joint interface.

In D-96 AG brazed joints made using 100-μm-thick TICUSIL® braze preforms, the alumina side of the joint interface consisted of a polycrystalline TiO layer, with particle sizes ranging from 100 to 200 nm (Figure 5.45). Within this TiO layer, isolated Ti_2O particles were also observed (Figure 5.46).

The average thickness of the TiO layer was ~210 nm. STEM-EDX was used to characterise the average composition of this TiO layer as 73.0Ti-1.6Cu-25.4O wt.%. Similarly, the average

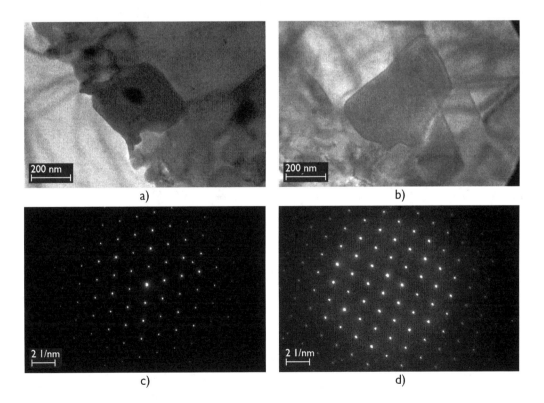

FIGURE 5.44 High magnification bright field TEM images of selected Ti_3Cu_3O particles in (a) D-96 AG and (b) D-100 AG brazed joints and corresponding selected area diffraction patterns in (c) D-96 AG and (b) D-100 AG brazed joints, with zone axes of [110].

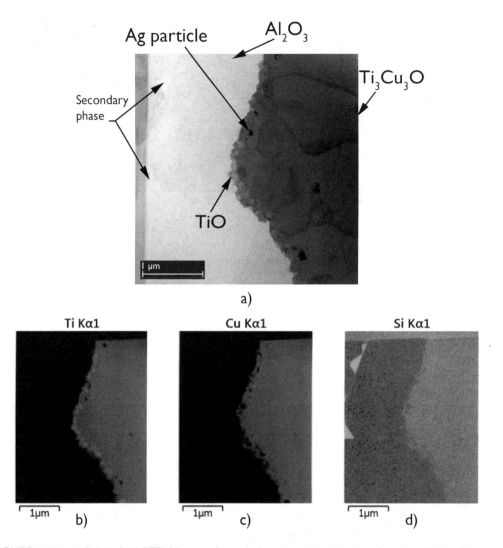

FIGURE 5.45 (a) Bright field STEM image of a typical region at the joint interface in a D-96 AG brazed joint made using a 100-μm-thick TICUSIL® braze preform and corresponding STEM-EDX maps showing distributions of (b) Ti, (c) Cu and (f) Si.

composition of the isolated Ti_2O particles was 83.0Ti-17.1O wt.%. Electron diffraction was used to confirm that the TiO layer was composed of face-centred cubic γ-TiO particles of the Fm-3m (225) space group with the lattice parameters $a = b = c = 4.21$ Å (Figure 5.47).

The diffraction pattern acquired resembled the NaCl-type structure and was consistent with the other diffraction patterns of the γ-TiO phase acquired at the interfaces of D-96 AG and D-100 AG brazed joints made using 50-μm-thick TICUSIL® braze preforms.

Despite these similarities, the thickness of the TiO layer increased by ~18% in joints made using 100-μm-thick TICUSIL® braze preforms as compared with that in joints made using 50-μm-thick TICUSIL® braze preforms. Furthermore, the overall continuity of the TiO layer had also improved and comprised a greater volume fraction of Ti_2O particles as a result of the increase in the TICUSIL® braze preform thickness (Figure 5.46).

In D-100 AG brazed joints made using 100-μm-thick TICUSIL® braze preforms, the alumina side of the joint interface consisted of a polycrystalline TiO layer, with particle sizes ranging from 100 to 200 nm. Within this TiO layer, isolated Ti_2O particles were again observed (Figure 5.48).

Microstructural Evolution 147

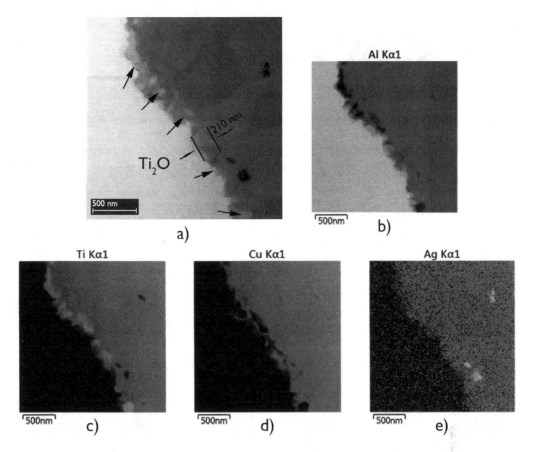

FIGURE 5.46 High magnification image of selected area in Figure 5.46a and corresponding STEM-EDX maps showing distributions of (b) Al, (c) Ti, (d) Cu and (e) Si.

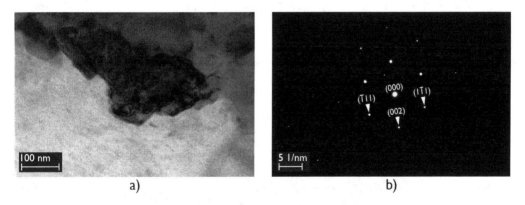

FIGURE 5.47 (a) High magnification bright field TEM image of a selected γ-TiO particle and (b) The corresponding selected area diffraction pattern acquired, with zone axes of [110].

The selected area diffraction patterns acquired confirmed that these phases (Figure 5.49) were similar in all of the D-96 AG and D-100 AG brazed joints made using 50- to 100-μm-thick TICUSIL® braze preforms.

Ag-rich braze infiltration was observed in the grain boundary regions of the alumina surface at the interfaces of the D-100 AG brazed joints made using 100-μm-thick TICUSIL® braze preforms

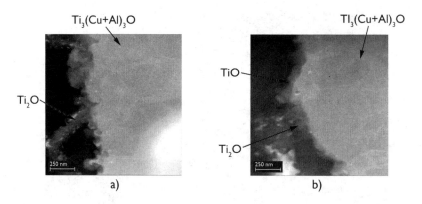

FIGURE 5.48 HAADF image of a typical region at the joint interface in a D-100 AG brazed joint made using a 100-μm-thick TICUSIL® braze preform (a) lower magnification and (b) higher magnification.

FIGURE 5.49 (a) High magnification bright field TEM image of a selected γ-TiO particle and (b) the corresponding selected area diffraction pattern acquired, with zone axes of [110].

(Figure 5.50). This appeared to be similar to the Ag-rich infiltration that was observed between the TiO layer and the $Ti_3(Cu + Al)_3O$ layer in D-100 AG brazed joints made using 50-μm-thick TICUSIL® braze preforms (Figure 5.41). These results suggested that the D-100 AG alumina surface may have comprised grinding-induced damage or residual porosity that provided intergranular pathways for braze infiltration. The composition of the braze that infiltrated the alumina surface was depleted in Ti; since Ti was consumed during reactions between the infiltrating braze alloy and the adjacent alumina grains. The relatively fluid Ag-rich phase that separated during the formation of the Cu-Ti compounds and was observed to flow towards the joint edges, may have also formed in part the braze composition that infiltrated the alumina surface. This phenomenon was not observed in the D-96 AG brazed joints which comprised the liquid phase sintered 96.0 wt.% Al_2O_3 alumina; however, Ag-rich particles were observed instead, in the $Ti_3(Cu + Al)_3O$ layer (Figure 5.45).

The compositions of the isolated Cu_4Ti_3 and Cu_4Ti phases and the Ag-rich phase that surrounded them, observed in the microstructures of the D-96 AG and D-100 AG brazed joints made using 100-μm-thick TICUSIL® braze preforms, were characterised using STEM-EDX. The composition of the Ag-rich phase was 95.6Ag-4.4Cu wt.% (Figure 5.51) while that of the Cu_4Ti_3 and Cu_4Ti phases were 60.1Cu-37.5Ti-2.4 Ag wt.% (Figure 5.52a) and 80.5Cu-17.7Ti-1.9Ag wt.% (Figure 5.52b), respectively. These compositions were similar to those characterised earlier in Section 5.2 using SEM-EDX which were 82.1Cu-15.3Ti-2.6Ag wt.% for the Cu_4Ti_3 phase and 62.5Cu-33.6Ti-3.9Ag wt.% for the Cu_4Ti phase.

Microstructural Evolution

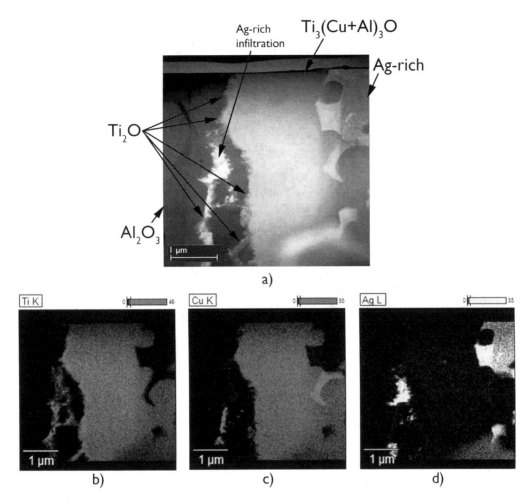

FIGURE 5.50 (a) HAADF TEM image of a typical region at the joint interface in a D-100 AG brazed joint made using a 100-μm-thick TICUSIL® braze preform and corresponding STEM-EDX maps showing distributions of (b) Ti, (c) Cu and (d) Ag.

Electron diffraction was used to confirm the crystal structures of the Cu_4Ti_3 and Cu_4Ti phases. The crystal structure of Cu_4Ti_3 was tetragonal of the I4/mmm (139) space group, with lattice parameters $a = b = 3.12$ Å and $c = 19.84$ Å (Figure 5.53). The crystal structure of Cu_4Ti was orthorhombic of the Pnma (62) space group, with lattice parameters $a = 4.53$ Å, $b = 12.93$ Å and $c = 4.34$ Å (Figure 5.54). These results were consistent with those reported in the literature (Dezellus et al., 2011; Pfeifer et al., 1968).

An increase in the TICUSIL® braze preform thickness from 50 to 100 μm did not affect the composition of the reaction layer in either of the D-96 or D-100 AG brazed joints. The TiO layer on the alumina side of the joint interface consisted of γ-TiO particles as the predominant phase, with isolated Ti_2O particles. However, the continuity of the TiO improved and its thickness increased by ~18%, from ~177 to ~210 nm, as the TICUSIL® braze preform thickness was increased from 50 to 100 μm. Furthermore, the volume fraction of the isolated Ti_2O particles in the TiO layer was also observed to have increased. A corresponding increase in the thickness of the $Ti_3(Cu + Al)_3O$ layer by 35.3% from 1.7 to 2.3 μm (see Section 5.3) could also be attributed to the increase in the TICUSIL® braze preform thickness from 50 to 100 μm. Due to a twofold increase in the relative Ti concentration factor, therefore, the continuity of the reaction layers improved and their thicknesses also increased while their compositions remained unchanged. While these results are in

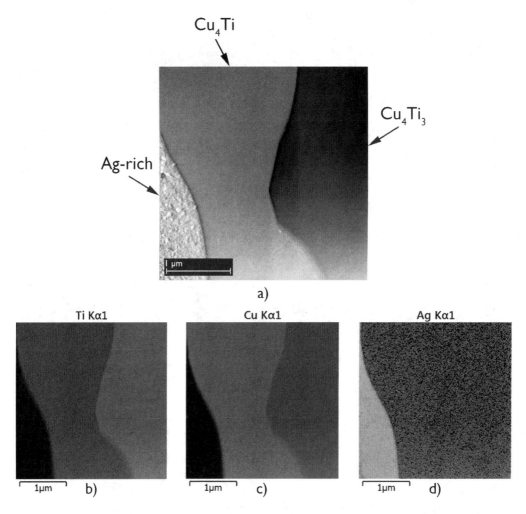

FIGURE 5.51 (a) HAADF TEM image of a typical region in the braze interlayer of a D-96 AG brazed joint made using a 100-μm-thick TICUSIL® braze preform and corresponding STEM-EDX maps showing distributions of (b) Ti, (c) Cu and (d) Ag. The selected area shows the edge of an isolated Cu_4Ti_3 and Cu_4Ti structure, surrounded by an Ag-rich phase.

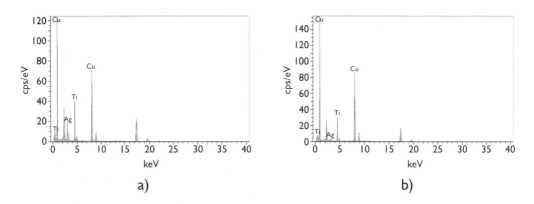

FIGURE 5.52 STEM-EDX spectrums acquired from the (a) Cu_4Ti_3 and (b) Cu_4Ti phases, corresponding to the selected area shown in Figure 5.51a.

Microstructural Evolution

FIGURE 5.53 (a) High magnification bright field TEM image of a selected Cu_4Ti_3 particle and (b) the corresponding selected area diffraction pattern acquired.

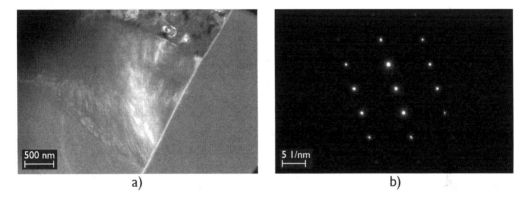

FIGURE 5.54 (a) High magnification bright field TEM image of a selected Cu_4Ti particle and (b) the corresponding selected area diffraction pattern acquired.

good agreement with the effect of Ti concentration as reported in the literature (see Section 2.2.2.3), it can be seen here that this effect can be controlled via the braze preform thickness.

The predominant phase in the TiO layer was characterised as γ-TiO; however, this phase is unstable at temperatures below ~1250°C according to the equilibrium O–Ti binary phase diagram (Figure 5.55) (Murray and Wriedt, 1987). At temperatures below ~450°C, α-TiO and Ti_2O_3 are likely to form. The nucleation of small quantities of other TiO reaction products at the brazing temperature, i.e. Ti_2O observed in the γ-TiO layer, is also favourable since the γ-TiO phase is not in a state of thermodynamic equilibrium (Ali et al., 2015; Valeeva et al., 2001).

Valeeva et al. (2001) found that the equilibrium phase transformation of γ-TiO to α-TiO required a slow cooling rate of 10°C/h. The significantly faster cooling rate of 10°C/min selected in this study, representative of typical cooling rates selected in the literature, may have inhibited this phase transformation, resulting instead in the formation of the metastable γ-TiO phase.

In a study by Kelkar et al. (1994), thermodynamic analysis was used to show that the high-temperature γ-TiO phase forms on the alumina side of the joint interface at relatively lower Ti activity levels, whereby it is in a local thermodynamic equilibrium with the Ti_3Cu_3O phase, which forms on the braze side of the joint interface. With relatively higher Ti activity levels, however, α-TiO would be expected to form on the alumina side of the joint interface, whereby it is in a local thermodynamic equilibrium with the Ti_4Cu_2O phase. The formation of these phases with the chemical activities of Ti at 945°C is shown in Figure 5.56. In this study, the formation of the γ-TiO phase on the alumina side

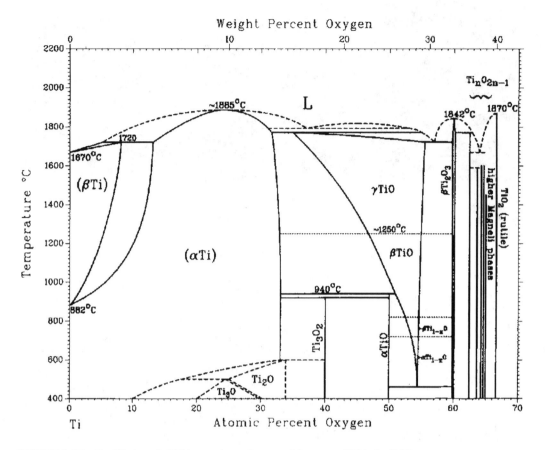

FIGURE 5.55 Equilibrium O–Ti binary phase diagram (Murray and Wriedt, 1987).

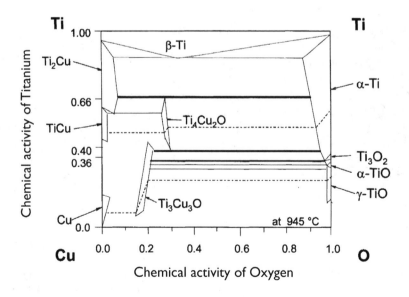

FIGURE 5.56 Titanium activity diagram corresponding to the Ti-Cu-O ternary at 945°C (Yang et al., 2000).

Microstructural Evolution

of the joint interface is coherent with that of the Ti_3Cu_3O phase on the braze side of the joint interface, and indicative of the relatively lower Ti activity levels in these brazing experiments.

5.3.3 Secondary Phase Interaction

Liquid phase sintered D-96 AG alumina, composed of 96.0 wt.% Al_2O_3 alumina, was found to contain 3.2 wt.% SiO_2 silica as the main secondary phase, as well as 0.55 wt.% MgO magnesia and 0.06 wt.% CaO calcia. Solid-state sintered D-100 AG alumina, composed of 99.7 wt.% Al_2O_3 alumina, was found to contain 0.30 wt.% SiO_2 silica, 0.03 wt.% MgO magnesia and 0.02 wt.% CaO calcia. This phase analysis was performed using EPMA, with average values based on 10 measurements (see Section 4.1).

The presence of secondary phase elements at the interfaces of D-96 AG and D-100 AG brazed joints was first studied using EPMA line scans. Elemental profiles from each scan performed across the interfaces of both the D-96 AG and D-100 AG brazed joints made using 100-μm-thick TICUSIL® braze preforms are shown in Figures 5.57b and 5.57d, respectively.

EPMA line scans performed across the interfaces of D-96 AG brazed joints showed negligible quantities of Mg and Ca; however, up to 10 at.% Si was observed in the vicinity of the Ti-rich reaction

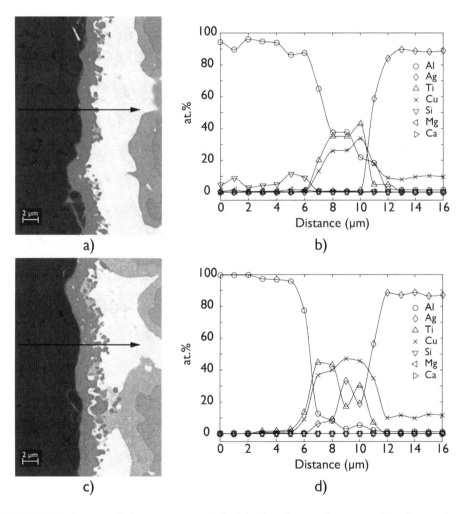

FIGURE 5.57 Backscattered electron images of the joint interfaces and corresponding elemental profiles (at.%) performed using electron probe microanalysis in (a)-(b) D-96 AG and (c)-(d) D-100 AG brazed joints made using 100-μm-thick TICUSIL® braze preforms. Line scans are indicated by arrows.

layers. A peak overlap in concentration of the elements: Ti, Cu and Si, was observed on the alumina side of the joint interface, where ~5 at.% Si was detected (Figures 5.57a and 5.57b). At the interfaces of D-100 AG brazed joints, negligible quantities of all Mg, Ca and Si were observed (Figures 5.57c and 5.57d). These results showed clear differences between the interfacial chemistries of joints made using D-96 AG and D-100 AG alumina ceramics, i.e. with significant quantities of Si at the interfaces of D-96 AG brazed joints. The presence of Si was not observed earlier, when characterising the reaction layers of these joints using SEM and TEM techniques. Such a result would also be in contradiction with the literature (see Section 2.4.1). Therefore, it was postulated that Si was distributed only at specific locations on the interface.

In this study, the presence of Si at the interfaces of D-96 AG brazed joints was observed, using SEM-EDX techniques, to be non-uniform and discontinuous across the bond line. The presence of Si was observed only at the locations where the triple pocket grain boundary regions of the D-96 AG alumina surface intersected with the Ti-rich reaction layers, beneath the γ-TiO layer on the alumina side of the joint interface (Figure 5.58).

These locations of secondary phase interaction, characterised using backscattered electron imaging, showed the formation of a relatively darker phase as compared with the $Ti_3(Cu+Al)_3O$ layer which formed on the braze side of the joint interface. These results appeared to show that Ti and/or Cu from the braze alloy had penetrated approximately 300 to 500 nm deep into the triple pocket grain boundary regions of the faying surfaces of D-96 AG alumina, where these elements had interacted with the relatively lower atomic number elements, Si and/or Mg. This secondary phase interaction was observed at every location where the triple pocket grain boundary regions of the D-96 AG alumina surface intersected with the Ti-rich reaction layers (Figure 5.59). No such observations were made at the interfaces of the D-100 AG brazed joints.

Using the in-situ lift-out method performed using FIB milling (see Section 3.12), TEM specimens were prepared from the locations of the interfaces of D-96 AG brazed joints, where secondary phase interaction had been observed (Figure 5.60). The first TEM specimen was prepared from a D-96 AG brazed joint made using 50-μm-thick TICUSIL® braze preform and the second TEM specimen was prepared from a D-96 AG brazed joint made using a 100-μm-thick TICUSIL® braze preform.

In the first TEM specimen, STEM-EDX analysis showed that the secondary phase interaction had led to the formation of Ti_5Si_3, with composition 67.4Ti-26.9Si-5.7O wt.% (Figure 5.61). These Ti_5Si_3 compounds were observed to have formed between a triple pocket secondary phase grain, with composition 25.6Si-10.0Al-7.0Mg-57.4O wt.%, and the Ti-rich reaction layers; γ-TiO and $Ti_3(Cu+Al)_3O$. The resulting Ti_5Si_3 compounds, which appeared to have formed following the partial consumption of both the secondary phase grains and the Ti-rich reaction layers, were dispersed in an area that was approximately 1.0 μm wide. At the centre of this interaction region, several Ag particles that were 30 to 60 nm in diameter were also observed (Figures 5.62 and 5.63).

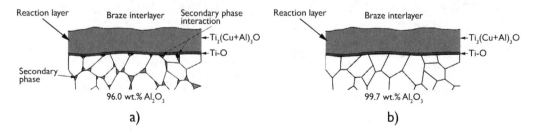

FIGURE 5.58 Schematics of the interfaces of (a) D-96 AG and (b) D-100 AG brazed joints. Secondary phase interaction occurred at the interfaces of brazed joints made using D-96 AG alumina at locations where the triple pocket grain boundary regions of the alumina surface intersected with the Ti-rich reaction layers. No secondary phase interaction was observed in D-100 AG brazed joints.

Microstructural Evolution

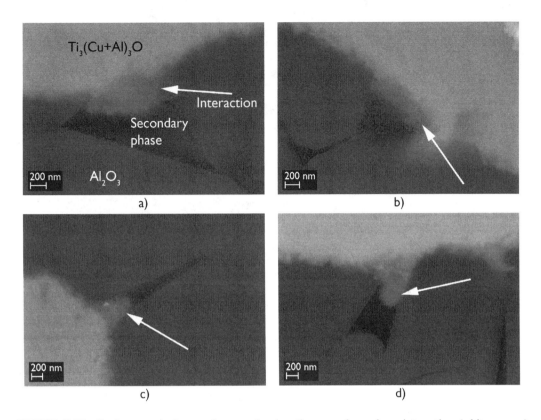

FIGURE 5.59 Backscattered electron images showing the secondary phase interaction (white arrows) observed in D-96 AG brazed joints, at locations where the triple pocket grain boundary regions of the alumina surface intersected with the Ti-rich reaction layers.

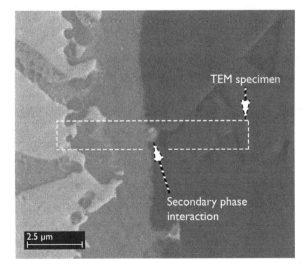

FIGURE 5.60 Ion beam image showing a region of the interface in a D-96 AG brazed joint made using a 50-μm-thick TICUSIL® braze preform, at which a TEM specimen was prepared using FIB milling according to the lift-out method. The dashed lines indicate the area from which a lamella was lifted out.

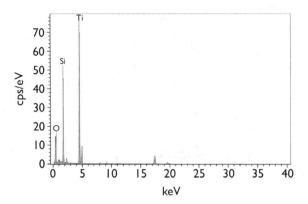

FIGURE 5.61 STEM-EDX spectrum acquired from Ti_5Si_3, corresponding to Figure 5.64.

Electron diffraction was used to confirm the hexagonal Ti_5Si_3 phase of the P63/mcm (193) space group with lattice parameters $a = b = 7.42$ Å and $c = 5.15$ Å (Figure 5.66). This was consistent with the selected area diffraction patterns of the Ti_5Si_3 phase as reported elsewhere in the literature (Lee et al., 2015).

The second TEM specimen from a similar interfacial region of another D-96 AG brazed joint, this time made using a 100-μm-thick TICUSIL® braze preform, showed a similar result, i.e. secondary phase interaction. Again, the formation of Ti_5Si_3, with composition 74.1Ti-23.3Si-2.6O wt.% was observed at locations where the triple pocket grain boundary regions of the D-96 AG alumina surface intersected with the Ti-rich reaction layers (Figure 5.64). At the centre of the 0.5 μm interaction region, small Ag particles were again observed.

These results showed that the formation of the Ti_5Si_3 phase resulted from the Si-rich secondary phase in D-96 alumina. With relatively higher amounts of Ti available to diffuse to the interface, in D-96 joints made using 150- and 250-μm-thick TICUSIL® braze preforms, it is likely that these Ti-Si compounds may have also formed in the same way.

The formation of Ti_5Si_3 at the interfaces of brazed joints made using D-96 alumina and TICUSIL® may have been due to the presence of SiO_2 at the triple pocket grain boundary regions of the alumina surface reacting with Ti, following the diffusion of Ti to the joint interfaces. A possible reaction mechanism is described in Equations 5.1 and 5.2. It is postulated that following the reduction of SiO_2 by Ti, the liberation of Si and its subsequent reaction with Ti led to the formation of Ti_5Si_3. These results were similar to the formation of Ti_5Si_3 reported in a study by Kang and Selverian (1992), in which a 3-μm-thick Ti coating was deposited onto the surface of a 95.0 wt.% Al_2O_3 alumina ceramic before the coated specimen was heated in a vacuum atmosphere to 980°C for 10 min, simulating a typical brazing experiment.

$$SiO_2 + 2Ti \rightarrow Si + 2TiO \tag{5.1}$$

$$3Si + 5Ti \rightarrow Ti_5Si_3 \tag{5.2}$$

The formation of Ti_5Si_3 has previously been reported to occur at the interfaces of joints made using Ti-containing braze alloys and other Si-based ceramic materials, such as SiC (Tamai and Naka, 1996; Liu et al., 2010) and Si_3N_4 (Asthana et al., 2013; Loehman et al., 1990; Naka et al., 1985; Suganuma et al., 1988; Singh et al., 2011; Singh et al., 2012; Tillman et al., 1996; Xian and Si, 1990). In these systems, a Ti_5Si_3 layer can typically form as a result of either free Si on the ceramic surface, or from Si that is liberated following the reduction of the SiC or Si_3N_4 ceramic surfaces by Ti.

Microstructural Evolution

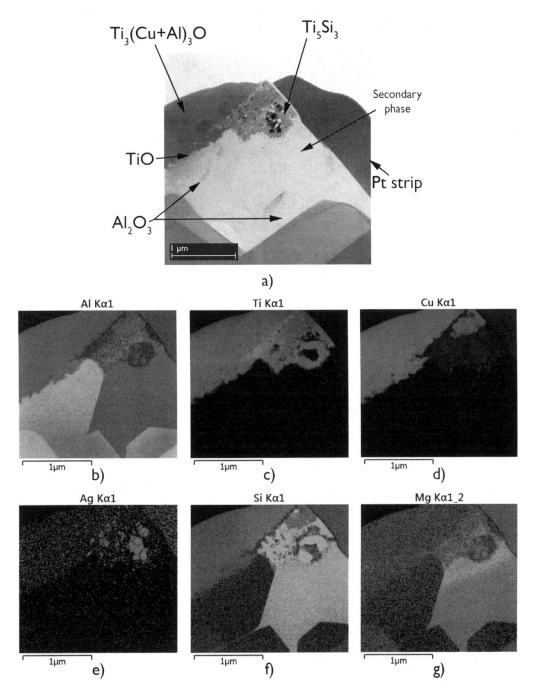

FIGURE 5.62 (a) Bright field STEM image of a typical region at the joint interface in a D-96 AG brazed joint made using a 50-μm-thick TICUSIL® braze preform, where secondary phase interaction had occurred (at locations where the triple pocket grain boundary regions of the D-96 AG alumina surface intersected with the Ti-rich reaction layers) and corresponding STEM-EDX maps showing distributions of (b) Al, (c) Ti, (d) Cu, (e) Ag, (f) Si and (g) Mg.

Asthana et al. (2013) observed, using TEM techniques, that the reaction layer which formed at the interfaces of Si_3N_4-to-Si_3N_4 brazed joints made using Cu-ABA (Cu-3Si-2Al-2.25Ti wt.%) consisted of TiN and Ti_5Si_3 reaction products. Tamai and Naka (1996) formed Si_3N_4-to-Si_3N_4 and SiC-to-SiC brazed joints using a 56Ag-38Cu-5Ti wt.% braze alloy. The interfaces of these

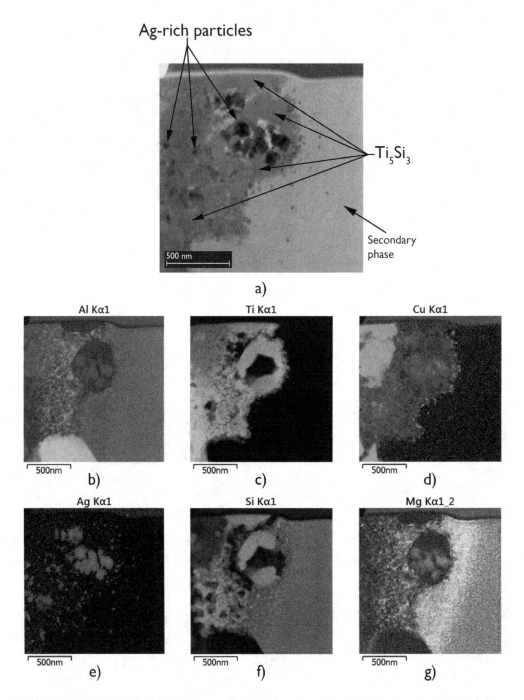

FIGURE 5.63 (a) Bright field STEM image of a typical region at the joint interface in a D-96 AG brazed joint made using a 50-μm-thick TICUSIL® braze preform, where secondary phase interaction had occurred (higher magnification image of Figure 5.64) and corresponding STEM-EDX maps showing distributions of (b) Al, (c) Ti, (d) Cu, (e) Ag, (f) Si and (g) Mg.

Si_3N_4-to-Si_3N_4 brazed joints consisted of TiN and Ti_5Si_3 reaction products while those of the SiC-to-SiC brazed joints consisted of TiC and Ti_5Si_3 reaction products. Tillman et al. (1996) formed Si_3N_4-to-Si_3N_4 and Si_3N_4-to-stainless steel brazed joints using TICUSIL®, and in both sets of these joints, the reaction layer, characterised using XRD, was found to consist of TiN and Ti_5Si_3 reaction

Microstructural Evolution

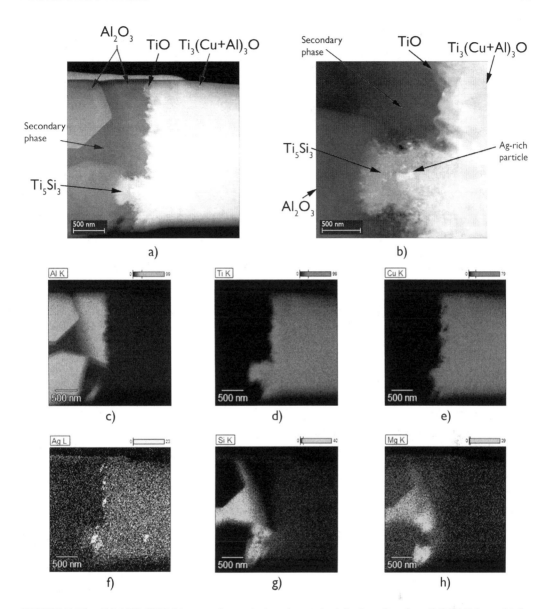

FIGURE 5.64 HAADF STEM image of a typical region at the joint interface in a D-96 AG brazed joint made using a 100-μm-thick TICUSIL® braze preform, where secondary phase interaction had occurred: (a) lower magnification and (b) higher magnification, and corresponding STEM-EDX maps showing distributions of (c) Al, (d) Ti, (e) Cu, (f) Ag, (g) Si and (h) Mg.

products. Lastly, Liu et al. (2010) observed, using TEM techniques, that the reaction layer which formed at the interfaces of SiC-to-SiC brazed joints made using Cusil ABA® consisted of a TiC layer and Ti$_5$Si$_3$ layer.

In all of these systems, preventing the formation of a Ti$_5$Si$_3$ layer on the ceramic side of the joint interface (which can act as a diffusion barrier to Ti), whilst also preventing the diffusion of Si into the braze interlayer (which can lead to brittle reaction products), has been reported to help minimise any degradation in joint strength (Urai and Naka, 1999; Xian and Si, 1990). The formation of a Ti$_5$Si$_3$ layer in these systems, therefore, has typically been found to adversely affect joint strength. In this study, however, Ti$_5$Si$_3$ was not observed as a continuous layer at the joint interface but instead as 0.5 to 1.0 μm wide interaction regions. The formation of Ti$_5$Si$_3$

was observed only at the locations where the triple pocket grain boundary regions of the D-96 alumina surface intersected with the Ti-rich reaction layers (Figure 5.65). At these locations, the Ti$_5$Si$_3$ compounds did not act as a diffusion barrier to Ti nor was any Si detected to have diffused into the braze interlayer.

In this study, the D-96 AG brazed joints were formed at a brazing temperature of 850°C. According to the equilibrium Ti-Si binary phase diagram, several Ti-Si compounds can form at this temperature (Figure 5.67). The composition of the Ti-Si phase which forms, however, can depend on the Ti and Si concentrations and their activity levels, i.e. relatively higher Ti activity levels are required to form Ti$_5$Si$_3$ rather than Ti$_2$Si (Nicholas, 1997).

In a similar study, Ali et al. (2015) used 50-μm-thick Cusil ABA® braze foils to produce several sets of alumina-to-alumina brazed joints using different grades of alumina. This included 95.0 wt.% Al$_2$O$_3$ alumina and 99.7 wt.% Al$_2$O$_3$ alumina ceramics. Brazing experiments were performed at 845°C for 15 min. It was reported that the alumina purity did not significantly affect the interfacial structure of the brazed joints. The secondary phase elements were observed in the reaction layers, with ~5 at.% Si in the Ti$_3$Cu$_3$O layer. Ali et al. (2015) suggested that Si and Al from the alumina ceramics could form solid solutions with the Ti$_3$Cu$_3$O layer (i.e. Ti$_3$(Cu + Al + Si)$_3$O. Isolated

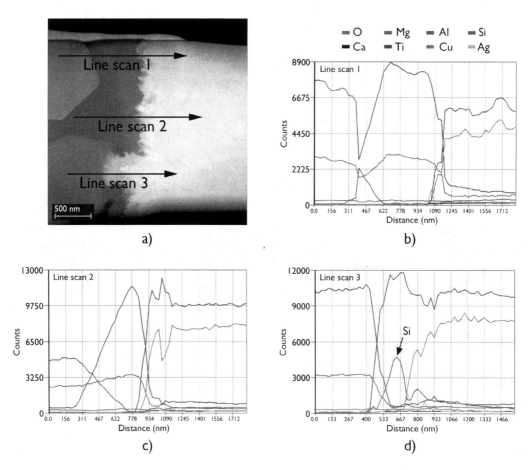

FIGURE 5.65 (a) HAADF STEM image of a typical region at the joint interface in a D-96 AG brazed joint made using a 100-μm-thick TICUSIL® braze preform, where secondary phase interaction led to the formation of Ti$_5$Si$_3$, (b) and (c) line scans 1 and 2 passed through the secondary phase grain, however, since this grain was separated from the TiO and Ti$_3$(Cu+Al)$_3$O reaction layers by an alumina grain, Ti$_5$Si$_3$ did not form, d) line scan 3 passed through Ti$_5$Si$_3$, which formed where the secondary phase grain was in contact with the TiO and Ti$_3$(Cu+Al)$_3$O reaction layers.

Microstructural Evolution

FIGURE 5.66 (a) High magnification bright field TEM image of a selected Ti_5Si_3 crystal and (b) the corresponding selected area diffraction pattern acquired with zone axes of [100].

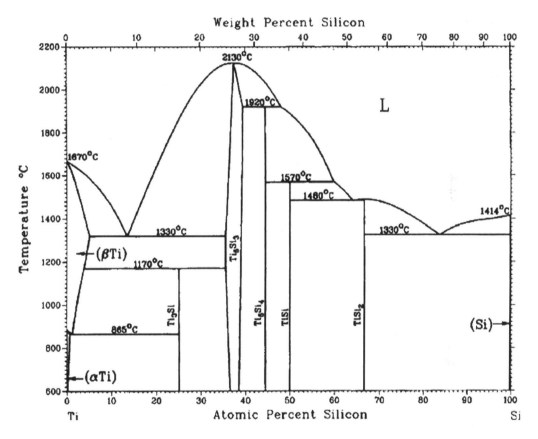

FIGURE 5.67 Equilibrium Ti-Si binary phase diagram (Massalski, 1990).

particles of the Ti_2Si phase, ~2 μm in size, were also observed on alumina alongside the titanium oxide particles, after the brazed joints were heat treated for 75 h.

While the observations by Ali et al. (2015) support the formation of Ti-Si compounds at the alumina/Ag-Cu-Ti joint interface, in this study Ti_5Si_3 was observed with regular spatial distribution only at the locations where the triple pocket grain boundary regions of the D-96 AG alumina surface intersected with the Ti rich reaction layers; Si was not observed within the $Ti_3(Cu + Al)_3O$

reaction layer. Therefore, the interfacial structures of the D-96 AG brazed joints were found to be markedly different to those of D-100 AG brazed joints (Figure 5.58). The effect of secondary phase interaction on the strength of D-96 AG brazed joints is discussed in Section 6.1.4.

5.4 MICROSTRUCTURES OF BRAZED JOINTS IN GROUND-AND-HEAT-TREATED CONDITION

Post-grinding heat treatment performed at 1550°C for 1 h prior to brazing was found to have affected the microstructures of both D-96 GHT and D-100 GHT brazed joints as compared to D-96 AG and D-100 AG brazed joints.

5.4.1 D-96 GHT Brazed Joints

D-96 GHT brazed joints made using 50-µm-thick TICUSIL® braze preforms had an average brazed joint thickness of 43 µm and an average reaction layer thickness of 1.5 µm (Figure 5.68a). The brazed joint microstructure consisted of the same phases that were observed in the D-96 AG and D-100 AG brazed joints made using 50-µm-thick TICUSIL® braze preforms (Figure 5.32a). The braze interlayer also consisted of the typical Ag-rich and Cu-rich phases distributed in a eutectic-like structure. The reaction layer composed of $Ti_3(Cu + Al)_3O$ in these D-96 GHT brazed joints was found to be have a similar thickness to that of the $Ti_3(Cu + Al)_3O$ layer observed in D-96 AG brazed joints made using 50-µm-thick TICUSIL® braze preforms. However, its structure appeared to be slightly broken at several isolated regions of the joint interface (Figure 5.68b).

The average brazed joint thickness of D-96 GHT brazed joints made using 50-µm-thick TICUSIL® braze preforms was 66% greater than that in the D-96 AG brazed joints made using 50-µm-thick TICUSIL® braze preforms. Furthermore, in these D-96 GHT brazed joints, the brazed joint thickness was found to be somewhat non-uniform. The most notable difference between the microstructures of D-96 GHT and D-96 AG brazed joints made using 50-µm-thick TICUSIL® braze preforms was the vacant grain boundary regions observed within ~10 µm of the D-96 GHT alumina surface. This was consistent with earlier observations where the post-grinding heat treatment was suggested to have led to the migration of the secondary phase in the alumina surface (see Section 4.5). During brazing, these vacant regions, characteristic of the triple pockets grain boundary regions otherwise accommodated by the secondary phase (see Figure 4.11), were instead found to have been infiltrated by the braze alloy (Figure 5.69).

D-96 GHT brazed joints made using 100-µm-thick TICUSIL® braze preforms had an average brazed joint thickness of 54 µm and an average reaction layer thickness of 2.3 µm. The brazed

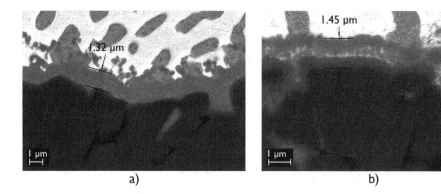

FIGURE 5.68 Backscattered electron images of the joint interface in a typical D-96 GHT brazed joint made using a 50-µm-thick TICUSIL® braze preform at (a) region where the $Ti_3(Cu + Al)_3O$ was continuous and (b) region where the $Ti_3(Cu + Al)_3O$ appeared to be broken.

Microstructural Evolution

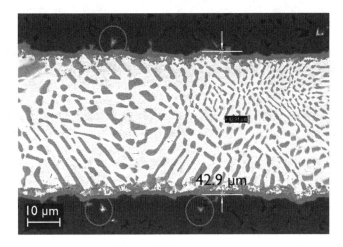

FIGURE 5.69 Backscattered electron image showing the microstructure of a typical D-96 GHT brazed joint made using a 50-μm-thick TICUSIL® braze preform. Circles indicate braze infiltration into vacant triple pocket grain boundary regions of the alumina surface which may have been created during the post-grinding heat treatment, prior to brazing.

joint microstructure consisted of the same phases observed in D-96 AG and D-100 AG brazed joints made using 100-μm-thick TICUSIL® braze preforms (Figure 5.32b), i.e., the braze interlayer consisted of the isolated Cu_4Ti_3 and Cu_4Ti phases and Ag-Cu eutectic-like transition regions; the reaction layer was characterised as $Ti_3(Cu + Al)_3O$.

The thicknesses of the $Ti_3(Cu + Al)_3O$ layers in D-96 GHT and D-96 AG brazed joints made using 100-μm-thick TICUSIL® braze preforms were similar; however, the average brazed joint thicknesses differed. In these D-96 GHT brazed joints, the brazed joint thickness was highly non-uniform, with minimum and maximum brazed joint thicknesses of 37 and 85 μm, respectively, at each of the joint edges (Figure 5.70). The average brazed joint thickness of these D-96 GHT brazed joints was also ~38% thicker than that of the D-96 AG brazed joints made using the same TICUSIL® braze preform thickness of 100 μm.

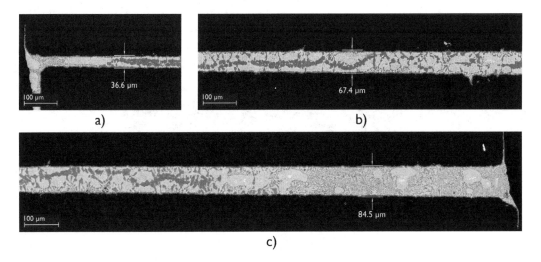

FIGURE 5.70 Backscattered electron images showing the microstructure of a typical D-96 GHT brazed joint made using a 100-μm-thick TICUSIL® braze preform, (a) left edge, (b) central region and (c) right edge, of the brazed joint cross-section.

A second set of D-96 GHT brazed joints made using 100-μm-thick TICUSIL® braze preforms showed the same high degree of non-uniformity in the brazed joint thickness. In these D-96 GHT brazed joints, the average brazed joint thickness was found to be 69 μm, with minimum and maximum brazed joint thicknesses of 21 and 85 μm, respectively, at each of the joint edges. These results confirmed that the post-grinding heat treatment had induced non-uniformity in the brazed joint thicknesses. This may have been caused by braze infiltration into the vacancies of the D-96 GHT alumina surface as a result of the migration of the secondary phase. The migration of the secondary phase may have also been influenced by the orientation of the faying surfaces of the short test bars during the heat treatment (see Figure 3.5b). As a result, braze infiltration may have also been non-uniform. It should be noted that the overall effect of the post-grinding heat treatment was found not to adversely affect the average flexural strength of the D-96 GHT standard alumina test bars, in which a 5.2% increase in strength was observed (see Section 4.5).

Braze infiltration in a fissured D-96 GHT alumina surface was observed in the cross-sectional microstructures of the brazed joints made using 100-μm-thick TICUSIL® braze preforms (Figure 5.71). This was similar to the braze infiltration which was observed in D-96 GHT brazed joints made using 50-μm-thick TICUSIL® braze preforms.

During the post-grinding heat treatment, which was performed at the near-sintering temperature of 1,550 °C, the secondary phase in the form of a liquid may have retracted away from the alumina

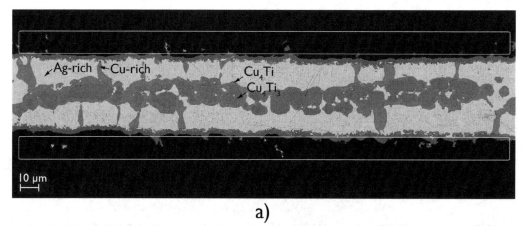

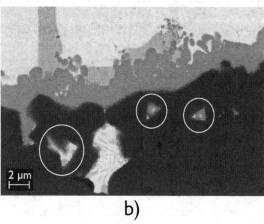

FIGURE 5.71 Backscattered electron images of (a) the brazed joint microstructure and (b) the joint interface, in a typical D-96 GHT brazed joint made using a 100-μm-thick TICUSIL® braze preform. Rectangles in (a) and circles in (b) indicate regions where braze infiltration occurred, into the vacant triple pocket grain boundary regions of the alumina surface, which may have been created during the post-grinding heat treatment.

surface filling crevices formed during grinding or otherwise. The liquid phase may have moved into grinding-induced defects, filling or blunting cracks in the D-96 GHT alumina surface (Moorhead and Simpson, 1993). Lange (1970) showed, using dye penetrant inspection, that cracks induced by thermally shocking alumina specimens could be healed during a heat treatment procedure through the migration of the liquid phase. This mechanism has also been reported elsewhere in the literature (Kirchner et al., 1970; Mizuhara and Mally, 1985).

Post-grinding heat treatment created vacancies in the intergranular and triple pocket grain boundary regions to produce a fissured D-96 GHT alumina surface. As a result, during brazing, the braze alloy was found to have infiltrated into these regions, occupying them ~10 μm into the D-96 GHT alumina surface. Migration of the finite secondary phase into damaged regions of the alumina surface, therefore, may have created new vacancies.

D-96 GHT brazed joints made using 150-μm-thick TICUSIL® braze preforms had an average brazed joint thickness of 86 μm and an average reaction layer thickness of 2.9 μm. The brazed joint microstructure consisted of the same phases observed in D-96 AG and D-100 AG brazed joints made using 150-μm-thick TICUSIL® braze preforms (Figure 5.32c), i.e., the braze interlayer consisted of a discontinuous multi-layered Cu-Ti structure composed of Ti, $CuTi_2$, CuTi, Cu_4Ti_3 and Cu_4Ti alongside the formation of the isolated Cu_4Ti_3 and Cu_4Ti phases. The typical Ag-rich and Cu-rich phases were also observed throughout the braze interlayer and the reaction layer was characterised as $Ti_3(Cu + Al)_3O$ (Figure 5.72).

The thicknesses of the $Ti_3(Cu + Al)_3O$ layers in the D-96 GHT and D-96 AG brazed joints made using 150-μm-thick TICUSIL® braze preforms were similar; however, the average brazed joint thicknesses again differed. The average brazed joint thickness of these D-96 GHT brazed joints made using 100-μm-thick TICUSIL® braze preforms was 31% greater than that of the D-96 AG brazed joints made using the same TICUSIL® braze preform thickness of 100 μm. Furthermore, the brazed joint thickness of these D-96 GHT brazed joints appeared once again to be non-uniform.

In D-96 GHT brazed joints made using 150-μm-thick TICUSIL® braze preforms, braze infiltration, ~20 μm into the intergranular regions of the alumina surface, was observed. This was similar to the braze infiltration observed in D-96 GHT brazed joints made using 50- and 100-μm-thick TICUSIL® braze preforms.

These results indicated that the post-grinding heat treatment could affect the microstructures of D-96 GHT brazed joints (Table 5.8), as compared to the microstructures of D-96 AG brazed joints despite the same brazing conditions (Table 5.7). The post-grinding heat treatment did not appear to affect the composition or thickness of the reaction layer, characterised as $Ti_3(Cu + Al)_3O$ in all of the D-96 AG and D-96 GHT brazed joints made using 50- to 150-μm-thick TICUSIL® braze preforms (Figure 5.73a).

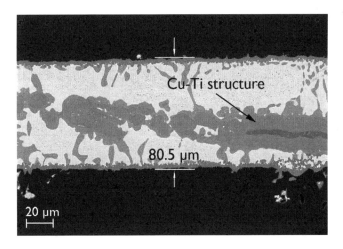

FIGURE 5.72 Backscattered electron image showing the microstructure of a typical D-96 GHT brazed joint made using a 150-μm-thick TICUSIL® braze preform.

TABLE 5.8
Properties of D-96 GHT Brazed Joints Made Using 50- to 150-μm-Thick TICUSIL® Braze Preforms

Sample Set	Alumina Purity, (wt.% Al$_2$O$_3$)	Surface Condition	Braze Preform Thickness (μm)	Specimens Characterised	Reaction Layer Thickness (μm)	Average Brazed Joint Thickness (μm)
5	D-96	GHT	50	1	1.5 ± 0.1	43 ± 1
6	D-96	GHT	100	1	2.3 ± 0.1	54 ± 4
7	D-96	GHT	100	1	2.4 ± 0.1	43 ± 3
8	D-96	GHT	150	1	2.9 ± 0.1	86 ± 4

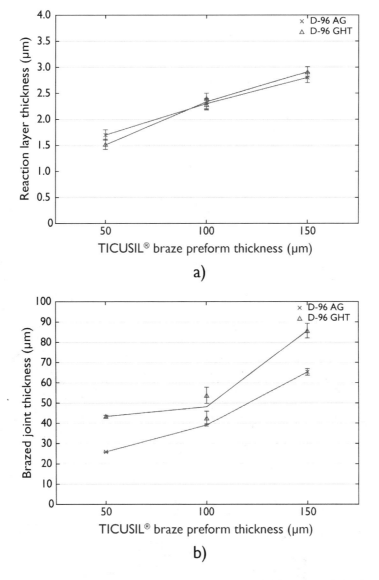

FIGURE 5.73 Comparisons between (a) the average reaction layer thicknesses and (b) the average brazed joint thicknesses, in D-96 AG and D-96 GHT brazed joints made using 50- to 150-μm-thick TICUSIL® braze preforms.

Microstructural Evolution

However, the uniformity of the D-96 GHT brazed joint thicknesses appeared to have been significantly affected, particularly when 100-µm-thick TICUSIL® braze preforms were used (Figure 5.73b). The post-grinding heat treatment had created vacant intergranular regions to an average depth of ~20 µm in the D-96 GHT alumina surface. In these regions, characteristic of where the secondary phase had migrated away from, braze infiltration was observed.

5.4.2 D-100 GHT Brazed Joints

D-100 GHT brazed joints made using 50-µm-thick TICUSIL® braze preforms had an average brazed joint thickness of 38 µm and an average reaction layer thickness of 2.0 µm (Figure 5.74a). The brazed joint microstructure consisted of the same phases observed in the D-96 AG and D-100 AG brazed joints made using 50-µm-thick TICUSIL® braze preforms (Figure 5.32a). The braze interlayer consisted of Ag-rich and Cu-rich phases which were distributed in a eutectic-like structure; the reaction layer was characterised as $Ti_3(Cu + Al)_3O$.

The thickness of the $Ti_3(Cu+Al)_3O$ layer in D-100 GHT brazed joints made using 50-µm-thick TICUSIL® braze preforms was 25% greater than that of the $Ti_3(Cu+Al)_3O$ layer in D-100 AG brazed joints made using the same TICUSIL® braze preform thickness of 50 µm. Furthermore, the average brazed joint thickness was observed to be non-uniform, with minimum and maximum thicknesses of 16 and 67 µm, respectively, at each of the joint edges. The average brazed joint thickness of these D-100 GHT brazed joints was ~80% greater as compared with that of the D-100 AG brazed joints. This result showed that the non-uniformity in the brazed joint thickness, induced by the post-grinding heat treatment, was unlikely to be associated with alumina purity. Braze infiltration is discussed further in Section 5.5.3.

D-100 GHT brazed joints made using 100-µm-thick TICUSIL® braze preforms had an average brazed joint thickness of 34 µm and an average reaction layer thickness of 2.9 µm (Figure 5.74b). The brazed joint microstructure consisted of the same phases observed in D-96 AG and D-100 AG brazed joints made using 100-µm-thick TICUSIL® braze preforms (Figure 5.32b). The braze interlayer consisted of isolated Cu_4Ti_3 and Cu_4Ti phases and Ag-Cu eutectic-like transition regions; the reaction layer was characterised as $Ti_3(Cu+Al)_3O$. The average brazed joint and reaction layer thicknesses of D-100 GHT and D-100 AG brazed joints made using 100-µm-thick TICUSIL® braze preforms were similar.

D-100 GHT brazed joints made using 150-µm-thick TICUSIL® braze preforms had an average brazed joint thickness of 94 µm and an average reaction layer thickness of 3.3 µm (Figure 5.74c). The brazed joint microstructure consisted of the same phases observed in D-96 AG and D-100 AG brazed joints made using 150-µm-thick TICUSIL® braze preforms (Figure 5.32c). The braze interlayer consisted of a discontinuous multi-layered Cu-Ti structure composed of Ti, $CuTi_2$, CuTi, Cu_4Ti_3 and Cu_4Ti alongside the formation of the isolated Cu_4Ti_3 and Cu_4Ti phases. The typical Ag-rich and Cu-rich phases were also observed throughout the braze interlayer and the reaction layer was characterised as $Ti_3(Cu + Al)_3O$ (Figure 5.72).

The thicknesses of the $Ti_3(Cu + Al)_3O$ layers in D-100 GHT and D-100 AG brazed joints made using 150-µm-thick TICUSIL® braze preforms were similar. However, the average brazed joint

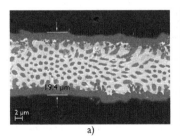
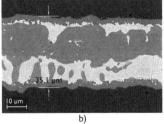
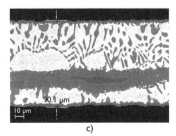

a) b) c)

FIGURE 5.74 Backscattered electron images showing the microstructures of typical D-100 GHT brazed joints made using (a) 50 µm, (b) 100 µm and (c) 150-µm-thick TICUSIL® braze preforms.

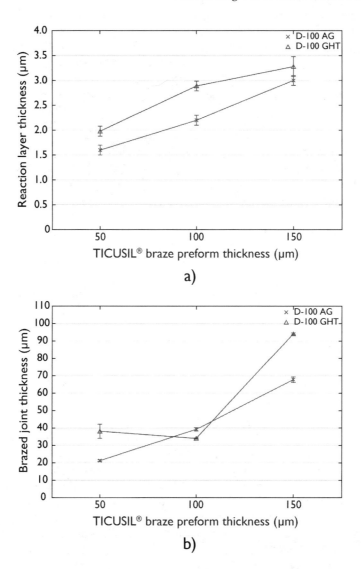

FIGURE 5.75 Comparisons between (a) the average reaction layer thicknesses and (b) the average brazed joint thicknesses, in D-100 AG and D-100 GHT brazed joints made using 50- to 150-μm-thick TICUSIL® braze preforms.

thickness of 94 μm in these D-100 GHT brazed joints was ~39% greater than that in the corresponding D-100 AG brazed joints. Furthermore, the brazed joint thickness of these D-100 GHT brazed joints was again found to be somewhat non-uniform.

These results showed that the microstructural evolution which occurred in the braze interlayer of D-96 AG and D-100 AG brazed joints as the TICUSIL® braze preform thickness was increased from 50 to 150 μm occurred independently of whether D-96 and D-100 alumina ceramics were brazed in as-ground or in ground-and-heat-treated conditions. Therefore, this microstructural evolution was likely to be a result of the braze preform thicknesses and the brazing conditions selected.

Post-grinding heat treatment led to an increase in the average brazed joint thickness in both D-96 GHT and D-100 GHT brazed joints, as compared to the average brazed joint thicknesses of D-96 AG and D-100 AG brazed joints, respectively. The post-grinding heat treatment also led to a high degree of non-uniformity in the brazed joint thicknesses of joints made using D-96 GHT and D-100

TABLE 5.9
Properties of D-100 GHT Brazed Joints Made Using 50- to 150-μm-thick TICUSIL® Braze Preforms

Sample Set	Alumina Purity, (wt.% Al$_2$O$_3$)	Surface Condition	Braze Preform Thickness (μm)	Specimens Characterised	Reaction Layer Thickness (μm)	Average Brazed Joint Thickness (μm)
13	D-100	GHT	50	1	2.0 ± 0.1	38 ± 4
14	D-100	GHT	100	1	2.9 ± 0.1	34 ± 1
15	D-100	GHT	150	1	3.3 ± 0.2	94 ± 1

GHT alumina, using 50- to 150-μm-thick TICUSIL® braze preforms (Tables 5.8 and 5.9). This non-uniformity in the brazed joint thickness, therefore, was also independent of alumina purity.

While in D-96 GHT brazed joints, the post-grinding heat treatment created fissures in the alumina surface, where braze infiltration was observed, this did not occur in D-100 GHT brazed joints. The lack of any secondary phase in solid-state sintered D-100 alumina, therefore, may have prevented any secondary phase migration and subsequent braze infiltration.

The average reaction layer thickness in D-100 GHT brazed joints was thicker than that in D-100 AG brazed joints (Figure 5.75). This increase in the average reaction layer thickness was observed in all of the D-100 GHT brazed joints made using 50- to 150-μm-thick TICUSIL® braze preforms and appeared to have occurred, therefore, due to the post-grinding heat treatment. No such increase in the average reaction layer thickness was observed in the D-96 GHT brazed joints as compared to that in the D-96 AG brazed joints. These results require further investigation.

In the literature, post-grinding heat treatment has been used to recover grinding damage in both liquid phase sintered and solid-state sintered alumina ceramics prior to brazing (see Table 2.18). In this study, post-grinding heat treatment was found to affect the microstructures of brazed joints made liquid phase sintered and solid-state sintered alumina ceramics differently. In Section 4.5, post-grinding heat treatment was found to have led to a small increase in the average flexural strength of the D-96 GHT standard alumina test bars, and a small decrease in the average flexural strength of the D-100 GHT standard alumina test bars. The effect of post-grinding heat treatment on the strengths of these brazed joints is discussed in Section 6.3.

In summary, the post-grinding heat treatment led to an increase in the average brazed joint thicknesses of both the D-96 GHT and D-100 GHT brazed joints as compared to the D-96 AG and D-100 AG brazed joints, respectively. A high degree of non-uniformity, as a result of the post-grinding heat treatment, in the thicknesses of both the D-96 GHT and D-100 GHT brazed joints was observed. The post-grinding heat treatment led to braze infiltration into the intergranular regions of the D-96 GHT alumina surface, however, this was not observed in the D-100 GHT brazed joints. Lastly, the post-grinding heat treatment led to an increase in the average reaction layer thickness in the D-100 GHT brazed joints as compared to the D-100 AG brazed joints, however, this was not observed in the D-96 GHT brazed joints. In this study, the post-grinding heat treatment, prior to brazing, was found to affect the microstructures of D-96 GHT and D-100 GHT brazed joints differently and, as a result, may also affect the strengths of both of these sets of joints differently (see Chapter 6).

5.4.3 Braze Infiltration

Electron probe microanalysis (EPMA) line scans performed across the interfaces of D-96 GHT brazed joints showed a lack of any significant Si concentration in the vicinity of the Ti-rich reaction layers (Figure 5.76). Therefore, the elemental profiles in the line scans performed across the interfaces of D-96 GHT brazed joints were markedly different from those of the line scans performed at the interfaces of D-96 AG brazed joints, in which ~5 at.% Si was observed at the joint interface (Figure 5.57).

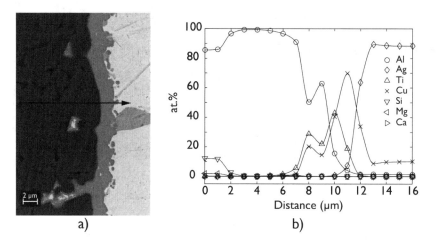

FIGURE 5.76 Backscattered electron image of the joint interface in a (a) D-96 GHT brazed joint made using a 100-μm-thick TICUSIL® braze preform and (b) corresponding elemental profile (at.%) of a line scan performed across the interface using electron probe microanalysis. Line scan is indicated by an arrow.

It is postulated that this lack of Si, at the interfaces of D-96 GHT brazed joints, occurred due to the post-grinding heat treatment whereby the secondary phase in part evaporated from the alumina surface and in part formed a liquid that flowed into defective areas of the alumina surface.

TEM techniques were used to characterise the regions of the D-96 GHT alumina surface in joints made using 100-μm-thick TICUSIL® braze preforms, where braze infiltration was observed. This included the locations where the triple pocket grain boundary regions of the D-96 GHT alumina surface intersected with the Ti-rich reaction layers, where Ti_5Si_3 was earlier characterised in D-96 AG brazed joints made using 100-μm-thick TICUSIL® braze preforms.

In D-96 GHT brazed joints made using 100-μm-thick TICUSIL® braze preforms, the reaction layer comprised a polycrystalline $Ti_3(Cu+Al)_3O$ layer with composition 43.9Ti-43.2Cu-5.2Al-7.7O wt.% on the alumina side of the joint interface. Electron diffraction was used to confirm the crystal structure of this phase as the cubic Ti_3Cu_3O phase of the Fd-3m (227) space group. The alumina side of the joint interface consisted of a continuous polycrystalline TiO layer composed of 72.5Ti-1.6Cu-25.9O wt.%, with particle sizes ranging from 150 to 200 nm (Figure 5.77a). Within this TiO layer, isolated Ti_2O particles composed of 82.0Ti-2.7Al-1.5Cu-13.7O wt.% were also observed (Figure 5.77b).

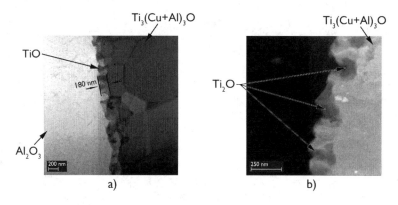

FIGURE 5.77 (a) Bright field STEM image of a typical region at joint interface in a D-96 GHT brazed joint made using a 100-μm-thick TICUSIL® braze preform showing the TiO layer and (b) HAADF image showing isolated Ti_2O particles in the TiO layer.

Microstructural Evolution 171

These results confirmed that the reaction layers which formed in both the D-100 AG and D-100 GHT brazed joints made using 100-μm-thick TICUSIL® braze preforms were similar.

At several locations where the triple pocket grain boundary regions of the D-96 GHT alumina intersected with the Ti-rich reaction layers, where braze infiltration was observed, the secondary phase had appeared to have migrated away (Figure 5.71). The triple pocket grain boundary region was instead completely accommodated by the braze alloy which led to the formation of polycrystalline Ti$_3$(Cu + Al)$_3$O structures, as Ti had reacted with the adjacent alumina grains. The perimeter of

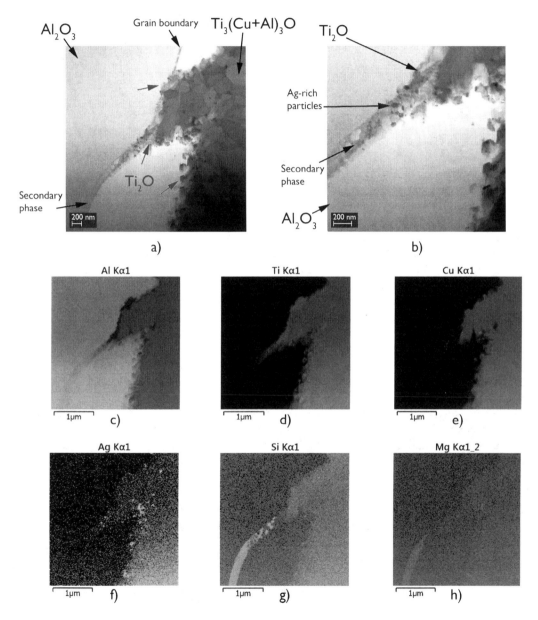

FIGURE 5.78 Bright field STEM images of typical regions in the D-96 GHT alumina surface, in a brazed joint made using a 100-μm-thick TICUSIL® braze preform, where braze infiltration occupied a triple pocket grain boundary region, (a) lower magnification and (b) higher magnification, and STEM-EDX maps showing distributions of (c) Al, (d) Ti, (e) Cu, (f) Ag, (g) Si and (h) Mg. Si and Mg are observed along the alumina grain boundaries having vacated the triple pocket grain boundary region.

the Ti$_3$(Cu + Al)$_3$O structures, surrounded by alumina grains, consisted of a TiO layer with isolated Ti$_2$O particles at the alumina/Ti$_3$(Cu + Al)$_3$O interfaces. Ag-rich particles were also observed, randomly scattered at the edges of the Ti$_3$(Cu + Al)$_3$O structures (Figure 5.78).

These results showed that despite the diffusion of Ti from the TICUSIL® braze alloy into the intergranular regions of the D-96 GHT alumina surface, no Ti-Si compounds were formed. This was inconsistent with the formation of Ti$_5$Si$_3$, which occurred at the interfaces of D-96 AG brazed joints, at similar locations where the triple pocket grain boundary regions of the D-96 AG alumina surface intersected with the Ti-rich reaction layers (see Section 5.4.3).

STEM-EDX maps performed in these triple pocket grain boundary regions of the D-96 GHT alumina surface, where braze infiltration was observed, confirmed the lack of any significant secondary phases. Instead, adjacent to these regions, the secondary phase elements Si and Mg were observed along the alumina grain boundaries (Figure 5.78). It is postulated, therefore, that the post-grinding heat treatment led to the migration of secondary phases along the alumina grain boundaries, creating vacancies in the triple pocket grain boundary regions of the D-96 GHT alumina surface, where braze infiltration subsequently occurred.

5.5 SUMMARY

The microstructures of alumina-to-alumina brazed joints made using D-96 and D-100 alumina ceramics in as-ground and ground and heat-treated conditions were studied. The effect of an increase in the TICUSIL® braze preform thickness on the microstructures of these joints was evaluated.

TICUSIL® braze foil is manufactured by a cladding process, whereby a Ti ribbon is sandwiched in an Ag-Cu eutectic before being rapidly solidified. The distribution of Ti in the 50- and 100-μm-thick braze foils was found to be somewhat non-uniform and consisted of Ti-depleted regions. However, the distribution of Ti in the 150- and 250-μm-thick foils considerably improved and was found to be uniform.

At the liquidus temperature of 900°C, TICUSIL® with the composition of 68.8Ag-26.7Cu-4.5Ti wt.% lies within the miscibility gap of the Ag-Cu-Ti system. Phase analysis revealed a two-phase melt forms, with 91% of an Ag-rich (Ag-23Cu-2Ti wt.%) liquid and 9% of a Ti-rich (Ag-63Cu-27Ti wt.%) liquid. At temperatures below 858°C, this two-phase melt enters into the $(L + Cu_4Ti_3 + Ag)$ domain; hence, Cu$_4$Ti$_3$ + Ag is observed tied to the Ti ribbon in as-received TICUSIL® braze foil. This showed that Ti has a higher chemical affinity with Cu than with Ag. Hence, at 1150°C, Ti has a solubility of ~67 wt.% in Cu and only ~3.7 wt.% in Ag.

The relative Ti concentration factor increased proportionally with the TICUSIL® braze preform thickness applied, i.e. as the TICUSIL® braze preform thickness was increased from 50 to 100 μm, the amount of Ti that would be available to diffuse to the joint interfaces increased twofold. With a limited alumina surface area, an increase in the TICUSIL® braze preform thickness, following the formation of stable reaction layers at the interface, led to excess solid Ti being retained in the braze interlayer.

As the TICUSIL® braze preform thickness increased from 50 to 250 μm, the microstructural evolutions observed in D-96 AG and D-100 AG brazed joints were similar and independent of the alumina purity. Each set of brazed joints comprised similar reaction layers which consisted of a γ-TiO layer with isolated Ti$_2$O particles on the alumina side of the joint interface and a Ti$_3$Cu$_3$O layer on the braze side of the joint interface.

Brazed joints made using 50-μm-thick TICUSIL® braze preforms consisted of a 26-μm-thick Ag-Cu eutectic-like braze interlayer. Ti completely diffused to the joint interfaces to form a 1.7-μm-thick reaction layer. As the TICUSIL® braze preform was increased from 50 to 100 μm, the brazed joint thickness increased from 26 to 39 μm. Despite a corresponding increase in the reaction layer thickness from 1.7 to 2.3 μm, excess Ti in the braze interlayer led to the formation of isolated Cu$_4$Ti$_3$ and Cu$_4$Ti phases surrounded by Ag-rich and Cu-rich phases in a eutectic-like distribution. Due to the formation of these Cu-Ti phases, the separation of an Ag-rich phase led to an estimated volume of braze outflow of ~46 vol.%.

Microstructural Evolution

Brazed joints made using 150-μm-thick TICUSIL® braze preforms were found to be 60-μm-thick and comprised a discontinuous multi-layered Cu-Ti structure in addition to the phases observed in joints made using 100-μm-thick TICUSIL® braze preforms. This multi-layered Cu-Ti structure consisted of a Ti core surrounded by $CuTi_2$, $CuTi$, Cu_4Ti_3 and Cu_4Ti layers, which accommodated relatively more Ti and consumed less Cu than the predominant Cu_4Ti_3 and Cu_4Ti phases observed in joints made using 100-μm-thick TICUSIL® braze preforms. Due to a relatively lower volume fraction of the Ag-rich phase, which separated as a result of the Cu-Ti phase formation, the estimated volume of braze outflow decreased to ~40 vol.%. Despite the excessive amounts of Ti retained in the braze interlayer, the reaction layer thickness increased from 2.3 to 2.8 μm as compared to joints made using 100-μm-thick TICUSIL® braze preforms.

Brazed joints made using 250-μm-thick TICUSIL® braze preforms were found to be 182-μm-thick and comprised the multi-layered Cu-Ti structure, as observed in joints made using 150-μm-thick TICUSIL® braze preforms, except that it was a continuous structure across the entire bond line. Furthermore, the isolated Cu_4Ti_3 and Cu_4Ti phases, which featured in joints made using both the 100- and 150-μm-thick TICUSIL® braze preforms, were no longer observed. Due to a further decrease in the volume fraction of the Ag-rich phase, which was found to separate as a result of the Cu-Ti phase formation, and the significantly thicker braze interlayer, the braze outflow was completely inhibited. Instead, Ag-rich phase separation was accommodated locally within the brazed joint microstructure. Despite the excessive amounts of Ti retained in the braze interlayer, the reaction layer thickness increased from 2.8 to 3.2 μm as compared to joints made using 150-μm-thick TICUSIL® braze preforms. The microstructures of joints made using 250-μm-thick TICUSIL® braze preforms most closely resembled that of the as-received TICUSIL® braze foils.

Cu-Ti phase formation in the microstructures of joints made using 100- to 250-μm-thick TICUSIL® braze preforms may have occurred due to two competing and simultaneous processes: (i) the dissolution of Ti in the surrounding Ag-Cu eutectic phase in amounts exceeding the solubility limit of Ti in the Ag-Cu braze alloy, and (ii) solid-state diffusion of Ti between the excess solid Ti in the braze interlayer and Cu in the surrounding Ag-Cu braze alloy. The brazing conditions of 850 °C for 10 min were sufficient to enable the complete diffusion of Ti to the joint interfaces when 50-μm-thick TICUSIL® braze preforms were selected. However, as the TICUSIL® braze preform thickness was increased further, the ~3 μm thick reaction layers provided a sufficient diffusion barrier such that the reduced Ti activity levels meant excess Ti in the braze interlayer preferentially interacted with the surrounding Ag-Cu eutectic to form the Cu-Ti phases.

The intergranular and triple pocket grain boundary regions of the liquid phase sintered D-96 AG alumina ceramic consisted of a secondary phase with composition 25.6Si-10.0Al-7.0Mg-57.4O wt.%. The formation of Ti_5Si_3 was observed at the interfaces of D-96 AG brazed joints at locations where the triple pocket grain boundary regions of the D-96 AG alumina surface intersected with the Ti-rich reaction layers. No such secondary phase interaction was observed in the D-100 AG brazed joints since D-100 alumina was solid-state sintered and did not consist of any significant amounts of secondary phases.

As the TICUSIL® braze preform thickness was increased from 50 to 250 μm the microstructures of D-96 GHT and D-100 GHT brazed joints were similar to the microstructures of D-96 AG and D-100 AG brazed joints. This showed that the microstructural evolution, which occurred as a result of an increase in the TICUSIL® braze preform thickness, was independent of the alumina purity and surface condition.

The post-grinding heat treatment led to an increase in the average reaction layer thickness in D-100 GHT brazed joints as compared to that in D-100 AG brazed joints; however, in D-96 GHT brazed joints this was not observed. Post-grinding heat treatment led to an increase in the average brazed joint thickness in both the D-96 GHT and D-100 GHT brazed joints, as compared to the average brazed joint thicknesses of the D-96 AG and D-100 AG brazed joints, respectively. A high degree of non-uniformity in the average brazed joint thickness was observed in all of the D-96 GHT and D-100 GHT brazed joints.

Post-grinding heat treatment created a fissured D-96 GHT alumina surface due to the migration of the secondary phase away from the triple pocket grain boundary regions, ~20 μm into the D-96 GHT alumina surface. This secondary phase migration may have occurred during the heat treatment, whereby the secondary phase may have evaporated from the alumina surface or have been re-distributed into defective areas of the alumina surface.

In D-96 GHT brazed joints, braze infiltration was observed whereby the vacated triple pocket grain boundary regions of the alumina surface were found to be accommodated by the braze alloy. These regions were characteristic of where the secondary phase would otherwise be observed in D-96 AG alumina. In the braze infiltrated regions, Ti from the braze alloy had reacted with the adjacent alumina grains to form similar reaction products to those observed at the joint interfaces. While the diffusion of Ti to the triple pocket grain boundary regions of the D-96 AG alumina surface led to the formation of the Ti_5Si_3 phase, in the braze infiltrated regions of the D-96 GHT alumina surface, Ti-Si phases were no longer observed.

These results showed that in the alumina/Ag-Cu-Ti system, the braze preform thickness is an important variable in the AMB process since it can induce a microstructural evolution that is a result of excess Ti, independent of the grade of alumina selected. Similarly, these results also showed that post-grinding heat treatment can affect brazed joints made using different grades of alumina somewhat differently. The effect of post-grinding heat treatment appears to be relatively complex, and requires further investigation.

5.6 CONCLUSIONS

1. A microstructural evolution occurred in alumina-to-alumina brazed joints, dimensioned to configuration c of ASTM C1161-13 and formed at 850°C for 10 min, as the TICUSIL® braze preform thickness was increased from 50 to 250 μm. This microstructural evolution was due to an increase in the amount of Ti which was available to diffuse to the joint interfaces and occurred independently of the grade of alumina selected.
2. The reaction layer, which consisted of a γ-TiO layer with isolated Ti_2O particles on the alumina side of the joint interface and a Ti_3Cu_3O layer on the braze side of the joint interface, increased in thickness from 1.7 to 3.2 μm as the braze preform thickness was increased from 50 to 250 μm.
3. The braze interlayer in joints made using 50-μm-thick TICUSIL® braze preforms consisted of an Ag-Cu eutectic and was 26-μm-thick. Ti had completely diffused to the joint interfaces.
4. In brazed joints made using 100-μm-thick TICUSIL® braze preforms, the braze interlayer was 39-μm-thick and consisted of isolated Cu_4Ti_3 and Cu_4Ti compounds surrounded by an Ag-Cu eutectic. The consumption of Cu in the formation of these Cu-Ti phases led to 46 vol.% Ag-rich braze outflow.
5. In brazed joints made using 150- and 250-μm-thick TICUSIL® braze preforms, the braze interlayers were 60- and 182-μm-thick, respectively. In addition to isolated Cu_4Ti_3 and Cu_4Ti compounds surrounded by an Ag-Cu eutectic, a multi-layered Cu-Ti structure consisting of a Ti core surrounded by $CuTi_2$, CuTi, Cu_4Ti_3 and Cu_4Ti also formed. This structure was observed to be discontinuous in joints made using 150-μm-thick TICUSIL® braze preforms, and continuous in joints made using 250-μm-thick TICUSIL® braze preforms. As a result, the Ag-rich braze outflow decreased from 40 vol. % to 0 vol. %.
6. The secondary phase, composed of 25.6Si-10.0Al-7.0Mg-57.4O wt.% in D-96 AG alumina, led to the formation of Ti_5Si_3 at regions of the joint interface where the triple pocket grain boundary regions of the alumina surface intersected with the Ti-rich reaction layers.
7. Post-grinding heat treatment led to a high degree of non-uniformity in the thicknesses of all brazed joints. A fissured D-96 GHT alumina surface, caused by the migration of the secondary phase was observed, which resulted in intergranular braze infiltration ~20 μm into the D-96 GHT alumina surface.

6 Joint Performance

6.1 STRENGTHS OF BRAZED JOINTS IN AS-GROUND CONDITION

As the TICUSIL® braze preform thickness was increased from 50 to 150 μm, the average strengths of both D-96 AG and D-100 AG brazed joints increased. With a further increase in the TICUSIL® braze preform thickness to 250 μm, the average strength of D-96 AG brazed joints decreased, while the average strength of D-100 AG brazed joints was not affected.

6.1.1 50-μm-Thick TICUSIL® Braze Preforms

The average joint strengths of D-96 AG and D-100 AG brazed joints made using 50-μm-thick TICUSIL® braze preforms were found to be 136 and 163 MPa, respectively. The failures in both of these sets of D-96 AG and D-100 AG brazed joints initiated at the joint interfaces as shown in Figures 6.1 and 6.2, respectively.

Fractography revealed two main failure modes in both D-96 AG and D-100 AG brazed joints made using 50-μm-thick TICUSIL® braze preforms: low-strength failures and high-strength failures (Table 6.1). The fracture paths in low-strength failures had propagated through the reaction layer at one of the two interfaces in each joint (Figures 6.1b and 6.1d in D-96 AG brazed joints and Figure 6.2c in D-100 AG brazed joints). Similarly, the fracture paths in high-strength failures had initiated at the

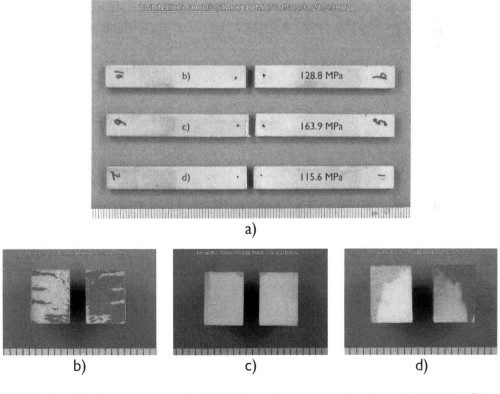

FIGURE 6.1 (a) Failures in sample set 1, D-96 AG brazed joints made using 50-μm-thick TICUSIL® braze preforms and, corresponding fracture surfaces showing (b) low, (c) high and (d) low, strength failures.

175

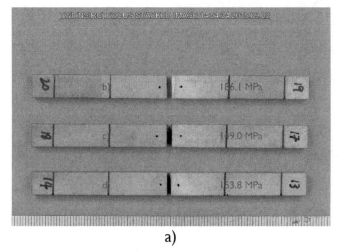

 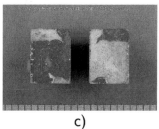 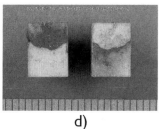

FIGURE 6.2 (a) Failures in sample set 9, D-100 AG brazed joints made using 50-μm-thick TICUSIL® braze preforms and, corresponding fracture surfaces showing (b) high, (c) low and (d) high, strength failures.

TABLE 6.1
Key for Failure Modes

Failure Location	Type	Failure Strength σ_f	% of Parent Strength σ_p
Failures at/close to joint interfaces	Low	$\sigma_f < 150\,\text{MPa}$	$\sigma_f < 60\%\ \sigma_p$
	High	$\sigma_f > 150\,\text{MPa}$	$\sigma_f > 60\%\ \sigma_p$
Failures in the ceramic	Defect	$\sigma_f < 150\,\text{MPa}$	$\sigma_f < 60\%\ \sigma_p$
	Optimum	$\sigma_f < 230\,\text{MPa}$	$\sigma_f < 90\%\ \sigma_p$
	Ideal	$\sigma_f > 230\,\text{MPa}$	$\sigma_f > 90\%\ \sigma_p$

reaction layer and propagated either through the braze interlayer (Figure 6.2c in D-100 AG brazed joints) or the alumina ceramic, close to the reaction layer (Figure 6.1c in D-96 AG brazed joints and Figure 6.2b in D-100 AG brazed joints). Both failure modes occurred at or close to the joint interfaces.

Despite the complete diffusion of Ti to the joint interfaces, the average reaction layer thickness of 1.7 μm in joints made using 50-μm-thick TICUSIL® braze preforms was relatively thin as compared with the reaction layers observed in joints made using 100- and 150-μm-thick TICUSIL® braze preforms. Joint strength can be correlated to the thickness of the reaction layer, which is a product of chemical reactions at the joint interface (Hao et al., 1994; Ning et al., 2003). Therefore, the joint interfaces may have been relatively weak.

Carim and Mohr (1997) showed that the ductile Ag-Cu braze interlayer is responsible for accommodating both thermally induced residual stresses and applied stresses via plastic deformation. The microstructures of both D-96 AG and D-100 AG brazed joints made using 50-μm-thick TICUSIL®

Joint Performance 177

braze preforms consisted of an Ag-Cu eutectic-like braze interlayer. Thermally induced residual stresses during cooling and applied stress during four-point bend testing, therefore, may not have been fully transferred to the braze interlayer due to the relatively insufficient chemical bonding at the joint interfaces. Hence, stresses in the vicinity of the joint interface may have initiated failures in the brittle reaction layer.

The percentage braze outflow estimated to have occurred in D-96 AG brazed joints made using 50-μm-thick TICUSIL® braze preforms was estimated to be 29 vol.%. This was based on the assumption that the 7 mm × 5 mm × 0.05 mm braze preform had completely filled the joint gap between the two 8 mm × 6 mm faying alumina surfaces to produce a uniform brazed joint thickness measured to be 26 μm. In some of the D-96 AG and D-100 AG brazed joints made using 50-μm-thick TICUSIL® braze preforms, incomplete joint filling was observed at the joint edges (see Appendix 2). This may have induced a notched-type effect during the four-point bend test in some of the joints, resulting in the two distinctly different failure modes observed at the joint interfaces, which therefore also included low-strength failures. The failure location and strength criteria by which the failure modes observed were divided in this study, are shown in Table 6.1.

6.1.2 100-μm-Thick TICUSIL® Braze Preforms

As the TICUSIL® braze preform thickness was increased from 50 to 100 μm, the average joint strengths of D-96 AG brazed joints increased by 75% from 136 to 238 MPa while the average joint strengths of D-100 AG brazed joints increased by 22% from 163 to 200 MPa. The failures in these D-96 AG brazed joints initiated in the alumina ceramic in some cases away from the joint interface (Figure 6.3), and failures in these D-100 AG brazed joints were mixed-mode and initiated both at the joint interface and in the alumina ceramic (Figure 6.4).

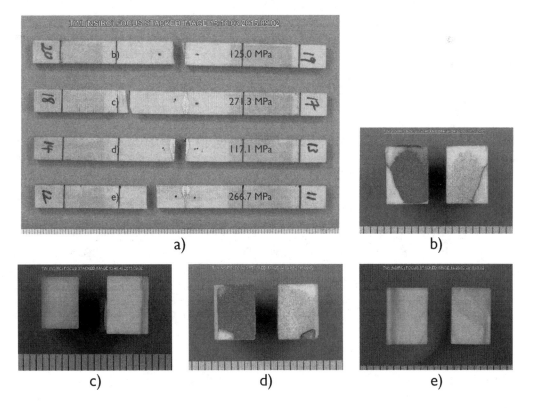

FIGURE 6.3 (a) Failures in sample set 2, D-96 AG brazed joints made using 100-μm-thick TICUSIL® braze preforms and, corresponding fracture surfaces showing (b) low, (c) ideal, (d) low and (e) ideal, strength failures.

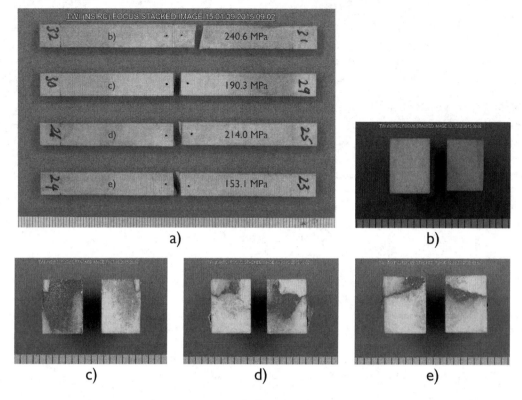

FIGURE 6.4 (a) Failures in sample set 10, D-100 AG brazed joints made using 100-μm-thick TICUSIL® braze preforms and, corresponding fracture surfaces showing (b) ideal, (c) high, (d) high and (e) high, strength failures.

Fractography revealed that the two low-strength failures that occurred at the joint interfaces of D-96 AG brazed joints made using 100-μm-thick TICUSIL® braze preforms were due to defects in the alumina ceramic (Figure 6.3b) and the improper pre-placement of the braze preform (Figure 6.3d). In both of these low-strength failures, the fracture paths propagated through the reaction layer, which is structurally the weakest constituent of the brazed joint microstructure. These failures were not characteristic, however, of the failure mode in D-96 AG brazed joints made using 100-μm-thick TICUSIL® braze preforms.

Two ideal strength failures occurred in D-96 AG brazed joints made using 100-μm-thick TICUSIL® braze preforms, at ~14 mm away from the joint interface (Figure 6.3c and 6.3e). The strengths of these joints were found to be 271 and 267 MPa, respectively, which were similar to the average flexural strength of 252 MPa achieved by D-96 AG standard alumina test bars. Therefore, D-96 AG brazed joints made using 100-μm-thick TICUSIL® braze preforms appeared to exhibit ideal joint strength, since failures occurred in the parent ceramic at strengths which represented the monolithic strength of D-96 AG alumina.

High-strength failures were the dominant failure mode in D-100 AG brazed joints made using 100-μm-thick TICUSIL® braze preforms, whereby fractures were initiated in the alumina ceramic on one side of the joint interface and propagated through the braze interlayer into the alumina ceramic on the other side of the joint interface (Figure 6.4c, 6.4d and 6.4e). A single ideal strength failure at 241 MPa was also observed, and this was similar to the average flexural strength of 250 MPa achieved by D-100 AG standard alumina test bars. Therefore, the performances of both D-96 AG and D-100 AG brazed joints were significantly improved with an increase in the TICUSIL® braze preform thickness from 50 to 100 μm. This was postulated to have occurred through a combination of several strengthening mechanisms.

Joint Performance

As the TICUSIL® braze preform thickness was increased from 50 to 100 µm, the average reaction layer thicknesses also increased, from 1.7 to 2.3 µm in D-96 AG brazed joints and from 1.6 to 2.2 µm in D-100 AG brazed joints. This indicated enhanced reaction kinetics in joints made using 100-µm-thick TICUSIL® braze preforms as compared with joints made using 50-µm-thick TICUSIL® braze preforms. Therefore, greater reactions at the interfaces of joints made using 100-µm-thick TICUSIL® braze preforms may have led to the increase in joint strengths.

As the TICUSIL® braze preform thickness was increased from 50 to 100 µm, the average brazed joint thicknesses also increased, from 26 to 39 µm in D-96 AG brazed joints and from 21 to 39 µm in D-100 AG brazed joints. The corresponding increase in the thicknesses of the braze interlayer, in both sets of joints, may have affected the extent to which thermally induced residual stresses were accommodated (see Section 2.1.6).

The microstructural evolutions in D-96 AG and D-100 AG brazed joints, which occurred due to an increase in the TICUSIL® braze preform thickness from 50 to 100 µm, led to the formation of isolated Cu_4Ti_3 and Cu_4Ti phases in the braze interlayer and the separation of an Ag-rich phase which flowed towards the joint edges (see Figure 5.32b). The hardness distribution in joints made using 100-µm-thick TICUSIL® braze preforms may have been relatively more complex than the uniform Ag-Cu braze interlayer in joints made using 50-µm-thick TICUSIL® braze preforms.

Several studies in the literature have reported that the strength of alumina-to-alumina and alumina-to-metal brazed joints can be improved by adding alumina particles or carbon fibres to create a composite-type Ag-Cu-Ti braze alloy (Yang et al., 2002, 2005; Zhu and Chung, 1997). These studies have found that particulate or fibre additions made to an Ag-Cu-Ti braze alloy, which react with Ti to form Ti-rich compounds in the braze interlayer, can significantly reduce the CTE mismatch and thereby the thermally induced residual stresses at the joint interfaces. The presence of relatively harder phases that are dispersed in the braze interlayer has been suggested to reinforce the brazed joint strength. The consumption of any Ti, in the braze interlayer, which has not diffused to the joint interfaces can decrease the microhardness of the surrounding matrix phase. This provides greater plasticity in accommodating both thermally induced residual stresses and applied stresses. The total volume of fibre or particulate additions, however, required to be balanced by the brazed joint thickness, which can be controlled by the volume of braze selected. This is to ensure that there is sufficient Ti available such that the formation of Ti-rich compounds in the braze interlayer does not inhibit interfacial reactions required in forming adequate chemical bonds at the joint interfaces, i.e. an optimum reaction layer thickness. While this could be achieved by simply reducing the total volume of fibre or particulate additions, the purpose of these additions which is to reduce the CTE of the braze interlayer may not be fully realised. Zhu and Chung (1997) formed alumina-to-stainless steel brazed joints using a 63.0Ag-34.25Cu-1.0Sn-1.75Ti wt.% braze alloy and found that by controlling the brazed joint thickness, the addition of 12 vol.% carbon fibre additions provided an optimum joint strength as compared with joints where no carbon fibre additions were made (Figure 6.5).

In D-96 AG and D-100 AG brazed joints made using 100-µm-thick TICUSIL® braze preforms, the formation of isolated Cu_4Ti_3 and Cu_4Ti phases may have reduced the overall CTE of the braze interlayer, thereby leading to an overall improvement in joint strength. The $Ti_3(Cu + Al)_3O$ layer, also rich in Cu and Ti, has an intermediate CTE ($15.1 \times 10^{-6}/°C$) relative to the Ag-Cu braze interlayer and alumina and hence, it can grade the CTE mismatch between them (see Section 2.1.5).

Unlike with particulate or fibre additions made to an Ag-Cu-Ti braze alloy, an increase in the braze preform thickness from 50 to 100 µm provided lower CTE phases in the braze interlayer without affecting the reaction layer thickness, which was observed to increase by ~36%. Similarly, the brazed joint thickness did not need to be controlled other than via the braze preform dimensions. In addition, the consumption of Ti in the braze interlayer led to the separation of an Ag-rich phase which was likely to have provided a relatively ductile matrix. Therefore, these results appear to have achieved the objectives set out in the design of composite Ag-Cu-Ti braze alloys without the

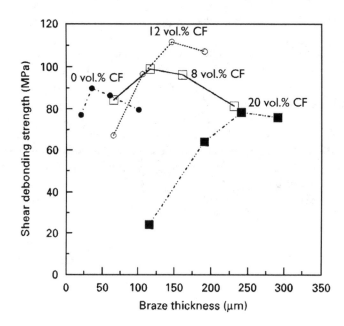

FIGURE 6.5 The effect of carbon fibre additions (vol.%) and brazed joint thickness on the strength of alumina-to-stainless steel brazed joints (Zhu and Chung, 1997).

modifications to the brazed joint microstucture through particulate or fibre additions. The consumption of Ti in the braze interlayer in the precipitation of Cu-Ti phases was associated to the excess Ti, following the formation of stable Ti-rich reaction layers at the joint interfaces.

The separation of an Ag-rich phase which flowed towards the joint edges in D-96 AG and D-100 AG brazed joints made using 100-μm-thick TICUSIL® braze preforms, led to ~46 vol.% braze outflow which was rich in Ag. As well as from better joint filling than in joints made using 50-μm-thick TICUSIL® braze preforms, the presence of an Ag-rich phase at the joint edges may have increased the ductility of the braze interlayer at the joint edges leading to the improvement in joint strength.

In ceramic-to-ceramic and ceramic-to-metal brazed joints, high stresses can develop at the free edges of the joint interfaces due to either a change in temperature or mechanical loading (Dobedoe, 1997). During four-point bend testing of brazed butt-joints, the free edges of the joint interface are particularly significant. High tensile stresses near these free edges can cause failure at the joint interface. Suganuma et al. (1985) showed, using finite-element analysis, that the stress contours developed in a cylindrical Si_3N_4-to-steel brazed butt-joint were significantly higher at the free edges of the joint (Figure 6.6). Therefore, Ag-rich braze outflow may have improved the strengths of joints made using 100-μm-thick TICUSIL® braze preforms by providing greater ductility at the joint edges.

The average joint strengths of D-96 AG brazed joints made using 100-μm-thick TICUSIL® braze preforms were 20% greater than the average joint strengths of D-100 AG brazed joints made using the same TICUSIL® braze preform thickness. This was despite the braze interlayers of these joints comprising similar microstructures. However, the average flexural strengths of D-96 AG and D-100 AG standard alumina test bars were similar; D-96 AG standard alumina test bars achieved 252 MPa and D-100 AG standard alumina test bars achieved 250 MPa. Both D-96 AG and D-100 AG standard and short alumina test bars were ground in the same way, according to ASTM C1161-13. Aside from slight differences in surface roughness, the only other significant difference between these two grades of alumina was the presence of a Si-rich secondary phase in D-96 AG alumina, which was liquid phase sintered. D-100 AG alumina was solid-state sintered and did not consist of any secondary phases. Therefore, the superior strength of D-96 AG brazed joints made using 100-μm-thick

Joint Performance 181

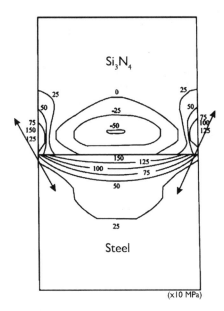

FIGURE 6.6 Stress contours developed in cylindrical Si$_3$N$_4$-to-steel brazed butt-joint showing high stresses near the free edges of the joint. Arrows indicate both the location and direction of the maximum tensile stress (Suganuma et al., 1985).

TICUSIL® braze preforms may have been achieved by additional chemical bonding evidenced by the formation and spatial distribution of Ti$_5$Si$_3$ reaction products (see Section 5.4.3).

6.1.3 150-μM-THICK TICUSIL® BRAZE PREFORMS

As the TICUSIL® braze preform thickness was increased from 100 to 150 μm, the average joint strengths of D-96 AG brazed joints increased by 13%, from 238 to 270 MPa, and the average joint strengths of D-100 AG brazed joints increased by 6%, from 200 to 212 MPa. The failures in the D-96 AG brazed joints predominantly occurred in the alumina ceramic (Figure 6.8), while the failures in the D-100 AG brazed joints were mixed-mode and occurred at both the joint interface and in the alumina ceramic (Figure 6.9).

Fractography revealed that in D-96 AG brazed joints made using 150-μm-thick TICUSIL® braze preforms, two high-strength failures were initiated in the reaction layer at the free edges of the joints (Figure 6.8d and 6.8e). As the TICUSIL® braze preform thickness was increased from 100 to 150 μm, the average brazed joint thickness in D-96 AG brazed joints also increased, by 67% from 39 to 65 μm. This increase in the average brazed joint thickness, therefore, may have led to an increase in the magnitude of residual stresses, particularly at the free edges of the joints.

Serizawa et al. (2001) used finite-element analysis to study the effect of joint thickness on the development of residual stresses at the joint interface of a ceramic-to-ceramic brazed butt-joint (Figure 6.7). The residual stresses at the joint edges were found to increase with an increase in the brazed joint thickness from 12.5 and 50 μm. The dimensions of the brazed joints in their model were exactly half that of those used in this study. The effect of an increase in the brazed joint thickness from 39 to 65 μm in this study, therefore, may have been similar to the effect of an increase in the brazed joint thickness from 20 to 33 μm in their FE model. Relatively high residual stresses at the edges of D-96 AG brazed joints made using 150-μm-thick TICUSIL® braze preforms may have resulted in failures initiating at the joint edges. Furthermore, these failures would be expected to have initiated at the reaction layer – a brittle constituent of the brazed joint microstructure. The ~42% Ag-rich braze outflow may have been insufficient in accommodating these high stresses at the joint edges.

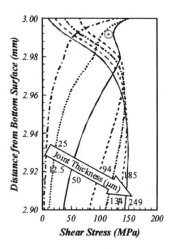

FIGURE 6.7 The effect of joint thickness on the development of residual stresses in ceramic-to-ceramic brazed butt-joints (Serizawa et al., 2001).

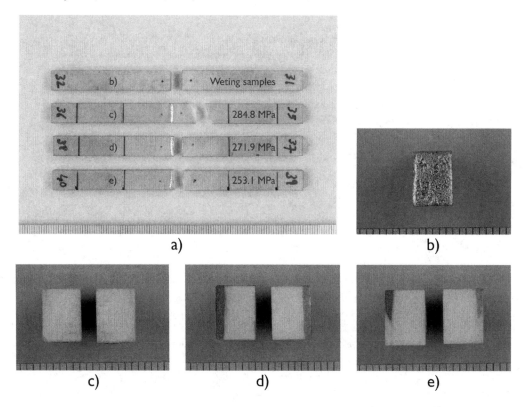

FIGURE 6.8 (a) failures in sample set 3, D-96 AG brazed joints made using 150-µm-thick TICUSIL® braze preforms and, corresponding fracture surfaces showing (b) wetting sample and, (c) ideal, (d) high and (e) high, strength failures.

The high-strength failures of 271.9 and 253.1 MPa in D-96 AG brazed joints made using 150-µm-thick TICUSIL® braze preforms achieved similar strengths to the single ideal failure of 284.4 MPa, which occurred in the alumina ceramic ~11 mm away from the joint interface in the same set of joints (Figure 6.8c). The strengths achieved by these high-strength failures were similar to the average flexural strength of 252 MPa achieved by D-96 AG standard alumina test bars.

Joint Performance

Despite failures initiating at the joint edges in the high-strength failures, the main fracture paths propagated through the alumina ceramic where residual stresses may have been lower. In this way, the resulting joint strengths were similar, despite the two distinctly different failure modes.

Failures in D-100 AG brazed joints made using 150-µm-thick TICUSIL® braze preforms were either high-strength failures or optimum failures. High-strength failures initiated in the alumina ceramic on one side of the joint interface and propagated through the braze interlayer into the alumina ceramic on the other side of the joint interface. Two high-strength failures were observed; these joints failed at 249.4 and 171.5 MPa (Figure 6.9c and 6.9e). Optimum strength failures, which occurred in the alumina ceramic, were observed in two joints which achieved strengths of 181.5 and 202.3 MPa (Figure 6.9b and 6.9d). These optimum strength failures were observed in joints which achieved strengths that were ~23% lower than the average flexural strength of D-100 AG standard alumina test bars, and did not occur at any considerable distances away from the joint interfaces. These results showed that failure locations can be misleading in determining joint performances, if failure loads are not also considered (see Section 2.5.1).

An increase in the TICUSIL® braze preform thickness from 100 to 150 µm had led to improvements in the strengths of both D-96 AG and D-100 AG brazed joints. The resulting 13% increase in the average strength of D-96 AG brazed joints from 238 to 270 MPa was ideal, limited only by the strength of the parent alumina ceramic. While the increase in the average strength of D-100 AG brazed joints from 200 to 212 MPa was not characteristic of the ideal failure mode; a 6% increase was still observed.

As the TICUSIL® braze preform thickness was increased from 100 to 150 µm, the average reaction layer thicknesses increased, from 2.3 to 2.8 µm in D-96 AG brazed joints, and from 2.2 to

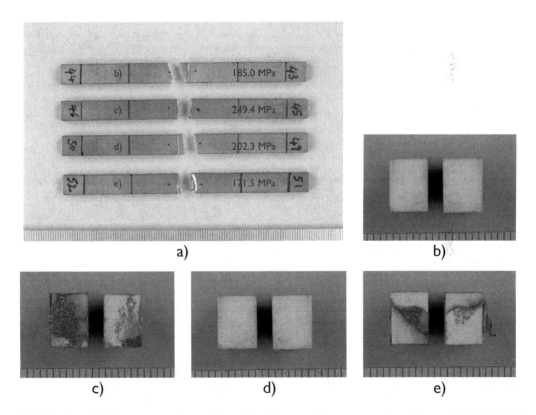

FIGURE 6.9 (a) Failures in sample set 11, D-100 AG brazed joints made using 150-µm-thick TICUSIL® braze preforms and, corresponding fracture surfaces showing (b) optimum, (c) high, (d) optimum and (e) high, strength failures.

3.0 µm in D-100 AG brazed joints. These indications of improved interfacial bonding may have contributed to the increase in joint strengths observed in both sets of joints.

The microstructural evolutions in D-96 AG and D-100 AG brazed joints, which occurred due to an increase in the TICUSIL® braze preform thickness from 100 to 150 µm, included the formation of a discontinuous multi-layered Cu-Ti structure (see Figure 5.32c). The total volume fraction of Ti retained in the braze interlayer of joints made using 150-µm-thick TICUSIL® braze preforms was higher than in joints made using 100-µm-thick TICUSIL® braze preforms. The reduced CTE of the braze interlayer in both sets of these joints, due to the presence of the multi-layered Cu-Ti structure and Cu-Ti phases respectively, may have led to a decrease in thermally induced residual stresses at the joint interfaces, thereby leading to the increase in joint strengths observed.

The average brazed joint thicknesses of D-100 AG brazed joints increased by 67%, from 39 to 65 µm, as the TICUSIL® braze preform thickness was increased from 100 to 150 µm. A corresponding and similar increase in the average brazed joint thicknesses of D-96 AG brazed joints was also observed. In both sets of joints, therefore, the development of residual stresses at the joint interfaces may have been greater in joints made using 150-µm-thick TICUSIL® braze preforms than in joints made using 100-µm-thick TICUSIL® braze preforms (Serizawa et al., 2001), albeit somewhat reduced by the formation of the relatively lower CTE Cu-Ti phases in the braze interlayers.

The two high-strength failures observed in D-100 AG brazed joints made using 150-µm-thick TICUSIL® braze preforms consisted of fractures which propagated through the braze interlayer. This suggested that chemical bonding at the joint interfaces was sufficient to enable load transfer to the braze interlayer and that the relatively thick reaction layer may not have been the cause of these failures. The braze interlayer could not accommodate this load; however, despite the increased brazed joint thickness as otherwise failures may have been deflected into the alumina ceramics. The presence of Ti and Cu-Ti phases in the braze interlayer may have been excessive, reducing the overall ductility of the braze interlayer. The relatively high volume fraction of brittle Cu-Ti phases may have initiated failures in the braze interlayer. The presence of these Cu-Ti-rich phases was evident in the fracture surfaces of these joints (Figure 6.9c and 6.9e).

The two optimum strength failures observed in D-100 AG brazed joints made using 150-µm-thick TICUSIL® braze preforms were not characteristic of the parent D-100 AG alumina strength. While the microstructures of all D-100 AG brazed joints made using 150-µm-thick TICUSIL® braze preforms were expected to be similar, the high-strength failures observed were unlikely to have initiated simply due to a weakened braze interlayer. Similarly, the joints in which optimum strength failures were observed achieved strengths that were ~23% lower than the parent D-100 AG alumina strength. Therefore, these optimum strength failures were also unlikely to have initiated simply due to ideal joints having been formed; instead failures appeared to have initiated from defects in surfaces of the alumina ceramics.

Similar failure strengths were observed in both high-strength failures and optimum strength failures observed in D-100 AG brazed joints made using 150-µm-thick TICUSIL® braze preforms. Despite the two distinctly different failure modes, high-strength failures initiated at the joint interface while optimum strength failures initiated close to the joint interface. These results suggested that the joint interfaces of D-100 AG brazed joints made using 150-µm-thick TICUSIL® braze preforms were highly stressed, such that two failure modes coincided. Therefore, these optimum strength failures may have been affected by defects in the surfaces of the alumina ceramics.

6.1.4 Secondary Phase Interaction

The average flexural strength of D-96 AG standard alumina test bars (252 MPa) and the average flexural strength of D-100 AG standard alumina test bars (250 MPa) were similar. D-96 AG and D-100 AG brazed joints made using 150-µm-thick TICUSIL® braze preforms were also similar in terms of their brazed joint microstructures, their average reaction layer thicknesses and their average braze interlayer thicknesses. Despite these similarities, the average joint strength of these D-96 AG brazed joints (270 MPa) was 27% greater than that of these D-100 AG brazed joints (212 MPa). Therefore, the mechanism responsible for the superior strengths of these D-96 AG brazed joints, relative to these D-100 AG brazed joints, was likely associated to differences between the alumina ceramics.

The only significant difference between the brazed joints made using these two grades of alumina, which were both ground in the same way according to ASTM C1161-13, was the presence of a Si-rich secondary phase in D-96 AG alumina, which was liquid phase sintered. D-100 AG alumina was solid-state sintered and did not consist of any secondary phases. Therefore, the formation of Ti_5Si_3 reaction products at the interfaces of D-96 AG brazed joints may have provided the improvement in joint strength observed, which could be explained based on two possible mechanisms.

The regular spatial distribution of the Ti_5Si_3 at locations of the interface where the triple pocket grain boundary regions of the D-96 AG alumina surface intersected with the Ti-rich reaction layers may have led to an improvement in the gradation of the CTE across the joint interface.

The $Ti_3(Cu + Al)_3O$ layer with a CTE value of $15.2 \times 10^{-6}.°C^{-1}$ introduces a gradual transition in the CTE across the joint interface between alumina, which has a relatively lower CTE value of 8.1 to $8.5 \times 10^{-6}.°C^{-1}$ and the braze interlayer, in which the Ag-rich and Cu-rich phases have CTE values of $19.2 \times 10^{-6}.°C^{-1}$ and $22.0 \times 10^{-6}.°C^{-1}$, respectively (Table 2.2). The CTE of a Ti_5Si_3 crystal lattice exhibits high anisotropy, with CTE values of $5.9 \times 10^{-6}.°C^{-1}$ along the *a*-axis and $16.9 \times 10^{-6}.°C^{-1}$ along the *c*-axis (Rodrigues et al., 2006). The formation of the Ti_5Si_3 reaction products may have made the residual stress distribution at the joint interface relatively complex. Secondary phase interaction may have further graded the CTE transition across the joint interface leading to improvements in both the residual stress state of the joints and the joint strength (Figure 6.10).

A second potential mechanism may be that the regular spatial distribution of Ti_5Si_3 compounds at locations of the interface, where the triple pocket grain boundary regions on the D-96 AG alumina surface intersected with the Ti-rich reaction layers, provided a nanostructured interlocking mechanism. This may be similar to the way in which ~20-nm-thick Ti_5Si_3 dendrites were suggested to improve the strengths of glass-to-Ti-6Al-4V joints. Oku et al. (2001) proposed that strong bonding could be achieved between a thin Ti_5Si_3 layer and a Ti-6Al-4V alloy that was joined to a bioactive glass coating. A 150-nm-thick Ti_5Si_3 layer on the Ti-6Al-4V alloy side of the joint interface was suggested to provide strain relaxation, while ~20-nm-thick-Ti_5Si_3 particles which grew as dendrites into the glass were suggested to provide mechanical interlocking which contributed to good adhesion.

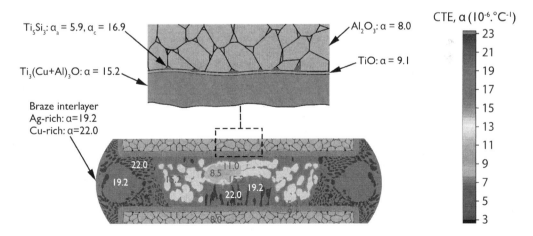

FIGURE 6.10 CTE values, α, ($\times 10^{-6}.°C^{-1}$) of phases in D-96 AG brazed joints made using 150-μm-thick TICUSIL® braze preforms.

Furthermore, the formation of these Ti_5Si_3 compounds at the interfaces of D-96 AG brazed joints may have provided additional chemical bonding, enabling relatively higher levels of both, thermally induced residual stresses and applied stresses, to be transferred into the relatively ductile matrix of the braze interlayer. Secondary phase interaction, therefore, may have provided a nanostructured interlocking mechanism which led to the improvement in joint strength observed.

The effect of Ti-Si reaction products on the strength of alumina-to-alumina or alumina-to-metal brazed joints made using Ag-Cu-Ti braze alloys could not be found in the literature. The formation of Ti_5Si_3 reaction products has been commonly observed, however, at the interfaces of Si_3N_4-to-Si_3N_4 and SiC-to-SiC brazed joints made using Ag-Cu-Ti braze alloys (see Section 5.4.3). In these systems, the formation of Ti-Si reaction products at the joint interfaces occurs following the reactions that are necessary to achieve chemical bonding with the surfaces of the silicon carbide and silicon nitride ceramics. While finely distributed Ti_5Si_3 particles (<2 μm) can be beneficial in achieving strong joints, larger Ti_5Si_3 particles (>2 μm), or a continuous Ti_5Si_3 layer, have been shown to lead to a degradation in joint strength (Urai and Masaaki, 1999; Xian and Si, 1990).

The degradation in strength observed in Si_3N_4-to-Si_3N_4 and SiC-to-SiC brazed joints due to excessive quantities of Ti_5Si_3 appears to be similar to the way in which an excessively thick $Ti_3(Cu+Al)_3O$ layer can also degrade the strength of alumina-to-alumina brazed joints made using Ag-Cu-Ti braze alloys (see Sections 2.3.1.3 and 2.3.2.3). This is primarily due to the brittle nature of these reaction product phases at the joint interface. The Ti_5Si_3 reaction products observed at the interfaces of D-96 AG brazed joints were ~200 nm in size and were isolated along the joint interface. Therefore, the additional chemical bonding in the formation of Ti_5Si_3 reaction products may have been beneficial to joint strength. Furthermore, due to the sizes of these Ti_5Si_3 reaction products, it is unlikely that joint strength would be adversely affected by their presence at the joint interface creating a diffusion barrier between Ti and the alumina surface. The regular spatial distribution of Ti_5Si_3 at locations of the interface where the triple pocket grain boundary regions of the alumina surface intersected with the Ti-rich reaction layers may have provided a nanostructured interlocking mechanism, which led to the improvement in joint strength observed.

6.1.5 250-μm-Thick TICUSIL® Braze Preforms

As the TICUSIL® braze preform thickness was increased from 150 to 250 μm, the average joint strengths of D-96 AG brazed joints decreased by 13%, from 270 to 235 MPa, while the average joint strengths of D-100 AG brazed joints remained at 212 MPa. The failures in these D-96 AG brazed joints predominantly initiated at the joint interface (Figure 6.11), while the failures in these D-100 AG brazed joints were mixed-mode and initiated at either the joint interface or in the alumina ceramic (Figure 6.12).

Fractography revealed that all of the high-strength failures in these D-96 AG brazed joints made using 250-μm-thick TICUSIL® braze preforms had consistently initiated in the sub-surface of the alumina ceramic, on one side of the brazed joint, before propagating through the reaction layer on the same side of the brazed joint (Figure 6.11).

Two ideal strength failures were observed in D-100 AG brazed joints made using 250-μm-thick TICUSIL® braze preforms (Figures 6.12b and 6.12e). These joints achieved strengths of 234 and 283 MPa respectively, similar to the average strength of 250 MPa achieved by the D-100 AG standard alumina test bars.

As the TICUSIL® braze preform thickness was increased from 150 to 250 μm, the average reaction layer thicknesses increased from 2.8 to 3.2 μm in D-96 AG brazed joints and from 3.0 to 3.2 μm in D-100 AG brazed joints. The thicknesses of the reaction layers in both sets of these joints, made using 250-μm-thick TICUSIL® braze preforms, may have been excessively thick. However, a single high-strength failure was observed in the D-100 AG brazed joints, in which the fracture paths propagated through the braze interlayer (Figure 6.12d). Furthermore, in the D-100 AG brazed joints, no failures had initiated at the joint interfaces. This suggests that chemical bonding at the interfaces of these D-100 AG brazed joints was sufficient to enable load transfer to the braze interlayer.

Joint Performance

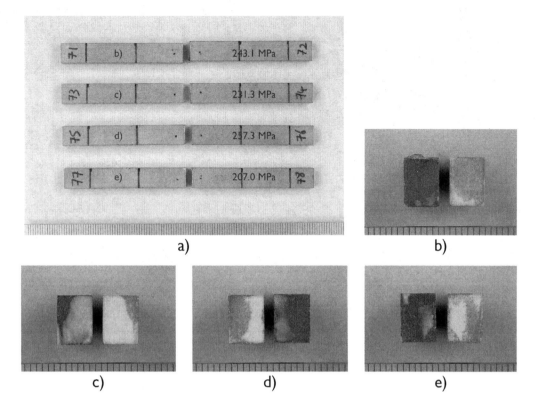

FIGURE 6.11 (a) Failures in sample set 4, D-96 AG brazed joints made using 250-μm-thick TICUSIL® braze preforms and, corresponding fracture surfaces showing (b) high and, (c) high, (d) high and (e) high, strength failures.

The brazed joint microstructures, average reaction layer thicknesses and average braze interlayer thicknesses were all similar in these D-96 AG and D-100 AG brazed joints. Therefore, it is unlikely that an excessively thick reaction layer was the main cause of the failures observed in D-96 AG brazed joints made using 250-μm-thick TICUSIL® braze preforms. However, since the reaction layer phase is a relatively weakest constituent of the brazed joint microstructure, failures that initiated from the alumina surface may have propagated into the reaction layer.

The lack of any braze outflow in both D-96 AG and D-100 AG brazed joints made using 250-μm-thick TICUSIL® braze preforms may have reduced the ductility of the joint edges (see Figure 5.34d). As the TICUSIL® braze preform thickness was increased from 150 to 250 μm, the average brazed joint thicknesses increased from 65 to 182 μm in D-96 AG brazed joints and from 68 to 191 μm in D-100 AG brazed joints. According to the finite-element analysis model in Serizawa et al. (2001), these brazed joint thicknesses would be expected to reduce the residual stresses at the joint edges and, instead, increase residual stresses towards the centre of the joint (Figure 6.7). Therefore, the Ag-Cu eutectic-like structure at the edges of these joints may have actually been sufficient to accommodate the relatively lower levels of residual stress.

Excessive growth of Ti_5Si_3 reaction products at the interfaces of D-96 AG brazed joints made using 250-μm-thick TICUSIL® braze preforms may have embrittled the joint interface. As the TICUSIL® braze preform thickness was increased from 150 to 250 μm, the 14% increase in the average reaction layer thickness in the D-96 AG brazed joints may have provided additional Ti for larger Ti_5Si_3 particles to form. This could explain why failures did not occur at the interfaces of D-100 AG brazed joints, since D-100 AG alumina was solid-state sintered and did not consist of any Si-rich secondary phases. Any reduced strengthening mechanisms caused by an excessively thick reaction

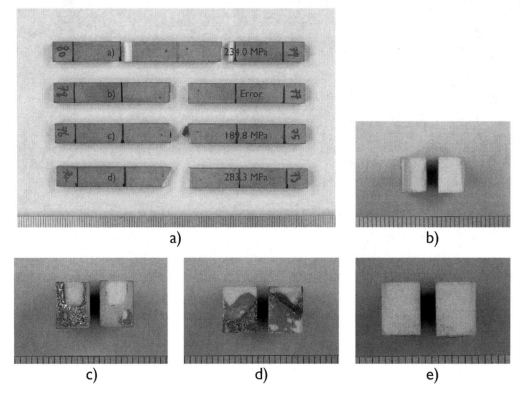

FIGURE 6.12 (a) Failures in sample set 12, D-100 AG brazed joints made using 250-μm-thick TICUSIL® braze preforms and, corresponding fracture surfaces showing (b) ideal, (c) error in brazing fixture, (d) high and (e) ideal, strength failures.

layer or brazed joint thickness may have been counterbalanced by a reduction in the CTE of the braze interlayer provided by the continuous Cu-Ti structure observed in all joints made using 250-μm-thick TICUSIL® braze preforms. Therefore, the failure mechanisms observed in the D-96 AG brazed joints were not observed in the D-100 AG brazed joints and are likely to be associated, therefore, to differences in the properties of the alumina ceramics – the Si-rich secondary phase in D-96 AG alumina.

The properties of D-96 AG and D-100 AG brazed joints made using 50- to 250-μm-thick TICUSIL® braze preforms are summarised in Table 6.2 and Figure 6.13.

6.2 NANOINDENTATION

Nanoindentation testing techniques were used to evaluate changes in the nanohardness distribution as a result of the microstructural evolution in both D-96 AG and D-100 AG brazed joints, as the TICUSIL® braze preform thickness was increased from 50 to 150 μm (see Figure 5.34). Nanoindentation was used to investigate whether the increase in joint strength observed, which corresponded to this microstructural evolution, could be correlated to the formation of the relatively harder Cu-Ti phases in the braze interlayer which may have reinforced the strength of the joints, and the relatively ductile Ag-rich phase which may have better accommodated stress particularly at the joint edges. The reinforcement of alumina-to-alumina brazed joints by the presence of Ti-rich phases in the braze interlayer has been suggested in typical the literature concerning composite Ag-Cu-Ti braze alloys with particulate or fibre additions (Yang et al., 2002, 2005; Zhu and Chung, 1997) (see Section 6.1.2).

In this study, nanoindentation testing techniques were selected over conventional hardness testing techniques since the latter method is impractical for performing targeted indents in μm-sized

TABLE 6.2
Properties of D-96 AG and D-100 AG Brazed Joints Made Using 50- to 250-μm-Thick TICUSIL® Braze Preforms

Sample Set	Alumina Purity, (wt.% Al$_2$O$_3$) and Surface Condition	Braze Preform Thickness (μm)	Specimens Characterised	Reaction Layer Thickness (μm)	Average Brazed Joint Thickness (μm)	Estimated Volume of Braze Outflow (%)	Specimens Mechanically Tested	Joint Strength (MPa)	Failure Location
1	D-96 AG	50	1	1.7 ± 0.1	26 ± 1	29	3	136 ± 14	Interface
2	D-96 AG	100	3	2.3 ± 0.1	39 ± 1	46	4	238 ± 31	Ceramic
3	D-96 AG	150	1	2.8 ± 0.1	65 ± 2	40	4	270 ± 9	Ceramic
4	D-96 AG	250	1	3.2 ± 0.2	182 ± 3	0	4	235 ± 11	Interface
9	D-100 AG	50	1	1.6 ± 0.1	21 ± 1	42	3	163 ± 12	Interface
10	D-100 AG	100	2	2.2 ± 0.1	39 ± 1	46	4	200 ± 19	Interface
11	D-100 AG	150	1	3.0 ± 0.1	68 ± 2	40	4	212 ± 7	Mixed
12	D-100 AG	250	1	3.2 ± 0.2	191 ± 6	0	4	212 ± 22	Mixed

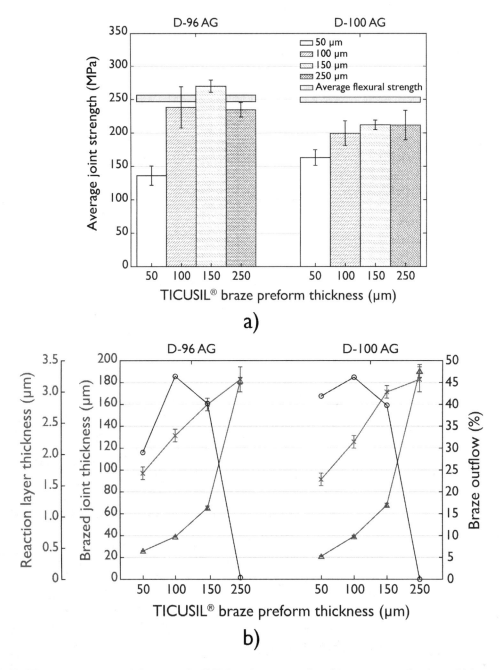

FIGURE 6.13 (a) Average joint strengths (MPa), and corresponding (b) average reaction layer thicknesses (μm), average brazed joint thicknesses (μm) and braze outflow (vol.%), in D-96 AG and D-100 AG brazed joints made using 50-, 100-, 150- and 250-μm-thick TICUSIL® braze preforms.

phases such as the constituents present in typical brazed joint microstructures, e.g. the reaction layer phase in joints made using 50- to 150-μm-thick TICUSIL® braze preforms increased from 1.7 to 2.8 μm. The pyramidal Berkovich indenter with a face angle of ~65.27°, which was used to perform the small-scale nanoindentation tests, had a tip radius of just 50 to 100 nm (Figure 6.14). Since the edges of this indenter were constructed to meet at a single point, the targeted indents could be performed more accurately.

Joint Performance

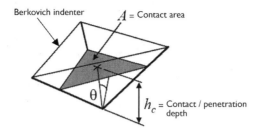

FIGURE 6.14 Indentation parameters for a Berkovich indenter (Fischer-Cripps, 2011).

The nm-to-μm-sized residual impressions produced in a typical nanoindentation test cannot be accurately measured using the same optical techniques that are used to adequately measure the relatively larger residual impressions produced in a typical conventional hardness test. Instead, the contact or penetration depth (h_c) of the indenter can be measured as a function of load and the known geometry of the indenter can be used to calculate the contact area (A) (Figure 6.14). The nanohardness values can be derived as follows:

$$A = 24.5 h_c^2 \qquad (6.1)$$

$$H = \frac{P}{24.5 h_c^2} \qquad (6.2)$$

where A is the contact area, h_c is the contact depth, H is the nanohardness and P is the maximum load (Fischer-Cripps, 2011).

The elastic modulus of the specimen can be determined from the unloading response of the load–displacement curve. In a typical nanoindentation test, the load is increased at the point where the indenter is in contact with the specimen and the displacement is zero. The load is then increased to a maximum before being released, at which point the specimen usually exhibits some elastic recovery due to the relaxation of elastic strains. This elastic recovery is somewhat limited, however, due to the plastic deformation induced by the indenter during the test (Figure 6.15). The elastic modulus can be calculated as follows (Fischer-Cripps, 2011):

$$E_r = \frac{1}{2} \frac{\sqrt{\pi}}{\sqrt{A}} \frac{dp}{dh} \qquad (6.3)$$

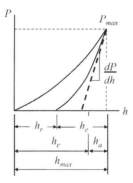

FIGURE 6.15 Load–displacement curve in a typical nanoindentation test, where P_{max} is the maximum load, h_{max} is the maximum penetration depth, h_r is the depth of the residual impression, h_e is the displacement associated with the elastic recovery, h_c is the contact depth and the unloading response $\frac{dP}{dh}$ is used to calculate the elastic modulus (Fischer-Cripps, 2011).

6.2.1 Calibration

Depth profiles were performed in two polished D-96 AG specimens, in which the load of a total of 355 indents was incrementally increased from 0.1 to 14 mN. The nanohardness of alumina was calculated from the contact depths of each indent, and the variation in nanohardness with contact depth was studied. Depth-controlled nanoindentation tests could then be performed to calibrated depths, once the nanohardness of alumina had been validated. The nanohardness of alumina was used as reference when testing other phases in the brazed joint microstructures.

The depth profiles performed in both alumina specimens were similar and showed that as the load increased from 0.1 to 14 mN, the contact depth also increased, from 2.1 to ~120 nm (Figure 6.16). At contact depths below ~40 nm; however, anomalously high nanohardness values of up to ~65 GPa were calculated. This indicated that these indents may have been too shallow to derive the nanohardness of alumina accurately. At contact depths greater than ~40 nm, the nanohardness of alumina reduced to ~30 GPa, which was similar to the nanohardness values of alumina found in the literature (Chakraborty et al., 2012; Dey et al., 2015; Stollberg et al., 2005).

In-situ scanning probe microscopy (SPM) imaging was used to confirm that the indents performed at contact depths less than 40 nm had not induced sufficient plastic deformation to create residual impressions (Figure 6.17a). Therefore, the anomalously high nanohardness may have resulted from indents that were within the elastic regime of alumina.

The onset of plastic deformation was first observed at contact depths of ~30 nm, where the resulting nanohardness values were ~35 GPa. However, in-situ SPM imaging found that this was only consistent with indents performed at, or near to, the alumina grain boundaries (Figure 6.17b and c). Plateaus observed in the depth profiles indicated that contact depths greater than ~40 nm could suitably derive a steady nanohardness value of ~30 GPa. In-situ SPM imaging showed that these indents had induced sufficient plastic deformation (Figure 6.17c). The nanohardness of alumina was evaluated as ~30 GPa, achievable at contact depths of at least 40 to 120 nm, independent of the indent locations.

With a further increase in the load range from 14 to 500 mN, the nanohardness of alumina decreased further, from ~30 GPa to ~16 GPa (Figure 6.18). This was similar to the Vickers hardness of 1590 Hv reported by the manufacturer, CoorsTek Ltd (Table 3.1). These results showed that nanohardness values are not absolute, since they can vary with the contact depth of the indenter. In comparing the nanohardness values of different phases, therefore, similar parameters should be used consistently.

The decrease in nanohardness observed with increasing contact depth may have occurred due to a phenomenon commonly referred to as the indentation size effect (Chakraborty et al., 2012). Several mechanisms have been used to explain the variation in nanohardness with contact depth, which also include the effects of fracture in ceramics (Bull, 2003). Chakraborty et al. (2012) suggested that this indentation size effect led to the nanohardness of alumina decreasing from 24 to 19 GPa with increasing indenter contact depth, from 50 to 850 nm. Therefore, the results obtained in this study were consistent with those in the literature.

The anisotropic nanohardness distribution in D-96 AG alumina may have been influenced by the presence of a secondary phase at the grain boundaries. This was shown by Dey et al. (2015), whereby the nanohardness of alumina was found to be higher at the centre of an alumina grain as compared with the nanohardness away from the centre of the alumina grain, and at the grain boundaries. This anisotropy in nanohardness was found to decrease, however, as the load increased from 1 to 10 mN. This occurred due to relatively larger indents, which were no longer able to specifically target the secondary phases at the grain boundaries. These results showed that in evaluating nanohardness, therefore, loads are required to be low enough to produce indents that are physically smaller than the phases targeted, but high enough to achieve contact depths that represent sufficient plastic deformation to overcome the indentation size effect.

Depth-controlled nanoindentation tests were used to evaluate the nanohardness values of the phases which formed in alumina-to-alumina brazed joints made using 50- to 150-μm-thick TICUSIL®

Joint Performance

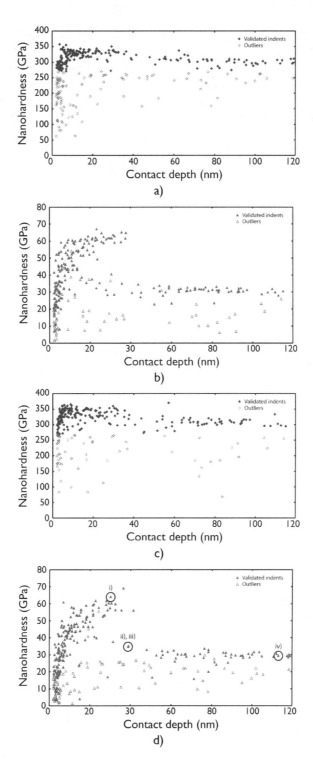

FIGURE 6.16 Depth profiles showing the variations in the reduced modulus (GPa) and nanohardness (GPa) of D-96 AG alumina with contact depth, in the load range 0.1–14 mN, in (a)-(b) specimen 1 and, (b)-(c) specimen 2, – Scanning probe microscope images for the indents circled and labelled as (i)-(iv) in (d) are shown in Figure 6.17. Validated indents and outliers were identified based on analysis of each corresponding load–displacement curve.

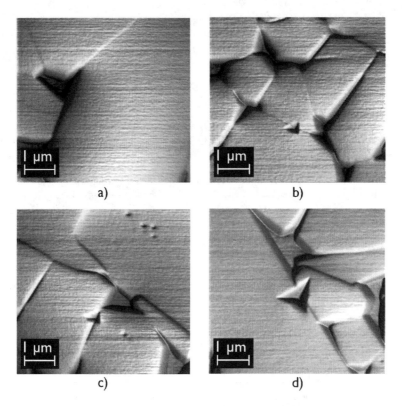

FIGURE 6.17 In-situ SPM images of selected indents as circled and labelled in Figure 6.16, (a) indent (i) at contact depth of 25 nm, (b) indent (ii) at contact depth of 30 nm, (c) indent (iii) at contact depth of 30 nm and (d) indent (iv) at contact depth of 115 nm.

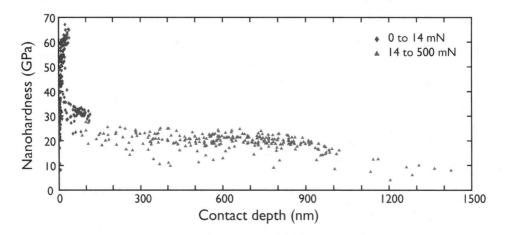

FIGURE 6.18 Depth profiles showing the nanohardness (GPa) variation with contact depth in the load ranges 0.1 to 14 mN and 14 to 500 mN, in D-96 AG alumina.

braze preforms. Displacement-controlled nanoindentation testing allowed for load application until the desired contact depths were achieved. This was preferred over load-controlled tests, which would require consideration of the mechanical properties of the relevant phases and as well as any variation in surface roughness of the specimen. Subsequently, in-situ SPM imaging was used to ensure that the residual impressions made by each indent had adequately targeted the relevant phases.

Joint Performance

Depth-controlled nanoindentation tests, set to a contact depth of 50 nm, were used to create grids of indents to evaluate the nanohardness distribution across the microstructures of brazed joints in a single experiment. The analyses of these results were limited to the reaction layer phase and phases in the braze interlayer, which were relatively more ductile than alumina. The portion of the grids which produced indents in alumina, were expected to lead to anomalously high nanohardness values, based on the results obtained from the depth profiles (Figures 6.16 and 6.17).

6.2.2 Targeted Indents in Alumina

Three depth-controlled indents, set to maximum penetration depths of 120 nm, were performed at the centres of two alumina grains adjacent to the reaction layer in a D-96 AG brazed joint made using a 50-µm-thick TICUSIL® braze preform. This resulted in an average maximum load of 8.3 mN and an average contact depth of 90.6 nm (Table 6.3). In-situ SPM imaging showed that these indents were within the grain boundaries of each alumina grain (Figure 6.19). The average nanohardness of alumina was calculated to be 28.6 GPa, which was consistent with the results obtained from the depth profiles in D-96 AG alumina (Figures 6.16 and 6.17).

6.2.3 Targeted Indents in the Reaction Layer

Six depth-controlled indents, set to maximum penetration depths of 50 nm, were performed at the centre of the reaction layer in a D-96 AG brazed joint made using a 50-µm-thick TICUSIL® braze preform. Figure 6.20 shows the load function used to perform these depth-controlled nanoindentation

TABLE 6.3
Properties of Targeted Indents Performed in D-96 AG Alumina Grains

Indent Number	Maximum Depth, h_{max} (nm)	Maximum Load, P_{max} (mN)	Contact Depth, h_c (nm)	Reduced Modulus, E_r (GPa)	Hardness, H (GPa)	Drift (nm/s)
1	124.1	8.3	91.5	312.3	28.4	0.010
2	124.3	8.5	90.2	308.8	29.6	0.065
3	123.7	8.0	90.0	294.2	27.9	0.064

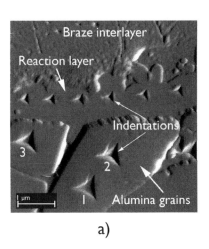
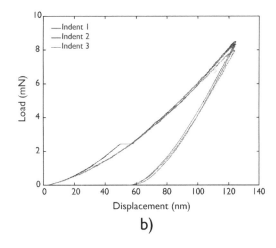

a)   b)

FIGURE 6.19 (a) In-situ SPM image showing three indents performed in alumina grains adjacent to the $Ti_3(Cu + Al)_3O$ reaction layer in a D-96 AG brazed joint made using a 50-µm-thick TICUSIL® braze preform, (b) corresponding load–displacement curves, showing good consistency.

tests. These indents were performed to maximum penetration depths of 50 nm which resulted in an average maximum load of 1.2 mN and an average contact depth of 53.9 nm (Table 6.4). In-situ SPM imaging showed that all of the indents had been successfully positioned within the 1.7-μm-thick reaction layer (Figure 6.19a). The corresponding load–displacement curves showed good consistency when superimposed (Figure 6.21), and the average nanohardness value of the reaction layer was calculated to be 14.7 GPa. The nanohardness of the reaction layer, characterised as $Ti_3(Cu + Al)_3O$, could not be found in the literature.

Two sets of 10 depth-controlled indents, set to maximum penetration depths of 50 nm, were performed at the centres of the reaction layers in D-96 AG brazed joints made using 100- and 150-μm-thick TICUSIL® braze preforms. This was to establish whether the 75% increase in joint strength from 136 to 238 MPa, as a result of an increase in the TICUSIL® braze preform thickness from 50 to 100 μm, and the 13% increase in joint strength from 238 to 270 MPa, as a result of an increase in the TICUSIL® braze preform thickness from 100 to 150 μm, was attributed to any variation in the mechanical properties of the reaction layers.

In-situ SPM imaging showed that all of the indents were within the 2.3-μm-thick reaction layer of the brazed joint made using a 100-μm-thick TICUSIL® braze preform (Figure 6.22), and within the 2.8 μm reaction layer of the brazed joint made using a 150-μm-thick TICUSIL® braze preform. The corresponding load–displacement curves, for both sets of indents, showed good consistency when superimposed (Figure 6.23).

The average nanohardness values of the reaction layers which formed in D-96 AG brazed joints made using both 100- and 150-μm-thick TICUSIL® braze preforms were 16.0 and 15.6 GPa, respectively (Table 6.5). These values were similar to the nanohardness value of 14.7 GPa evaluated from the reaction layer in the D-96 AG brazed joint made using a 50-μm-thick TICUSIL® braze preform. These results were consistent with the fact that the reaction layer phase in all three sets of joints was

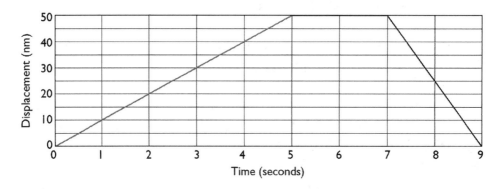

FIGURE 6.20 Load function used for depth-controlled nanoindentation experiments.

TABLE 6.4
Properties of Targeted Indents Performed in the $Ti_3(Cu + Al)_3O$ Reaction Layer

Indent Number	Maximum Depth, h_{max} (nm)	Maximum Load, P_{max} (mN)	Contact Depth, h_c (nm)	Reduced Modulus, E_r (GPa)	Hardness, H (GPa)	Drift (nm/s)
1	53.8	1.2	40.1	211.2	15.3	0.066
2	54.5	1.1	41.4	197.1	13.3	0.011
3	53.8	1.1	41.3	208.3	13.5	0.152
4	53.9	1.2	40.7	213.0	14.8	0.050
5	53.9	1.3	39.1	214.2	17.2	0.030
6	53.7	1.2	41.2	216.6	14.0	0.090

Joint Performance

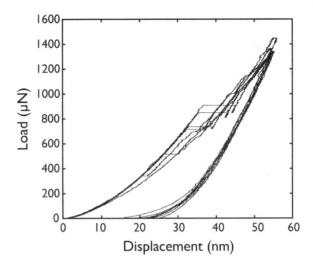

FIGURE 6.21 Load–displacement curves of six indents performed in the $Ti_3(Cu + Al)_3O$ reaction layer in a D-96 AG brazed joint made using a 50-μm-thick TICUSIL® braze preform.

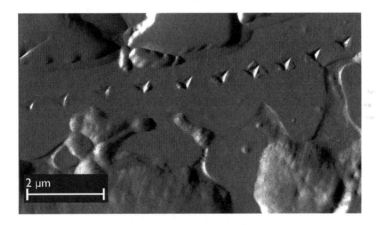

FIGURE 6.22 In-situ SPM image showing ten indents performed in the reaction layer in a D-96 AG brazed joint made using a 100-μm-thick TICUSIL® braze preform.

characterised as $Ti_3(Cu + Al)_3O$ (see Section 5.3). Therefore, the average nanohardness of the $Ti_3(Cu + Al)_3O$ layer in D-100 AG brazed joints was also likely to be ~15.4 GPa.

6.2.4 Targeted Indents in the Braze Interlayer

Ten depth-controlled indents, set to maximum penetration depths of 50 nm were performed at the centre of the Ag-rich phase, which formed in the braze interlayers of both D-96 AG and D-100 AG brazed joints made using 50- to 250-μm-thick TICUSIL® braze preforms. This resulted in an average maximum load of 239.5 μN and an average contact depth of 51.3 nm (Table 6.6).

Similarly, eight depth-controlled indents, set to maximum penetration depths of 50 nm, were performed at the centre of the Cu-rich phase, which also formed in the braze interlayers of both D-96 AG and D-100 AG brazed joints made using 50- to 250-μm-thick TICUSIL® braze preforms. This resulted in an average maximum load of 169.5 μN and an average contact depth of 51.5 nm (Table 6.6).

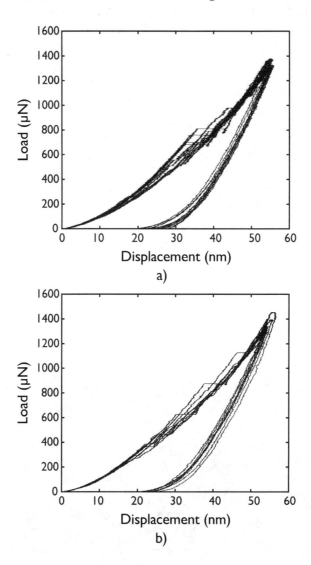

FIGURE 6.23 Load–displacement curves of sets of 10 indents performed in the $Ti_3(Cu + Al)_3O$ reaction layers in D-96 AG brazed joints made using (a) 100-μm- and, (b) 150-μm-thick TICUSIL® braze preforms.

TABLE 6.5
Properties of Targeted Indents Performed in the $Ti_3(Cu + Al)_3O$ Reaction Layers of D-96 AG Brazed Joints Made Using 50- to 150-μm-Thick TICUSIL® Braze Preforms

Alumina	Braze Preform Thickness (μm)	Maximum Depth, h_{max} (nm)	Maximum Load, P_{max} (mN)	Contact Depth, h_c (nm)	Reduced Modulus, E_r (GPa)	Hardness, H (GPa)	Drift (nm/s)
D-96 AG	50	53.9	1.2	40.6	210.1 ± 0.4	14.7 ± 0.2	0.067
D-96 AG	100	54.7	1.3	41.5	229.2 ± 1.5	16.0 ± 0.3	0.065
D-96 AG	150	55.2	1.3	42.3	230.7 ± 1.8	15.6 ± 0.2	0.079

TABLE 6.6
Properties of Targeted Indents Performed in the Ag-Rich and Cu-Rich Phases of the Braze Interlayer

Phase	Maximum Depth, h_{max} (nm)	Maximum Load, P_{max} (μN)	Contact Depth, h_c (nm)	Reduced Modulus, E_r (GPa)	Hardness, H (GPa)	Drift (nm/s)
Ag-rich	55.0	169.5	51.5	95.5	1.5	0.162
Cu-rich	54.9	239.4	51.3	132.1	2.1	0.093

In-situ SPM imaging showed that all of the indents were performed within the targeted Ag-rich and Cu-rich phases, respectively, away from any adjacent interfaces (Figure 6.24). The corresponding load–displacement curves, for both sets of indents, showed good consistency when superimposed (Figure 6.25).

The average nanohardness values of the Ag-rich and Cu-rich phases were found to be 1.5 GPa and 2.1 GPa, respectively. These results showed that the Ag-rich phase was relatively more ductile than the Cu-rich phase by ~40%. It is likely, therefore, that the Ag-rich phase can undergo relatively more plastic deformation than the Cu-rich phase, in accommodating thermally induced residual stresses and applied stresses in a brazed joint. This may also explain why a relatively higher dislocation density was observed in the Ag-rich phase as compared with the Cu-rich phase in joints made using 50-μm-thick TICUSIL® braze preforms (see Figure 5.45).

Due to the uncertainty surrounding the depth of these Ag-rich and Cu-rich phases in the cross-sectional microstructures of the brazed joints, depth profiles were performed in each phase to evaluate any variation in nanohardness with contact depth. Three sets of load/partial unload indents were performed, whereby depth profiles were evaluated, with a total of 60 cyclic indents at a single location in each phase. In each cycle, the maximum penetration depth was incrementally increased until a maximum penetration depth of 200 nm was achieved. A partial unload between each cycle enabled the nanohardness to be calculated, relative to the instantaneous contact depth. Figure 6.26 shows the load–displacement curves for one of the load/partial unload indents performed in the Ag-rich phase, which corresponds to one of the relatively larger residual impressions, as shown in Figure 6.24.

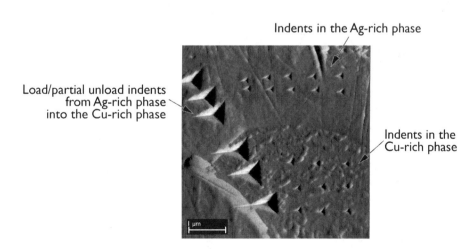

FIGURE 6.24 In-situ SPM image showing ten indents performed in the Ag-rich phase, and eight indents performed in the Cu-rich phase, in the braze interlayer of a D-96 AG brazed joint made using a 50-μm-thick TICUSIL® braze preform.

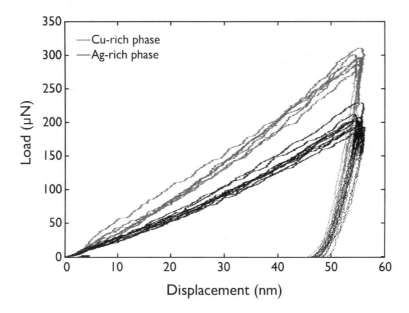

FIGURE 6.25 Load–displacement curves of ten indents performed in the Ag-rich phase, and eight indents performed in the Cu-rich phase, in the braze interlayer of a D-96 AG brazed joint made using a 50-μm-thick TICUSIL® braze preform.

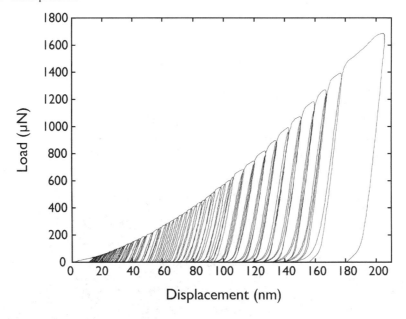

FIGURE 6.26 Load/partial unload–displacement curves of 60 cyclic indents performed during a single experiment and in one location in the Ag-rich phase, in the braze interlayer of a D-96 AG brazed joint made using a 50-μm-thick TICUSIL® braze preform.

The depth profiles performed in both the Ag-rich and Cu-rich phases in the braze interlayer of a D-96 AG brazed joint made using a 50-μm-thick TICUSIL® braze preform showed that the nanohardness values of both phases decreased by ~1 GPa as the contact depth increased from ~15 to ~170 nm (Figure 6.27). The Ag-rich phase was consistently more ductile than the Cu-rich phase.

From these results, the differences in the nanohardness values of the phases in the brazed joint microstructure could be compared. At contact depths of 40 nm, therefore, the nanohardness of

Joint Performance

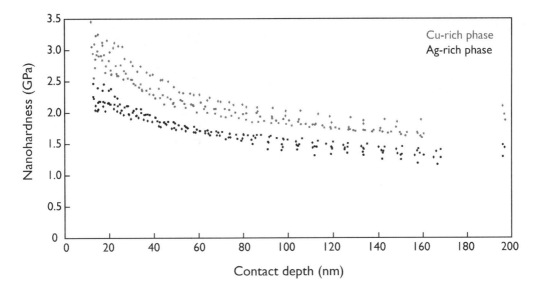

FIGURE 6.27 Depth profiles showing the nanohardness variation with contact depth of the Ag-rich and Cu-rich phases in the braze interlayer of a D-96 AG brazed joint made using a 50-μm-thick TICUSIL® braze preform.

alumina was ~30 GPa while the nanohardness of the $Ti_3(Cu + Al)_3O$ reaction layer was 15.4 GPa. These phases were relatively harder than the Ag-rich and Cu-rich phases in the braze interlayer, which had nanohardness values of 1.9 and 2.5 GPa, respectively, at contact depths of 40 nm (Figure 6.28).

6.2.5 Nanohardness Distribution Plots

A grid of 289 indents with 1.8 μm spacings, set to maximum penetration depths of 50 nm, were performed in a 30 μm × 25 μm selected area across the brazed joint microstructure of a D-96 AG brazed joint made using a 50-μm-thick TICUSIL® braze preform (Figure 6.29a). The results were used to plot the nanohardness distribution in the 26-μm-thick brazed joint, which consisted of a 1.7-μm-thick $Ti_3(Cu + Al)_3O$ reaction layer, and Ag-rich and Cu-rich phases in a eutectic-like structure (Figure 6.29b). The nanohardness distribution plot showed a good correlation with the microstructure of the

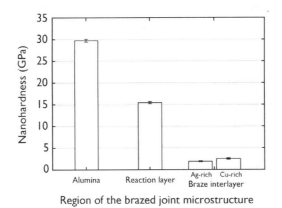

FIGURE 6.28 Comparison of the nanohardness (GPa) values of phases in a D-96 AG brazed joint made using a 50-μm-thick TICUSIL® braze preform, at contact depths of 40.0 nm.

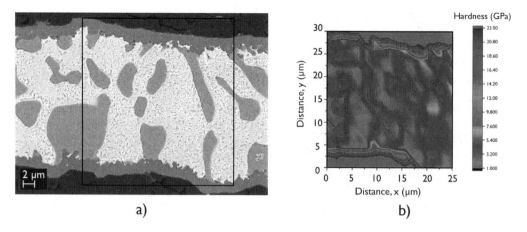

FIGURE 6.29 (a) Backscattered electron image showing the microstructure of a typical D-96 AG brazed joint made using a 50-μm-thick TICUSIL® braze preform and, (b) corresponding nanohardness distribution in the selected area (indicated by a rectangle).

joint, which was consistent throughout the joint including at the joint edges where braze outflow was observed (see Figure 5.34a).

A grid of 600 indents, set to maximum penetration depths of 50 nm, were performed in a 30 μm × 25 μm selected area in the brazed joint microstructure of a D-96 AG brazed joint made using a 100-μm-thick TICUSIL® braze preform (Figure 6.30a). Spacings of 1.8 μm in the x-direction and 0.9 μm in the y-direction were set between the indents. The results were used to plot the nanohardness distribution in the 39-μm-thick brazed joint, which consisted of a 2.3-μm-thick $Ti_3(Cu+Al)_3O$ reaction layer, and isolated Cu_4Ti_3 and Cu_4Ti phases that were surrounded by Ag-rich and Cu-rich phases (Figure 6.30b).

The nanohardness values of the Cu_4Ti_3 and Cu_4Ti phases were derived from the nanohardness distribution plot by analysing the load–displacement curves of selected indents at the centres of each phase (Figure 6.31). The corresponding load–displacement curves, for all sets of indents, showed good consistency when superimposed. The 8.5 GPa average nanohardness of Cu_4Ti was 26% greater than the 10.7 GPa average nanohardness of Cu_4Ti_3 (Table 6.7), which was consistent with reported differences between the nanohardness values of these phases (Chen et al., 2015).

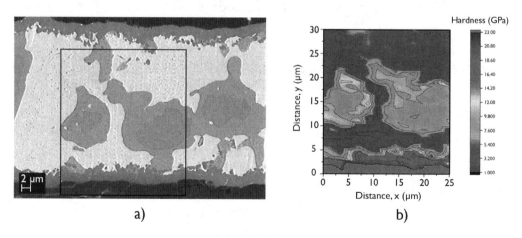

FIGURE 6.30 (a) Backscattered electron image showing the microstructure of a typical D-96 AG brazed joint made using a 100-μm-thick TICUSIL® braze preform and, (b) corresponding nanohardness distribution in the selected area (indicated by a rectangle).

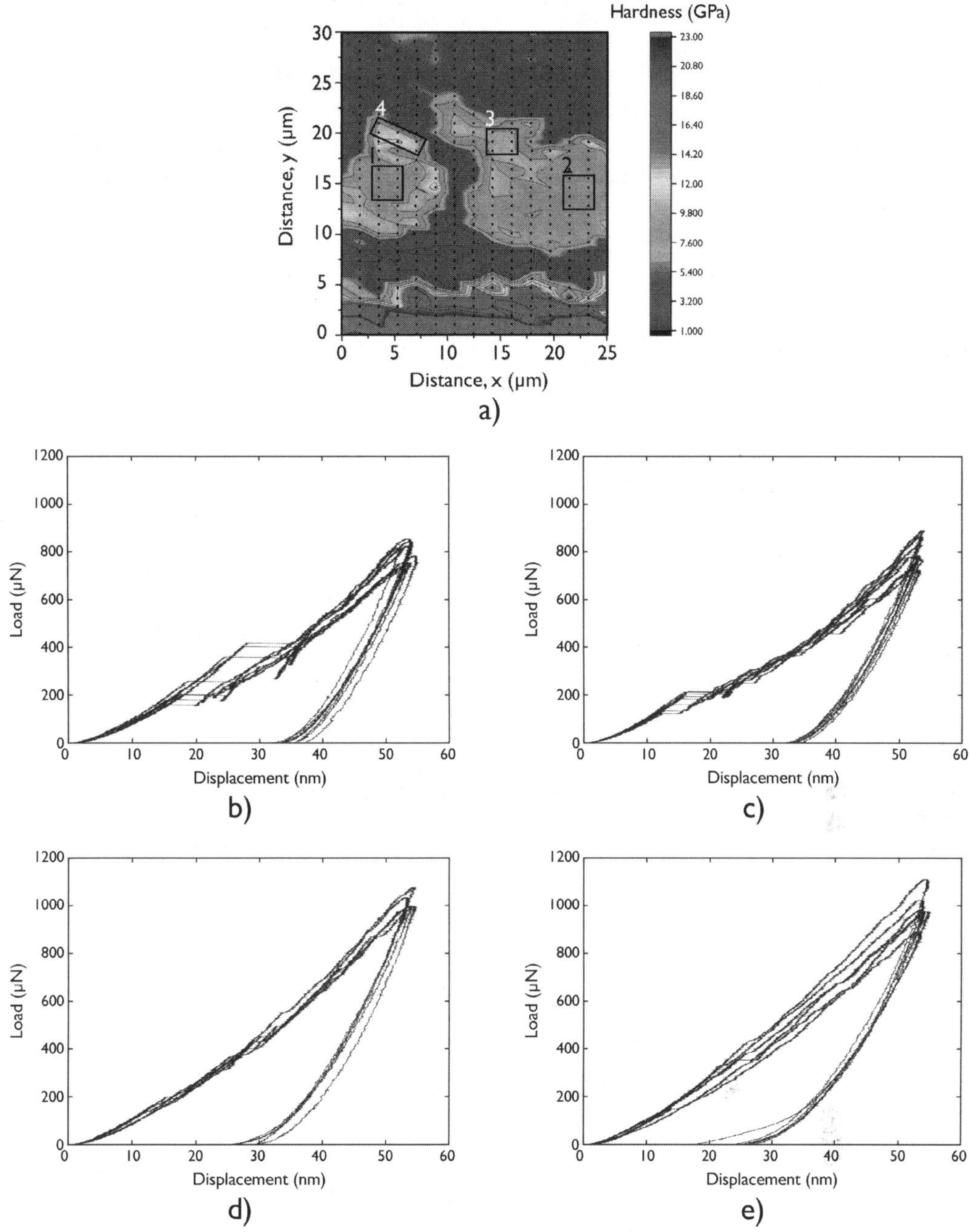

FIGURE 6.31 Nanohardness distribution plot, with indent locations, in a selected area of a D-96 AG brazed joint made using a 100-μm-thick TICUSIL® braze preform and, the corresponding load–displacement curves in (b) area 1 and (c) area 2, of sets of indents performed in the Cu_4Ti_3 phase and, (d) area 3 and (e) area 4, of sets of indents performed in the Cu_4Ti phase.

These results showed that the isolated Cu_4Ti_3 and Cu_4Ti phases, which formed in the braze interlayer of joints made using 100- and 150-μm-thick TICUSIL® braze preforms, were relatively harder than the surrounding Ag-rich and Cu-rich matrix. Literature studies have shown that relatively harder particulate phases in an otherwise ductile braze interlayer can, through reinforcement, lead to an improvement in joint strength (Yang et al., 2002, 2005; Zhu and Chung, 1997). Therefore, the formation of isolated

TABLE 6.7
Properties of Targeted Indents Performed in the Cu$_4$Ti$_3$ and Cu$_4$Ti Phases of the Braze Interlayer

Phase	Maximum Depth, h_{max} (nm)	Maximum Load, P_{max} (μN)	Contact Depth, h_c (nm)	Reduced Modulus, E_r (GPa)	Hardness, H (GPa)	Drift (nm/s)
Cu$_4$Ti	53.6	922.0	42.8	191.0 ± 5.6	10.7 ± 0.2	0.345
Cu$_4$Ti$_3$	44.5	778.7	53.4	192.0 ± 3.6	8.5 ± 0.2	0.318

Cu$_4$Ti$_3$ and Cu$_4$Ti phases in the reinforced braze interlayer of joints made using 100- and 150-μm-thick TICUSIL® braze preforms may have contributed to the increase in joint strengths observed.

A grid of 1681 indents, set to maximum penetration depths of 50 nm, were performed in a 60 μm × 60 μm selected area at the joint edge of a D-96 AG brazed joint made using a 100-μm-thick TICUSIL® braze preform (Figure 6.32). The absence of the Cu-rich phase at the joint edge resulted in a nanohardness distribution that was relatively ductile as compared with the nanohardness distribution in D-96 AG brazed joints made using 50-μm-thick TICUSIL® braze preforms (Figure 6.29b).

These results suggested that the separation of an Ag-rich phase which flowed towards the joint edges may have made the edges of joints made using 100- and 150-μm-thick TICUSIL® braze preforms relatively more ductile than that of joints made using 50-μm-thick TICUSIL® braze preforms (see Figure 5.24). Therefore, the ductile edges in these joints may have also contributed to the increase in joint strengths observed.

A grid of 1681 indents with 1.5-μm spacings, set to maximum penetration depths of 50 nm, were performed in a 60 μm × 60 μm selected area across the brazed joint microstructure of a D-96 AG brazed joint made using a 150-μm-thick TICUSIL® braze preform (Figure 6.33a). The results were used to plot the nanohardness distribution in a typical region of the 182-μm-thick brazed joint, which comprised a 3.2-μm-thick Ti$_3$(Cu + Al)$_3$O reaction layer, a multi-layered Cu-Ti structure composed of Ti, CuTi$_2$, CuTi, Cu$_4$Ti$_3$ and Cu$_4$Ti, isolated Cu$_4$Ti$_3$ and Cu$_4$Ti phases, and the typical Ag-rich and Cu-rich phases (see Figure 5.34c).

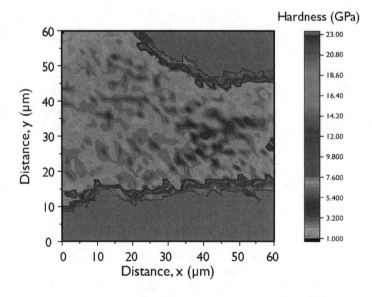

FIGURE 6.32 Nanohardness distribution plot in a selected area at the edge of a D-96 AG brazed joint made using a 100-μm-thick TICUSIL® braze preform, where Ag-rich braze outflow was observed.

Joint Performance

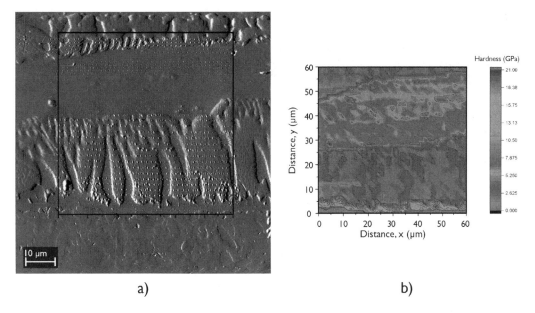

FIGURE 6.33 (a) In-situ SPM image showing 1681 indents performed in the brazed joint microstructure of a D-96 AG brazed joint made using a 150-μm-thick TICUSIL® braze preform and, (b) corresponding nanohardness distribution in selected area (indicated by a rectangle).

The nanohardness distribution plot showed that the multi-layered Cu-Ti structure was relatively harder than the surrounding Ag-rich and Cu-rich matrix (Figure 6.33b). Within the multi-layered Cu-Ti structure, however, the nanohardness distribution was found to be relatively complex. The average nanohardness at the outskirts of the multi-layered Cu-Ti structure, composed of Cu_4Ti and Cu_4Ti_3, was found to be between 7 and 9 GPa. The average nanohardness at the centre of the multi-layered Cu-Ti structure, composed of a Ti core surrounded by layers of $Cu-Ti_2$ and Cu-Ti, was found to decrease to between ~3 and 4 GPa. Therefore, the Cu_4Ti phase was the hardest Cu-Ti phase in the multi-layered Cu-Ti structure. The Cu_4Ti phase surrounded other Cu-Ti phases that were relatively less hard, and was itself surround by the relatively ductile Ag-rich and Cu-rich matrix (see Figure 5.25a).

Ductile regions of the braze interlayer can plastically deform to a accommodate stress, whereas interfacial reaction products are limited in their deformability which results in localised stress at the joint interface (Do Nascimento et al., 1999). Evaluating the nanohardness distribution of phases in the braze interlayer, therefore, is important; particularly in understanding the effect of the microstructural evolution found in this study on the improvement in joint strength observed.

Ghosh et al. (2012) performed load-controlled nanoindentation tests at a load of 100 mN to evaluate the nanohardness distribution in an alumina-to-alumina brazed joint made using a 150-μm-thick TICUSIL® braze foil. The nanohardness of the reaction layer was 5.7 ± 1.1 and the nanohardness of the braze interlayer was 1.8 ± 0.6 GPa (Figure 2.6). Differences in nanohardness between phases (which were not characterised) in the braze interlayer, were not evaluated. The difference in thickness between the ~30-μm-thick brazed joint and the 150-μm-thick TICUSIL® braze foil selected was suggested to have been due to braze infiltration into the porous alumina ceramic; however, this was not evident in the SEM micrographs shown. While the micrographs do not clearly show the phases formed in the braze interlayer, 7-μm-thick reaction layers were observed at the joint interfaces. The relatively high brazing temperature of 910°C may have enabled the complete diffusion of Ti to the joint interfaces. Nevertheless, the variations in nanohardness reported were similar with those in this study for joints made using 50-μm-thick TICUSIL® braze preforms, in which Ti was observed to have completely diffused to the joint interfaces (Figure 6.28).

The results of these nanoindentation tests suggest that the improvement in joint strength observed, in both D-96 AG and D-100 AG brazed joints, as a result of the microstructural evolution that occurred with an increase in the TICUSIL® braze perform thickness (see Figure 5.32), may be explained on the basis of two mechanisms: (i) the formation of Cu-Ti phases in the braze interlayer produces a nanohardness distribution that is relatively complex; relatively hard Cu-Ti phases are embedded in a relatively ductile Ag-Cu matrix which also comprises ductile Ag-rich phases. This provides a composite-type reinforced braze interlayer, (ii) the separation of an Ag-rich phase that flows towards the joint edges provides a strengthening effect; through the formation of relatively ductile joint edges. By these two mechanisms, the microstructural evolution may be correlated to the improvement in joint strengths observed.

6.3 STRENGTHS OF BRAZED JOINTS IN GROUND-AND-HEAT-TREATED CONDITION

As the TICUSIL® braze preform thickness was increased from 50 to 150 µm, the average strengths of D-96 GHT brazed joints were significantly affected by the post-grinding heat treatment, as compared with the average strengths of D-96 AG brazed joints. However, the average strengths of D-100 GHT brazed joints were similar to that of the D-100 AG brazed joints, as the TICUSIL® braze preform thickness was increased from 50 to 150 µm.

6.3.1 50-µm-Thick TICUSIL® Braze Preforms

The average strengths of D-96 GHT brazed joints made using 50-µm-thick TICUSIL® braze preforms increased by ~55% from 136 to 211 MPa due to the post-grinding heat treatment. All of the failures in the D-96 GHT brazed joints were initiated at the joint interfaces (Figure 6.34).

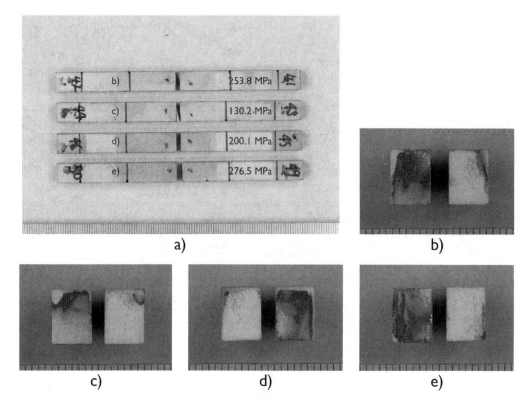

FIGURE 6.34 (a) Failures in sample set 5, D-96 GHT brazed joints made using 50-µm-thick TICUSIL® braze preforms and, corresponding fracture surfaces showing (b) high, (c) low, (d) high and (e) high, strength failures.

Fractography revealed that despite incomplete joint filling in the D-96 GHT brazed joints (which was also observed in D-96 AG and D-100 AG brazed joints made using 50-μm-thick TICUSIL® braze preforms), high-strength failures were the predominant failure mode.

The average reaction layer thickness of 1.5 μm, in D-96 GHT brazed joints made using 50-μm-thick TICUSIL® braze preforms, was similar to the average reaction layer thickness of 1.7 μm in D-96 GHT brazed joints made using the same TICUSIL® braze preform thickness of 50 μm. It was unlikely, therefore, that the increase in joint strength observed was due to stronger interfacial bonding. The fracture surfaces of each of the D-96 GHT brazed joints consistently showed that the fracture paths propagated through the sub-surface of the alumina ceramics (Figure 6.34b and 6.34e).

The average brazed joint thickness of 43 μm in D-96 GHT brazed joints made using 50-μm-thick TICUSIL® braze preforms was 66% greater than that of 26 μm in D-96 AG brazed joints made using the same TICUSIL® braze preform thickness of 50 μm. Therefore, the increase in joint strength observed may have been due to a relatively thicker Ag-Cu eutectic-like braze interlayer. The non-uniform brazed joint thickness observed in D-96 GHT brazed joints, however, may have limited this strengthening effect. Despite the predominant failure mode improving from low-strength failure to high-strength failure, following the post-grinding heat treatment in the D-96 GHT brazed joints, the failures were still found to have initiated at the joint interface.

Braze infiltration into the triple pocket grain boundary regions of the alumina surface was observed during microstructural examination of the D-96 GHT brazed joints (see Section 5.5.3). This may have weakened the alumina surface which would explain why the fractures observed in D-96 GHT brazed joints were initiated at the joint interface and propagated towards the reaction layer (through selected regions of the alumina surface). The fracture paths may have corresponded to weakened areas of the alumina surface, where braze infiltration was observed.

Post-grinding heat treatment had led to a 5.2% increase in the average flexural strength of the D-96 GHT standard alumina test bars, as compared with that of the D-96 AG standard alumina test bars. It was postulated that post-grinding heat treatment may have recovered some grinding damage in the D-96 GHT alumina surface (see Section 4.5). Therefore, post-grinding heat treatment may have led to the increase in joint strength observed in the D-96 GHT brazed joints, by facilitating recovery of grinding damage in the alumina surface, prior to brazing.

The average strengths of D-100 GHT brazed joints made using 50-μm-thick TICUSIL® braze preforms increased by ~18% from 163 to 192 MPa due to the post-grinding heat treatment. The predominant failure mode improved from low-strength failure to high-strength failure following the post-grinding heat treatment in the D-100 GHT brazed joints; however, the failures were still found to have initiated at the joint interface (Figure 6.35). The increase in joint strength observed may not have been related to the post-grinding heat treatment, which was found to decrease the flexural strength of the D-100 GHT standard alumina test bars by 8.4%. The average reaction layer thicknesses in D-100 GHT brazed joints were slightly thicker than that in D-100 AG brazed joints, for all braze preform thicknesses selected (see Figure 5.77a). It is likely that the improvement in joint strength observed, therefore, may have resulted from a slight improvement in interfacial bonding.

6.3.2 100-μm-Thick TICUSIL® Braze Preforms

D-96 GHT brazed joints made using 100-μm-thick TICUSIL® braze preforms achieved an average joint strength of 106 MPa. Despite the 5.2% increase in the average flexural strength of D-96 AG standard test bars following the post-grinding heat treatment, once brazed, a 56% reduction in the average joint strength was observed. All failures observed in these D-96 GHT brazed joints initiated at the joint interfaces (Figure 6.36). These results were not in accordance with those found in the literature, in which post-grinding heat treatment at near-sintering temperatures has been used to improve the strength of brazed joints made using alumina ceramics, prior to brazing (see Table 2.18).

The average brazed joint thickness of 54 μm in D-96 GHT brazed joints made using 100-μm-thick TICUSIL® braze preforms was ~38% greater than that of 39 μm in D-96 AG brazed joints made

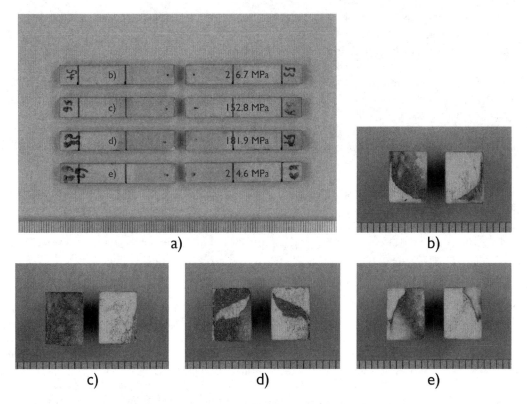

FIGURE 6.35 (a) Failures in sample set 13, D-100 GHT brazed joints made using 50-μm-thick TICUSIL® braze preforms and, corresponding fracture surfaces showing (b) high, (c) high, (d) high and (e) high, strength failures.

using the same TICUSIL® braze preform thickness of 100 μm. The brazed joint thickness of a typical D-96 GHT brazed joint made using a 100-μm-thick TICUSIL® braze preform exhibited a large degree of non-uniformity, with minimum and maximum brazed joint thicknesses of ~37 and ~85 μm, respectively (see Figure 5.72). Therefore, non-uniform stress distribution in these D-96 GHT brazed joints may have led to the degradation in joint strength observed.

Post-grinding heat treatment was postulated to have led to the retraction of the secondary phase away from the alumina surface, resulting in fissures at the triple pocket grain boundary regions of the alumina surface. While secondary phase migration may have healed areas of grinding induced damage, the resulting fissures may have acted as stress concentrators, initiating failures. The overriding effect, however, was found to have been a 5.2% increase in the average flexural strength of D-96 GHT standard alumina test bars (see Section 4.5). Once brazed, these vacant triple pocket regions in the alumina surface were observed to have been infiltrated by the braze alloy. As a result, thermally induced residual stresses may have structurally weakened the alumina surface (see Section 5.5.1). Furthermore, it was not known whether braze infiltration had occurred in all of the vacant triple pocket regions, as otherwise vacancies may have also acted as failure initiation sites. These competing mechanisms, therefore, may explain the variation in failure strengths observed in the D-96 GHT brazed joints, which included two low-strength failures at 96.9 and 121.2 MPa (Figure 6.36b and 6.36d) and a single high-strength failure at 150.7 MPa (Figure 6.36c). Fractography revealed that in both of these failure modes, fractures were initiated in the D-96 GHT alumina surface, albeit closer to the reaction layer in the low-strength failures and relatively deeper into the alumina surface for the high-strength failure. Both failure modes consistently indicated that the sub-surface region of D-96 GHT alumina may have been highly stressed and structurally weak.

Post-grinding heat treatment did not lead to a degradation in the strength of D-96 GHT brazed joints made using 50-μm-thick TICUSIL® braze preforms, in which similar braze infiltration was

Joint Performance

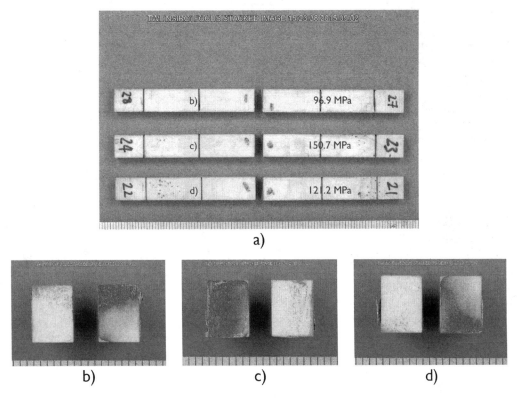

FIGURE 6.36 (a) Failures in sample set 6, D-96 GHT brazed joints made using 100-μm-thick TICUSIL® braze preforms and, corresponding fracture surfaces showing (b) low, (c) high and (d) low, strength failures.

also observed. However, these joints consisted of relatively uniform braze joint thicknesses, unlike the brazed joint thicknesses of D-96 GHT brazed joints made using 100-μm-thick TICUSIL® braze preforms, which was highly non-uniform. Therefore, the degradation in joint strength observed in D-96 GHT brazed joints made using 100-μm-thick TICUSIL® braze preforms may have been highly influenced by the non-uniform brazed joint thickness, rather than solely by the braze infiltration.

In order to confirm the degradation in joint strength observed, in D-96 GHT brazed joints made using 100-μm-thick TICUSIL® braze preforms, another set of brazed joints were produced under the same conditions. These joints also exhibited highly non-uniform brazed joint thicknesses, and achieved an average joint strength of 133 MPa. This represented a ~44% decrease in strength of the D-96 GHT brazed joints, once brazed, and following the post-grinding heat treatment as compared with D-96 GHT brazed joints made using 50-μm-thick TICUSIL® braze preforms. These results confirmed that D-96 GHT brazed joints made using 100-μm-thick TICUSIL® braze preforms were highly sensitive to the post-grinding heat treatment. The resulting degradation in the strength of these D-96 GHT brazed joints, therefore, may have been due to a combined effect of a non-uniform stress distribution in the joint coupled with the presence of either fissured regions of the alumina surface or regions in which braze infiltration had occurred. Further investigation is required to better understand these mechanisms, which led to the degradation in joint strength observed in these D-96 GHT brazed joints.

D-100 GHT brazed joints made using 100-μm-thick TICUSIL® braze preforms achieved an average joint strength of 190 MPa, which was similar to the 200 MPa achieved by D-100 AG brazed joints made using the same TICUSIL® braze preform thickness of 100 μm. Fractography revealed that the failures in these D-100 GHT brazed joints included two high-strength failures at 219.0 and 198.7 MPa (Figure 6.37b and 6.37d), a single low-strength failure at 121.1 MPa (Figure 6.37c) and a single optimum strength failure, at 223.0 MPa (Figure 6.37e). The similarities in the failure modes and joint strengths achieved by both D-100 GHT and D-100 AG sets of brazed joints made using 100-μm-thick TICUSIL® braze preforms suggests that the post-grinding heat treatment did not significantly affect the strengths of D-100 GHT brazed joints as much as the D-96 GHT brazed joints. These results were consistent with earlier findings which showed that the post-grinding heat treatment did not significantly affect the strengths of D-100 GHT brazed joints made using 50-μm-thick TICUSIL® braze preforms. Furthermore, the non-uniform brazed joint thicknesses of D-100 GHT brazed joints did not significantly affect joint strength. This supports the idea that the degradation of D-96 GHT brazed joints joints may have been more likely associated to the secondary phase migration, in which fissured regions and vacancies may have acted as failure initiation sites, rather than simply by the non-uniform brazed joint thickness.

The average brazed joint thickness of 34 μm in D-100 GHT brazed joints made using 100-μm-thick TICUSIL® braze preforms was similar to that of 39.1 μm in D-96 AG brazed joints made using the same TICUSIL® braze preform thickness of 100 μm. However, the average reaction layer thickness of 2.9 μm in D-100 GHT brazed joints made using 100-μm-thick TICUSIL® braze preforms was ~26% greater than the average reaction layer thickness of 2.3 μm in D-100 GHT brazed joints made using 50-μm-thick TICUSIL® braze preforms. Any expected improvement in strength as a result of improved interfacial bonding may have been counterbalanced by the effect of post-grinding heat treatment on the flexural strength of D-100 GHT alumina, which decreased by 8.4% (see Section 4.5). Overall, the effect of the post-grinding heat treatment appeared to be

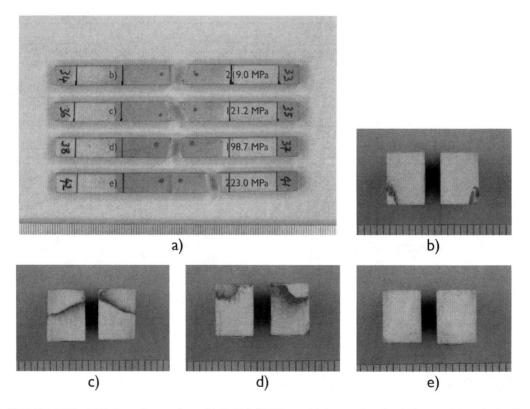

FIGURE 6.37 (a) Failures in sample set 14, D-100 GHT brazed joints made using 100-μm-thick TICUSIL® braze preforms and, corresponding fracture surfaces showing (b) high, (c) low, (d) high and (e) optimum, strength failures.

Joint Performance

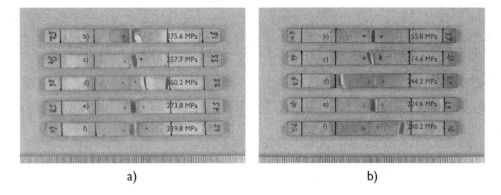

FIGURE 6.38 (a) Failures in D-96 GHT (sample set 8) and D-100 GHT (sample set 15) brazed joints made using 150-μm-thick TICUSIL® braze preforms. Failures in both sets of joints were high ($\sigma > 150$ MPa), optimum ($\sigma < 230$ MPa) or ideal ($\sigma > 230$ MPa) strength failures, and were all initiated in the alumina ceramic, close to or away from the joint interface.

more significant to the strength of the D-96 GHT brazed joints as compared to that of the D-100 GHT brazed joints.

6.3.3 150-μM-THICK TICUSIL® BRAZE PREFORMS

The average joint strengths of D-96 GHT and D-100 GHT brazed joints made using 150-μm-thick TICUSIL® braze preforms were found to be 261 and 216 MPa, respectively. Failures in both of these sets of joints initiated in the alumina ceramics either close to or away from the joint interfaces (Figure 6.38). The similarities in both the failure modes and joint strengths in these D-96 GHT and D-100 GHT brazed joints made using the same TICUSIL® braze preform thickness of 150 μm were similar to their as-ground counterparts. Therefore, the post-grinding heat treatment did not significantly affect the strengths of either D-96 GHT or D-100 GHT brazed joints when the TICUSIL® braze preform thickness of 150 μm was selected.

The use of 150-μm-thick TICUSIL® braze preforms led to ideal joint failures in both D-96 AG and D-100 AG brazed joints and this was postulated to have been due to several important strengthening mechanisms associated to the microstructural evolutions that were observed in the braze interlayer as the TICUSIL® braze preform thickness was increased (see Section 6.1.3). These strengthening mechanisms may have overridden any adverse effects of the post-grinding heat treatment on the strengths of both D-96 GHT and D-100 GHT brazed joints made using 150-μm-thick TICUSIL® braze preforms. This was more evident in these D-96 GHT brazed joints than in these D-100 GHT brazed joints, particularly since it was the D-96 GHT brazed joints that suffered from both non-uniform brazed joint thicknesses and weakened alumina surfaces as a result of the post-grinding heat treatment. Figure 6.39 summarises the effect of braze preform thickness on the strengths of D-96 GHT and D-100 GHT brazed joints as compared with the strengths of D-96 AG and D-100 AG brazed joints, formed under the same conditions. The properties of D-96 GHT and D-100 GHT brazed joints made using 50- to 150-μm-thick TICUSIL® braze preforms are summarised in Table 6.8.

6.4 SUMMARY

The strengths of alumina-to-alumina brazed joints made using D-96 and D-100 alumina ceramics in as-ground and ground-and-heat-treated conditions were studied. The effect of microstructural evolutions, due to an increase in the TICUSIL® braze preform thickness, on the mechanical and nanomechanical properties of alumina-to-alumina brazed joints were evaluated.

TABLE 6.8
Properties of D-96 GHT and D-100 GHT Brazed Joints Made Using 50- to 150-μm-Thick TICUSIL® Braze Preforms

Sample Set	Alumina Purity, (wt.% Al$_2$O$_3$) and Surface Condition	Braze Preform Thickness (μm)	Specimens Characterised	Reaction Layer Thickness (μm)	Average Brazed Joint Thickness (μm)	Specimens Mechanically Tested	Joint Strength (MPa)	Failure Location
5	D-96 GHT	50	1	1.5 ± 0.1	43 ± 1	4	211 ± 31	Interface
6	D-96 GHT	100	1	2.3 ± 0.1	54 ± 4	3	106 ± 20	Interface
7	D-96 GHT	100	1	2.4 ± 0.1	43 ± 3	4	133 ± 48	Mixed
8	D-96 GHT	150	1	2.9 ± 0.1	86 ± 4	5	261 ± 7	Ceramic
13	D-100 GHT	50	1	2.0 ± 0.1	38 ± 4	4	192 ± 15	Interface
14	D-100 GHT	100	1	2.9 ± 0.1	34 ± 1	4	190 ± 24	Mixed
15	D-100 GHT	150	1	3.3 ± 0.1	94 ± 1	5	216 ± 16	Ceramic

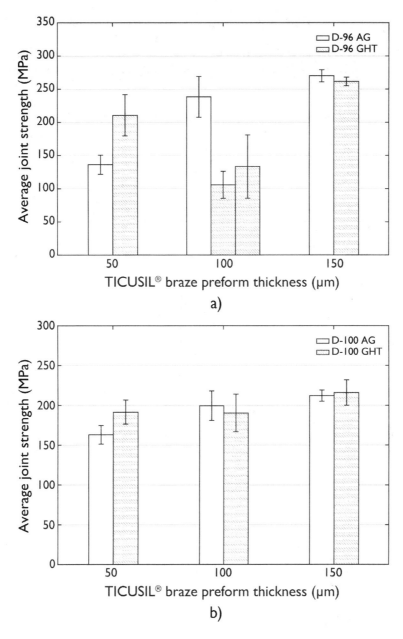

FIGURE 6.39 Comparisons between the average strengths of (a) D-96 AG and D-96 GHT brazed joints and, (b) D-100 AG and D-100 GHT brazed joints, made using 50- to 150-μm-thick TICUSIL® braze preforms.

Both D-96 AG and D-100 AG brazed joints made using 50-μm-thick TICUSIL® braze preforms predominantly failed at the joint interface, and achieved average strengths of 136 and 163 MPa, respectively. The ~1.7-μm-thick reaction layers represented relatively reduced chemical bonding at the interfaces despite the complete diffusion of Ti in the formation of a 26-μm-thick ductile Ag-Cu braze interlayer. Incomplete joint filling may have also introduced a notch-type effect at the joint edges.

The nanohardness of the Ag-rich and Cu-rich phases, which formed in the braze interlayer of all joints, were 1.9 and 2.5 GPa, respectively, at contact depths of 40 nm. Similarly, the nanohardness of alumina and the $Ti_3(Cu + Al)_3O$ reaction layer were ~30 and ~15.4 GPa, respectively at the same contact depths, which showed that these phases were considerably harder than the other constituents of the brazed joints.

When 100-μm-thick TICUSIL® braze preforms were used, the strengths of D-96 AG brazed joints increased by 75% to 238 MPa and the strengths of D-100 AG brazed joints also increased by 22% to 200 MPa. This was attributed to both a 35% increase in the reaction layer thickness to ~2.3 μm and a 51% increase in the brazed joint thickness to 39 μm.

The overall CTE of the braze interlayer in joints made using 100-μm-thick TICUSIL® braze preforms may have been reduced by the formation of the Cu_4Ti_3 and Cu_4Ti phases. This may have decreased the thermally induced residual stresses at the joint interface leading to the increased joint strengths observed.

The nanohardness of the Cu_4Ti and Cu_4Ti_3 phases were 8.5 and 10.7 GPa, respectively, at contact depths of 50 nm. The presence of these relatively harder phases dispersed in the braze interlayer may have reinforced the strength of the joints made using 100-μm-thick TICUSIL® braze preforms.

Brazed joints made using 100-μm-thick TICUSIL® braze preforms consisted of ~46 vol.% Ag-rich braze outflow. In addition to improved joint filling, the edges of these joints were relatively more ductile than the edges of joints made using 50-μm-thick TICUSIL® braze preforms.

D-96 AG brazed joints made using 100-μm-thick TICUSIL® braze preforms achieved average strengths of 238 MPa which was similar to the flexural strength of 252 MPa achieved by D-96 AG standard alumina test bars. These joints, in which failures predominantly occurred in the ceramic, outperformed D-100 AG brazed joints made using 100-μm-thick TICUSIL® braze preforms by ~20%. While the D-100 AG brazed joints achieved average strengths of 200 MPa, this was still ~20% lower than the flexural strength of 250 MPa achieved by D-100 AG standard alumina test bars.

The formation of Ti_5Si_3 compounds in D-96 AG brazed joints, at locations where the triple pocket grain boundary regions of the D-96 AG alumina surface intersected with the Ti-rich reaction layers, indicated additional chemical reactions had occurred. Since this was the only difference between D-96 AG and D-100 AG brazed joints, these compounds may have resulted in additional chemical bonding which provided a nanostructured interlocking mechanism that strengthened the joints.

The distribution and anisotropic CTE of the Ti_5Si_3 compounds may have made the residual stress distribution at the interfaces of D-96 AG brazed joints relatively complex. With a CTE lower than both the reaction layer and braze interlayer, however, these Ti_5Si_3 compounds may have further improved the existing gradual transition in CTE across the joint interface.

When 150-μm-thick TICUSIL® braze preforms were used, the average strength of D-96 AG brazed joints increased by 13% to 270 MPa, and the average strength of D-100 AG brazed joints increased by 6% to 212 MPa as compared with joints made using 100-μm-thick TICUSIL® braze preforms. This was attributed to the 2.8-μm-thick reaction layer and the ~40 vol.% Ag-rich braze outflow which may have enhanced ductility at the joint edges.

Despite the somewhat reduced CTE, due to the multi-layered Cu-Ti structure and isolated Cu-Ti phases in the braze interlayer, the interfaces of joints made using 150-μm-thick TICUSIL® braze preforms may have exhibited greater residual stresses due to the ~65 μm brazed joint thickness, thus weakening the D-100 AG brazed joints.

Despite the relatively thick braze interlayer in joints made using 150-μm-thick TICUSIL® braze preforms, the strengths of D-96 AG brazed joints were ideal, limited only by the monolithic strength of D-96 AG alumina. These joints may have benefitted from the thicker and reinforced braze interlayer and the highly ductile Ag-rich matrix, which could plastically deform to accommodate stress to a greater extent. The ideal joint strength may have also been achieved due to additional chemical bonding at the joint interface provided by the spatial distribution of the Ti_5Si_3 compounds in providing both, a gradation in CTE across the joint interface and a nanostructured interlocking mechanism. Due to this, greater load may have been transferable to the braze interlayer in these D-96 AG brazed joints.

The nanohardness of the multi-layered Cu-Ti structure in the braze interlayer of joints made using 150- and 250-μm-thick TICUSIL® braze preforms was between 7.0 to 9.0 GPa at its edges, and 3.0 to 4.0 GPa at its centre. Although the nanohardness distribution in the braze interlayer of these

Joint Performance

joints became relatively complex due to the presence of these Cu-Ti structures, this was found to be beneficial to the strengths of both of these D-96 AG and D-100 AG joints.

When 250-μm-thick TICUSIL® braze preforms were used, the resulting brazed joint thickness of ~182 μm meant that both of these sets of D-96 AG and D-100 AG brazed joints failed at relatively low loads. The joint interfaces may have been significantly weakened by thermally induced residual stresses despite ~3.2-μm-thick reaction layers in both sets of joints. These larger stresses may have overridden the combined strengthening mechanisms as discussed for joints made using 150-μm-thick TICUSIL® braze preforms. It appeared that additional strengthening provided by the Ti_5Si_3 in D-96 AG brazed joints may have reached its limit (see Figure 6.13a).

Ag-rich phase separation, which occurred due to Cu-Ti phase formation was accommodated locally in the braze interlayer of joints made using 250-μm-thick TICUSIL® braze preforms, and thus braze outflow was completely inhibited.

The average joint strengths of D-96 AG brazed joints decreased by 13% from 270 to 235 MPa as the TICUSIL® braze preform thickness was increased from 150 to 250 μm. Similarly, no further improvement in the average joint strength of D-100 AG brazed joints, which remained at ~212 MPa, was observed.

Post-grinding heat treatment led to a 55% increase in the strength of D-96 GHT brazed joints made using 50-μm-thick TICUSIL® braze preforms. A 66% increase in the brazed joint thickness, from 26 to 43 μm, may have accounted for this increase in strength. A fissured alumina surface caused by the post-grinding heat treatment led to failures in D-96 GHT brazed joints that propagated in the alumina surface, close to the joint interface. Braze infiltration into the vacant triple pocket grain boundary regions may have made the stress distribution in the vicinity of the joint interface relatively complex. The strengths of these joints were still significantly lower than those of the D-96 AG brazed joints made using 100- and 150-μm-thick TICUSIL® braze preforms.

D-96 GHT brazed joints made using 100-μm-thick TICUSIL® braze preforms consisted of severely non-uniform brazed joint thicknesses, which ranged from ~21 and ~85 μm in both sets of joints. Despite the 5.2% increase in the average flexural strength of D-96 GHT standard test bars following the post-grinding heat treatment, once brazed, a ~56% reduction in the average joint strength of these joints, was observed. A fissured alumina surface appeared to have caused the initiation of failures, which propagated ~20 μm in the alumina surface.

The strengthening mechanisms which contributed to the ideal strengths of D-96 AG brazed joints made using 150-μm-thick TICUSIL® braze preforms appeared to override the adverse effects of post-grinding heat treatment in D-96 GHT brazed joints made using the same TICUSIL® braze preform thickness of 150 μm.

Post-grinding heat treatment did not significantly affect the strengths of D-100 GHT brazed joints made using 150-μm-thick TICUSIL® braze preforms, which were statistically similar to their as-ground counterparts. However, while these D-100 GHT brazed joints achieved 216 MPa and failed in the ceramic, their as-ground conterparts, D-100 AG brazed joints made the same TICUSIL® braze preform thickness of 150 μm, achieved 212 MPa and failed at the joint interface.

The post-grinding heat treatment may have relieved compressive residual stresses, un-pinning cracks in D-100 GHT alumina, which led to a shift in the failure mode. This was similar to the effect of post-grinding heat treatment on the average flexural strength of D-100 GHT standard alumina test bars, which decreased by 8.4%, as compared to that of the D-100 AG standard alumina test bars.

The strengthening mechanisms which occurred as a result of the microstructural evolutions in both D-96 and D-100 brazed joints, in both as-ground (AG) and in ground-and-heat-treated (GHT) conditions, as the TICUSIL® braze preform thickness was increased from 50 to 150 μm, can be summarised as follows: (i) an increase in the braze interlayer thickness, which could plastically deform to accommodate stress to a greater extent, (ii) improved chemical bonding at the joint interfaces indicated by an increase in the reaction layer thickness, which could transfer both residual stress and applied stress into the braze interlayer to a greater extent, (iii) the formation of Cu-Ti phases in the braze interlayer which reduced its CTE and thereby the generation of thermally induced residual stresses at the joint interfaces, (iv) the presence of relatively harder Cu-Ti phases in the braze interlayer could re-inforce the strength of the brazed joints, (v) the formation of Cu-Ti phases in the braze interlayer produced relatively ductile Ag-rich phases in the braze interlayer, and (vi) the separation and subsequent outflow of an Ag-rich phase made the joints edges relatively ductile.

In addition to the strengthening mechanisms listed above, D-96 AG brazed joints outperformed D-100 AG brazed joints and achieved ideal performances when 150-μm-thick TICUSIL® braze preforms were selected. The only significant difference between the two grades of alumina was the presence of a Si-rich secondary phase in D-96 alumina. Therefore, the relatively inexpensive liquid phase sintered D-96 AG alumina ceramic provided ideal joint performance due to the formation of Ti_5Si_3 compounds as compared with D-100 AG alumina. These Ti_5Si_3 compounds were formed at locations where the triple pocket grain boundary regions of the D-96 AG alumina surface intersected with the Ti-rich reaction layers and due to their regular spatial distribution, may have strengthened these joints via (i) a gradation in CTE across the joint interface and (ii) a nanostructured interlocking mechanism. This result is required to be compared with other higher purity grades of alumina, such as those that comprise relatively lower levels of porosity than D-100 AG alumina.

6.5 CONCLUSIONS

1. Improvement in the strength of alumina-to-alumina brazed joints, dimensioned to configuration c of ASTM C1161-13 and formed at 850°C for 10 minutes, could be correlated to a microstructural evolution that occurred as the TICUSIL® braze preform thickness was increased from 50 to 150 μm.
2. As the TICUSIL® braze preform thickness was increased from 50 to 150 μm, a microstructural evolution was correlated to an increase in the average strength of D-96 AG brazed joints, which increased from 136 to 270 MPa. The failure mode was found to shift from the joint interface to the parent alumina ceramic. Similarly, the average strength of D-100 AG brazed joints increased from 163 to 212 MPa whereby mixed-mode failures were observed, predominantly at the joint interface, while the parent alumina strength was not achieved.
3. As the TICUSIL® braze preform thickness was increased from 50 to 150 μm, the average reaction layer thickness increased from 1.7 to 2.1 μm and the average braze interlayer thickness increased from 26 to 65 μm, in both D-96 and D-100 brazed joints. The composition of the reaction layer comprising γ-TiO and Ti_3Cu_3O was unaffected, while the braze interlayer changed from comprising an Ag-Cu eutectic to comprising Cu-Ti phases embedded in an Ag-Cu eutectic with greater Ag content, in the form of an Ag-rich phase which flowed towards the joint edges.
4. Nanoindentation testing showed that the Cu-Ti phases were relatively harder as compared with both the Ag-Cu eutectic and the Ag-rich phase, which were found to be relatively ductile. In addition of thicker reaction layers inferring improved chemical bonding, and thicker brazed joints affecting joint strength; the Cu-Ti phases may have reinforced the braze interlayer while the Ag-rich phase may have improved the ductility of the joints, particularly at the joint edges.
5. The strengths of D-96 AG brazed joints outperformed those of D-100 AG brazed joints, with ideal performances achieved using 150-μm-thick TICUSIL® braze preforms. The significant difference between the two grades of alumina was the presence of a Si-rich secondary phase in D-96 alumina. TEM analysis was used to confirm the formation of Ti_5Si_3 at regions where the triple pocket grain boundary regions of the 96.0 wt.% Al_2O_3 alumina surface intersected with the Ti-rich reaction layers. The regular spatial distribution of this phase may have provided a gradation in CTE across the joint interface as well as a nanostructured interlocking mechanism.
6. As the TICUSIL® braze preform thickness was increased from 150 to 250 μm, the average strength of the D-96 AG brazed joints decreased from 270 to 235 MPa, while that of D-100

AG brazed joints remained constant at ~212 MPa. Ag-rich phase separation was accommodated locally within the joints and no braze outflow was observed. Insufficient melting of the braze alloy, excessively thick brazed joints and reaction layers may have led to the degradation in strengths observed.

The findings from this work can be used to arrive at the following optimum conditions for alumina/Ag-Cu-Ti brazed joints:

Brazed joints comprising an alumina ceramic with a silica-rich secondary phase, and an Ag-Cu-Ti braze alloy that comprises at least 1.75 wt.% Ti whereby the surface area of the alumina ceramic, braze preform dimensions (including thickness), and other brazing conditions selected result in the brazed joint microstructures shown in Figure 5.32, preferably 5.32c. Optionally, the joint interface comprises the secondary phase interaction similar to that shown in Figure 6.10 in any of said brazed joint microstructures of Figure 5.32.

Appendix 1
Advanced Ceramics Definition

This work has focussed on the active metal brazing of alumina, an advanced ceramic material which exhibits excellent mechanical properties. Due to the recent developments in ceramic processing techniques coupled with differences in terminologies found in international standards, it is necessary to define the context in which advanced ceramic materials are used.

The definition of a 'ceramic' involves concepts that have developed over centuries, owing to the diverse range of applications in which raw ceramic materials have been used. Generally, ceramics are defined as inorganic and non-metallic materials which are moulded from a mass of raw material at room temperature, and subsequently gain their typical physical properties through high-temperature sintering (Verband der Keramischen Industrie, 2004).

Ceramics have undergone a rapid development in recent decades and have received extensive research attention for their suitability in industrial applications (Kang and Selverian, 1992; Rohde et al., 2009; Shiue et al., 2000). Their superior physical, chemical and mechanical properties have meant that ceramic materials are being used in a growing number of applications and fields (Kar et al., 2007). Their desirable properties are often reported to include excellent high-temperature strength, heat-, corrosion- and wear resistance at elevated temperatures (Hahn et al., 1998; Mandal et al., 2004b; Rohde et al., 2009).

In the literature concerning the brazing of ceramics, the terms 'structural ceramics' (Bang and Liu, 1994; Hahn et al., 1998), 'advanced ceramics' (Bang and Liu, 1994; Ghosh et al., 2012), 'engineering ceramics' (Fernie et al., 2009; Rohde et al., 2009) and 'technical ceramics' (Oyama and Stribe, 1998) are used somewhat interchangeably. Bang and Liu (1994) referred to structural ceramics as those ceramic materials in which improved processing techniques are used over traditional ceramics, and advanced ceramics as those ceramic materials which have enhanced mechanical properties. These differences in terminology, therefore, may have evolved as a result of technological developments in the processing of ceramic materials and as well as the applications in which ceramic materials were used.

In the 1950s, there was a strong drive to establish materials with the ability to retain good mechanical strength at high temperatures for use in aircraft gas turbines since higher temperature fuel combustion meant that higher operating efficiencies reduced fuel consumption and lower environmental pollution could all be achieved. In the 1970s, increased fuel costs led to an impetus and a surge in the interest of ceramic materials as there was a need to economise and increase the efficiency of energy-conversion applications (Moorhead and Keating, 1986). These industrial requirements initiated the development and classification of ceramic materials.

Binary inorganic compounds predominantly consisting of covalent bonds, which included a wide variety of metal oxides, nitrides, borides and carbides, were explored for these purposes and were termed as 'special ceramics'. This scientifically vague definition subsequently led to alternative designations such as 'high temperature ceramics', with both excellent structural and high-temperature properties, and 'engineering ceramics', which covered all ceramic materials which exhibited excellent mechanical properties. However, the increasing dominance of ceramics used in electrical (including magnetic and optical) applications led to the term 'functional ceramics'. Ceramic materials, whose mechanical properties were the main source of interest, were subsequently referred to as 'structural ceramics'. In the late 1990s, 'bioceramics' leapt forward as another important subdivision of the original special ceramics category, and since the increasing diversity of applications are not all entirely within the field of engineering, the whole range of materials were since referred to as 'advanced ceramics' (Thompson, 2003).

In May 2000, a UK Foresight panel defined 'advanced ceramics' as 'those functional or structural ceramics with materials excellence, used in high tech applications' (Foresight Materials Panel Advanced Ceramics Task Force, 2001). The Japanese standard JIS R 1600 defined the term 'fine ceramics', as ceramics which are 'produced with precisely controlled chemical compositions, microstructures, configurations and production processes to fulfil intended functions and which are composed mainly of non-metallic, inorganic substances' (Kyocera Corporation). However, the German ceramics industry referred to 'fine ceramics' and 'coarse ceramics' based on the grain sizes in the microstructures achieved during sintering. The Japanese definition of 'fine ceramics' was instead comparable to 'high performance ceramics' in the German standard DIN V ENV 12 212, which were defined as 'highly developed, high-strength ceramic materials, that are primarily non-metallic and inorganic and possess specific functional attributes'.

The definition of 'functional ceramics' in this German standard was consistent with the use of this term elsewhere in the international ceramics community. These ceramic materials were defined as high-performance ceramics that are used in a given application based on specific properties e.g. electric, magnetic or optical properties. However, the German standard also defined 'technical ceramics' in addition to high-performance ceramics and functional ceramics that includes ceramics used for structural, engineering and industrial applications and that are designed to withstand mechanical stresses, bending and pressure.

The rather cumbersome and imprecise nature of these definitions is principally a reflection of the diverse range of applications in which ceramic materials can be used. These terminologies and definitions have evolved together with the realisation, growth and development of international ceramic markets and applications. While these definitions do not exclude materials such as diamond and graphite, they are rather more precise than the definition of ceramics used in the popular work by Ashby and Jones (1980), in which ceramics were defined as 'materials that are neither metals nor polymers'.

In order to aid translation between worldwide communities, the international standards organisation brought together the commonalities in ISO 20507: 2014 in which it was stated that 'advanced ceramics', 'advanced technical ceramics' and 'fine ceramics' can all be used interchangeably in business, trade, scientific literature and ISO standards. These materials were defined as highly engineered, high-performance, predominantly non-metallic, inorganic, ceramic materials having specific functional attributes. For the purposes of this work, therefore, the term 'advanced ceramics' is deemed suitable.

According to a joint study conducted by VAMAS (Versailles Project on Advanced Materials and Standards; Early and Rook, 2004) and the ASTM Committee C-28 on Advanced Ceramics, and the USACA (United States Advanced Ceramics Association) 'advanced ceramics', which include structural, functional and bioceramics exhibit the following properties:

- Highly specialized by exploiting unique electric, magnetic, optical, mechanical, biological and environmental properties;
- Performing well under extreme conditions such as high temperature, high pressure, high stress, high radiation and high corrosive exposure;
- Predominantly inorganic – non-metallic;
- Relatively expensive, with properties and failure mechanisms not yet fully understood; capable of solving current manufacturing and use problems;
- High value-added products owing to their sophisticated processing technology;
- Not presently profitable in terms of return of investment (ROI), but which offer great promise for the future; and
- Positioned at the beginning of the development cycle and not yet widely used with respect to their potential (Heiman, 2010).

Appendix 2
Macro Images of Brazed Joints

Each of the four sides of the 15 sets of brazed joints (see Section 3.15) was inspected using a Cannon F-1 macro stand system prior to mechanical testing. Particular attention was paid to the alignment of the butt-joints and any braze outflow that occurred. Photographs of the as-brazed sets are shown below, with corresponding Ra values (μm) measured on the faying surfaces prior to brazing.

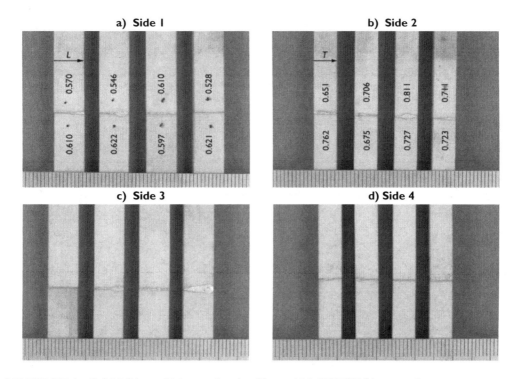

SAMPLE SET 1 D-96 AG brazed joints made using 50-μm-thick TICUSIL® braze preforms.

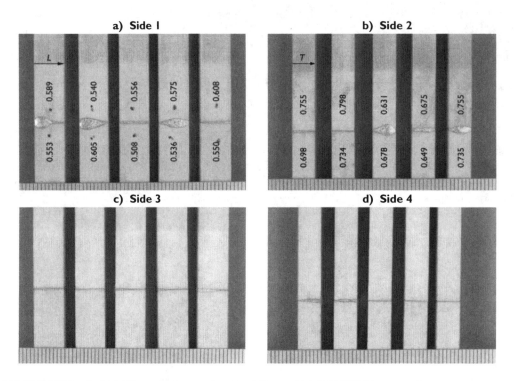

SAMPLE SET 2 D-96 AG brazed joints made using 100-μm thick TICUSIL® braze preforms.

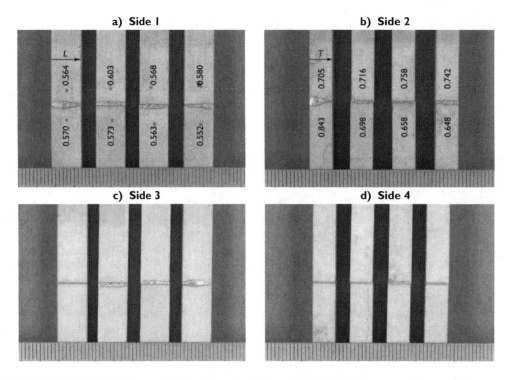

SAMPLE SET 3 D-96 AG brazed joints made using 150-μm thick TICUSIL® braze preforms.

Appendix 2

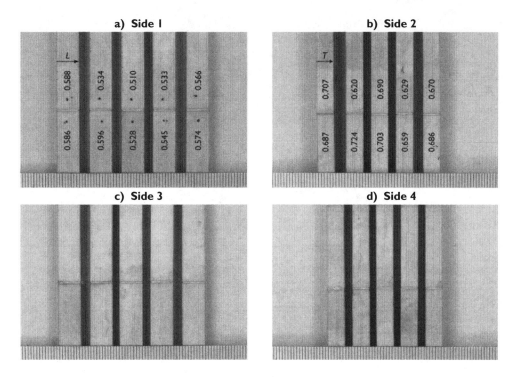

SAMPLE SET 4   D-96 AG brazed joints made using 250-μm thick TICUSIL® braze preforms.

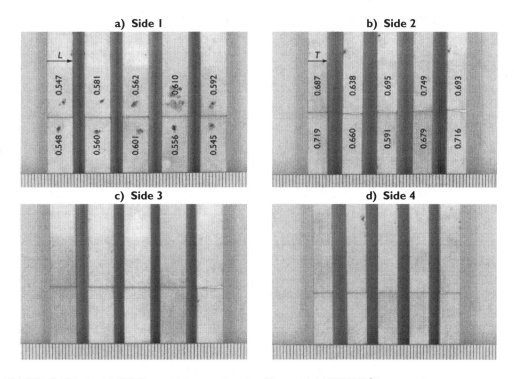

SAMPLE SET 5   D-96 GHT brazed joints made using 50-μm thick TICUSIL® braze preforms.

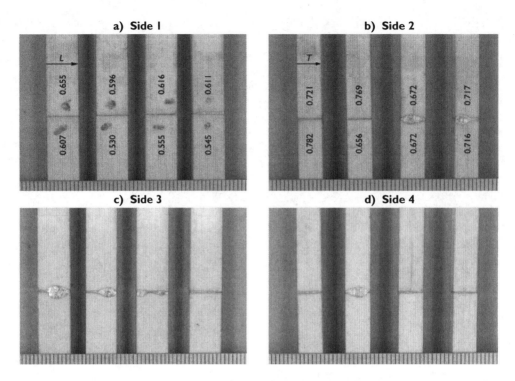

SAMPLE SET 6 D-96 GHT brazed joints made using 100-μm thick TICUSIL® braze preforms.

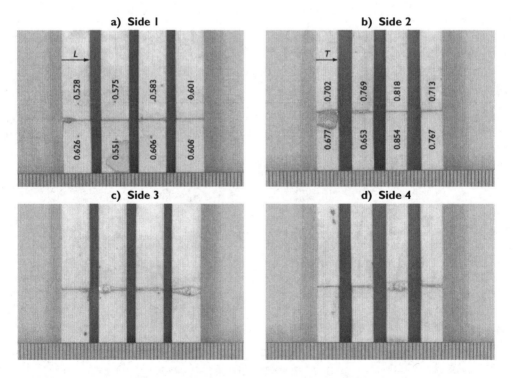

SAMPLE SET 7 D-96 GHT brazed joints made using 100-μm thick TICUSIL® braze preforms (specimen with significant braze outflow in (b) occurred due to improper pre-placement of perform).

Appendix 2

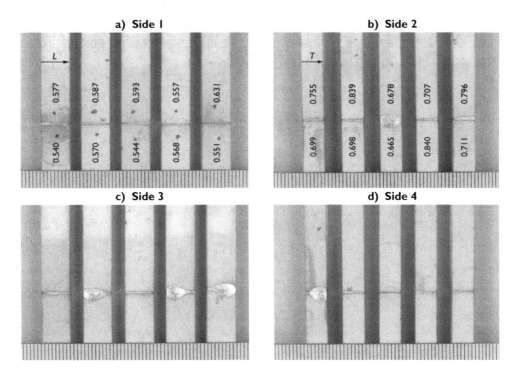

SAMPLE SET 8 D-96 GHT brazed joints made using 150-µm thick TICUSIL® braze preforms.

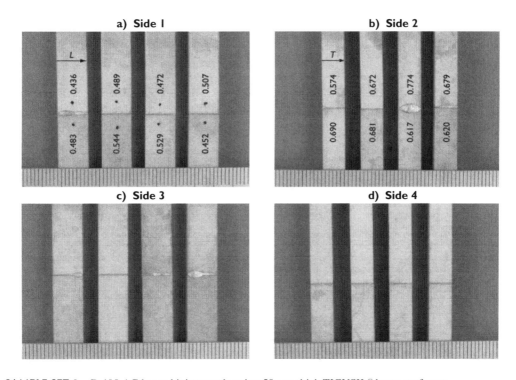

SAMPLE SET 9 D-100 AG brazed joints made using 50-µm thick TICUSIL® braze preforms.

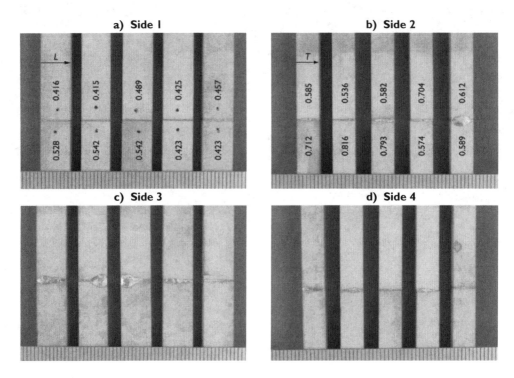

SAMPLE SET 10 D-100 AG brazed joints made using 100-μm thick TICUSIL® braze preforms.

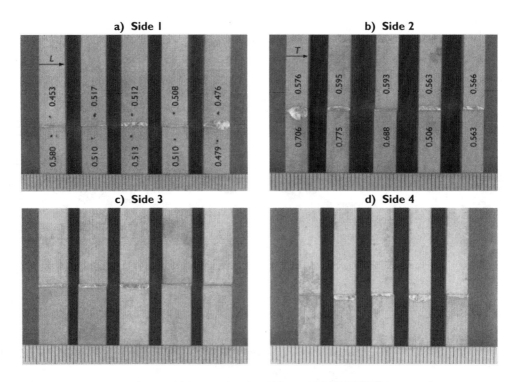

SAMPLE SET 11 D-100 AG brazed joints made using 150-μm thick TICUSIL® braze preforms.

Appendix 2

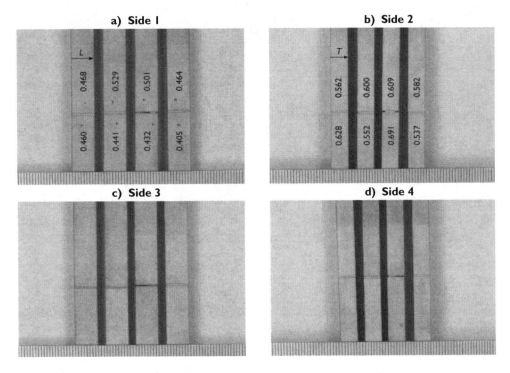

SAMPLE SET 12 D-100 AG brazed joints made using 250-μm thick TICUSIL® braze preforms.

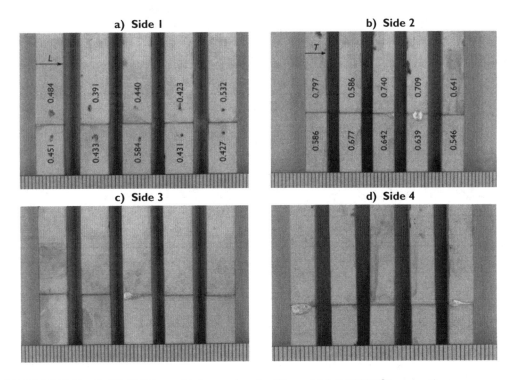

SAMPLE SET 13 D-100 GHT brazed joints made using 50-μm thick TICUSIL® braze preforms.

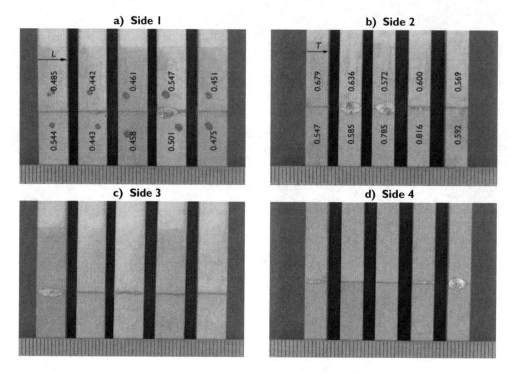

SAMPLE SET 14 D-100 GHT brazed joints made using 100-μm thick TICUSIL® braze preforms.

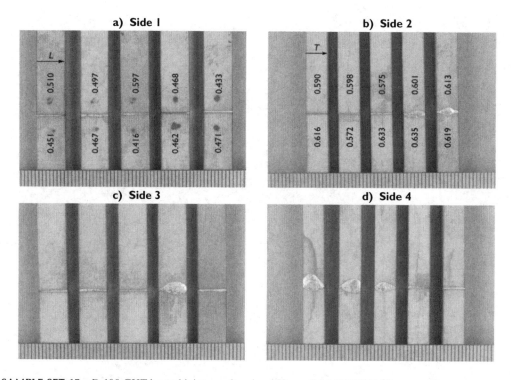

SAMPLE SET 15 D-100 GHT brazed joints made using 150-μm thick TICUSIL® braze preforms.

Appendix 3
Four-Point Bend Testing

The four-point bend testing of ceramic-to-ceramic and ceramic-to-metal brazed joints as a means to evaluate joint strength is not found in any recent or updated standard. The most recent standard which provides details of the four-point bend testing of ceramic-to-ceramic and ceramic-to-metal brazed joints is AWS C3.2M/C3.2:2008. However, several studies have adapted the ASTM C1161 standard, 'Standard Test Method for Flexural Strength of Advanced Ceramics at Ambient Temperature' to evaluate the strengths of brazed butt-joints, in four-point bending.

In this study, the ASTM C1161-13 standard was followed for the preparation and testing of standard alumina test bars. With the lengths of these standard test bars halved, short alumina test bars were brazed in butt-joint configurations and subsequently tested in four-point bend. Joint strengths of the brazed butt-joints were compared with the flexural strengths of the standard test bars. In order to achieve this, a brazing fixture was designed to support the butt-joint assemblies during brazing (see Appendix 5). The experimental method employed in this study (see Section 3) may benefit progression towards an updated standard for the four-point bend testing of ceramic-to-ceramic and ceramic-to-metal brazed joints.

Examples of studies in the literature that have adopted four-point bend testing for evaluating the strengths of ceramic-to-ceramic and ceramic-to-metal brazed joints, in butt-joint configurations are as follow:

References	Brazed Joint Material 1	Brazed Joint Material 2
Carter, E. (2004) *The Effect of Heat Treatment Time on the Structure and Mechanical Properties of Ag-Cu-Ti based Brazed Alumina Joints*. Sheffield Hallam University.	Alumina	Alumina
Cho, H.C. and Yu, J. (1992) Effects of brazing temperature on the fracture toughness of Al_2O_3/Ag-Cu-0.5Ti joints, *Scripta Metallurgica et Materialia*, 26, 797–802. (SENB Three-Point Bend Test).	Alumina	Alumina
Conquest, D.B. (2003) *Brazing of Alumina for High Temperature Applications*. University of Cambridge.	Alumina	Alumina
Hahn, S., Kim, M. and Kang, S. (1998) A study of the reliability of brazed Al_2O_3 joint systems, *IEEE Transactions on Components, Packaging, and Manufacturing Technology - Part C*, 21(3), 211–216.	Alumina	Alumina
Hanson, H.B., Ironside, K.I. and Fernie, J.A. (2000) Active metal brazing of zirconia. *Acta Materialia*, 48, 4673–4676.	ZrO_2	ZrO_2
Kim, T.W. (2001) Brazing parameters and mechanical properties of silicon nitride/stainless steel joint, *25th Annual Conference on Composites, Advanced Ceramics, Materials, and Structures: B*, 627–634.	Si_3N_4	Stainless steel
Kim, J.Y., Hardy, J.S. and Weil, K.S. (2006) High-temperature tolerance of the silver-copper oxide braze in reducing and oxidizing atmospheres, *Journal of Materials Research*, 21(6), 1434–1442.	Alumina	Alumina
Lewisohn, C.A. and Jones, R.H. (2001) Silicon carbide based joining materials for fusion energy and other high-temperature, structural applications, *25th Annual Conference on Composites, Advanced Ceramics, Materials, and Structures: B*, 621–625.	SiC	SiC
Lo, P.L., Chang, L.S. and Lu, Y.F. (2009) High strength alumina joints via transient liquid phase bonding. *Ceramics International*, 38, 3091–3095.	Alumina	Alumina

(*Continued*)

References	Brazed Joint Material 1	Brazed Joint Material 2
Moorhead, A.J. and Simpson Jr., W.A. (1993) Effect of surface preparation on strength of ceramic joints brazed with active filler metal, *Ceramic Transactions*, 35, 207–218.	Alumina	Alumina
Moorhead, A.J., Henson, H.M. and Henson, T.J. (1987) The role of interfacial reactions on the mechanical properties of ceramic brazements, in Pask, J.A. and Evans, A.G. (eds.) *Ceramic Microstructures '86: Role of interfaces (Materials Science Research, Vol. 21)*. New York: Plenum Press, pp. 949–958.	Alumina	Alumina
Ormston, D.R. (2001) *The Particulate Reinforcement of Active Braze Alloys for Joining Silicon Carbide*. University of Cambridge.	SiC	SiC
Palit, D. and Meier, A.M. (2006) Reaction kinetics and mechanical properties in the reactive brazing of copper to aluminium nitride, *Journal of Materials Science*, 41, 7197–7209.	AlN	AlN
Rana, A. (2006) *Comparative Techniques of Joining Alumina*. Sheffield Hallam Univeristy.	Alumina	Alumina
Sung, M.H. and Glaeser, A.M. (2005) *Reduced-Temperature Transient-Liquid-Phase Bonding of Alumina using a Ag-Cu-Based Brazing Alloy*. Lawrence Berkeley National Laboratory.	Alumina	Alumina
Turwitt, M., Elssner, G. and Petzow, G. (1987) On the fracture behaviour and microstructure of metal-to-ceramic joints, in Pask, J.A. and Evans, A.G. (eds.) *Ceramic Microstructures '86: Role of interfaces (Materials Science Research, Vol. 21)*. New York: Plenum Press, pp. 969–979.	Alumina	Alumina
Unal, O., Andersen, I.E. and Maghsoodi, S.I. (1997) A test method to measure shear strength of ceramic joints at high temperatures, *Journal of the American Ceramic Society*, 80(5), 1281–1284. (Asymmetric Four-Point Bend Test).	SiC	SiC
Xian, A.P. and Si, Z.Y. (1990) Joining of Si_3N_4 using Ag57Cu38Ti5 brazing filler metal, *Journal of Materials Science*, 25, 4483–4487.	Si_3N_4	Si_3N_4
Xiong, H., Chuangeng, W. and Zhenfeng, Z. (1998) Joining of Si_3N_4 to Si_3N_4 using rapidly-solidified CuNiTiB brazing filler foils, *Journal of Materials Processing Technology*, 75, 137–142.	Si_3N_4	Si_3N_4
Yadav, D.P., Kaul, R., Ganesh, P., Shiroman, R., Sridhar, R. and Kukreja, L.M. (2014) Study on vacuum brazing of high purity alumina for application in proton synchrotron, *Materials and Design*, 64, 415–422.	Alumina	Alumina

Appendix 4
Surface Roughness Measurements

Surface roughness measurements were performed in this study using a contact profilometer (see Section 3). A solid acrylic Perspex block was designed to support both the standard and short alumina test bars, prepared according to ASTM C1161-13. Surface roughness measurements were made on faying surfaces and outer surfaces of the test bars (see Figure 3.4). In the literature, the surface roughness (Ra values) of ceramic surfaces is typically measured prior to wetting and brazing experiments. Furthermore, several studies have incorporated the formation of brazed butt joints, subsequently tested in four-point bend.

In the literature, surfaces roughness measurements are usually performed in a single direction on the ceramic surface. In this study, surface roughness measurements were accurately performed in two directions; the *L*- and *T*-directions relative to the grinding directions of the alumina test bars, and the Ra values were found to vary accordingly.

In addition to the brazing fixture designed to support the butt-joint assemblies during brazing (see Appendix 5), a fixture to support the standard test bars during surface roughness measurements may also benefit progression towards an updated standard for the four-bend testing of ceramic-to-ceramic and ceramic-to-metal brazed joints.

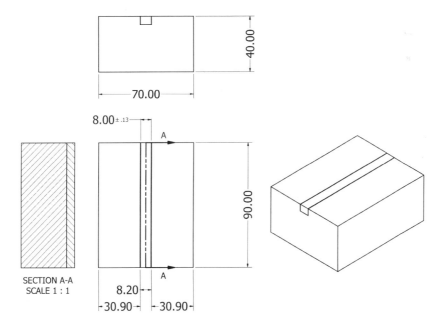

Acrylic Perspex mount designed and used in this study for supporting standard and short alumina test bars prepared according to ASTM C1161-13, during surface roughness measurements performed using a contact profilometer, prior to brazing.

Appendix 5
Brazing Fixture

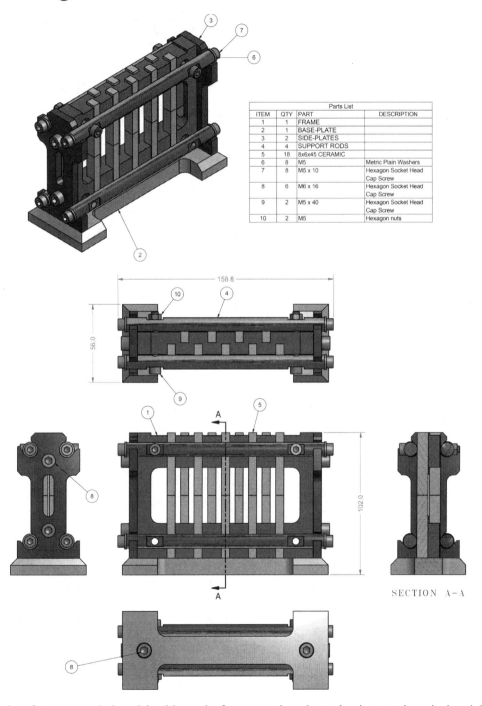

Brazing fixture was designed in this study for supporting short alumina test bars in butt-joint configurations during vacuum furnace brazing experiments (Courtesy of TWI Ltd).

Nomenclature

Å	Angstroms
α_1	Coefficient of thermal expansion
Ag	Silver
Ag-Cu-Ti	Silver-copper-titanium braze alloy
Al	Aluminium
Al_2O_3	Alumina
Ar	Argon
at.%	Atomic per cent
a_{Ti} or γ_{Ti}	Titanium activity coefficient
Atm	Standard atmospheric pressure (101325 Pa or 1.01325 bar)
Au	Gold
b	Specimen width
Ca	Calcium
CaO	Calcia
χ_{Ti}	Final mol. fraction of titanium in a melt
cm	Centimetre
Cr	Chromium
Cu	Copper
Cu-Ti	Copper-titanium
d	Specimen thickness
D	Diffusion coefficient
$\dfrac{\partial C}{\partial x}$	Concentration gradient across the distance x
$\dfrac{\partial C}{\partial t}$	Concentration gradient with respect to time t
°	Degrees
°C	Degrees Celsius
E_r	Reduced elastic modulus
exp	Exponential
g	Grams
G	Activation free energy
γ_{LV}	Interfacial energy at the liquid–vapour interface
γ_{SL}	Interfacial energy at the solid–liquid interface
γ_{SV}	Interfacial energy at the solid–vapour interface
GPa	Gigapascal
h	Height
h_c	Indenter contact depth
H	Hardness
H	Hydrogen
H^*	Temperature-dependent activation entropy
He	Helium
Hf	Hafnium
ΔH_i^∞	Partial enthalpy of mixing
H_v	Vicker's Hardness
J	Net flux of atoms

K	Kelvin
K_{IC}	Fracture toughness
k_p	Reaction layer thickness
kJ	Kilojoule
kT	Thermal energy of an atom at a given temperature T
l	Length
Δl	Change in length
L	Longitudinal direction
L	Outer span
L	Liquid
λ	Jump distance
m	Metre
'm'	Weibull modulus
Mbar	Millibar
Me-i	Solute i infinitely diluted in Me
Mg	Magnesium
MgO	Magnesia
Min	Minute
Mo	Molybdenum
MPa	Megapascal
mN	Millinewton
M_6O	Compound with six parts metal and one part oxygen
µm	Micrometre
µN	Micronewton
N	Nitrogen
Nb	Niobium
Ni	Nickel
nA	Nanoamp
nm	Nanometre
v	Successful jump frequency
v_0	Frequency of atomic vibrations
O	Oxygen
Ω	Ohm
P	Load
Pa	Pascal
pA	Pecoamp
P_f	Failure probability
Pt	Platinum
r	Increase in surface area
Ra	Arithmetic mean surface roughness
Rst_{outer}	Ra measured on outer surface of standard alumina test bar
Rsh_{outer}	Ra measured on outer surface of short alumina test bar
Rsh_{faying}	Ra measured on faying surface of short alumina test bar
®	Registered trade mark
s	Second
S^*	Temperature-independent activation enthalpy
Si	Silicon
SiO_2	Silica
SiC	Silicon Carbide
Si_3N_4	Silicon Nitride
Sn	Tin

Nomenclature

σ	Stress at failure
t	Time
T	Temperature
T	Transverse direction
Ti	Titanium
Ti-Cu-O	Titanium-copper-oxide
Ti-O	Titanium oxide
Ti-Si	Titanium-silicide
T_m	Melting temperature
ΔT	Change in temperature
θ_d	Dynamic contact angle
θ_f	Steady-state equilibrium contact angle
θ_i	Initial or intrinsic contact angle
θ_m	Macroscopic contact angle
θ_q	Quasi-steady contact angle
V	Vanadium
vol.%	Volume per cent
W	Watt
W	Tungsten
kW	Kilowatt
wt.%	Weight per cent
x	Distance or thickness
$\bar{x}$	Mean distance
X–X'	Section along path X–X'
Y	Yttria
Zr	Zirconium

References

Akselsen, O.M. (1992) Review, advances in brazing of ceramics, *Journal of Materials Science*, 27, 1989–2000.

Ali, M., Knowles, K.M., Mallinson, P.M. and Fernie, J.A. (2016) Interfacial reactions between sapphire and Ag-Cu-Ti-based active braze alloys, *Acta Materialia*, 103, 859–869.

Ali, M., Knowles, K.M., Mallison, P.M. and Fernie, J.A. (2015) Microstructural evolution and characterisation of interfacial phases in Al_2O_3/Ag-Cu-Ti/Al_2O_3 braze joints, *Acta Materialia*, 96, 143–158.

Andrieux, J., Dezellus, O., Bosselet, F. and Viala, J.C. (2009) Low-temperature interface reaction between titanium and the eutectic silver-copper brazing alloy, *Journal of Phase Equilibria and Diffusion*, 30(1), 40–45.

Andrieux, J., Dezellus, O., Bosselet, F., Sacerdote-Peronnet, M., Sigala, C., Chiraic and R., Viala, J.C. (2008) Details on the formation of Ti_2Cu_3 in the Ag-Cu-Ti system in the temperature range 790–860°C, *Journal of Phase Equilibria and Diffusion*, 29, 156–162.

Ashby, M. and Jones, D. (1980) *Engineering Materials: An Introduction to Their Properties and Applications*. New York: Pergamon.

Asthana, R. and Singh, M. (2008) Joining of partially sintered alumina to alumina, titanium, hastealloy and C-SiC composite using Ag-Cu brazes, *Journal of the European Ceramic Society*, 28, 617–631.

Asthana, R., Singh, M. and Martinez-Fernandez, J. (2013) Joining and interface characterization of in-situ reinforced silicon nitride, *Journal of Alloys and Compounds*, 552, 137–145.

Asthana, R. (2017) University of Wisconsin-Stout, WI, USA Private Communication.

ASTM Standards (2016) *F19-11: Standard Test Method for Tension and Vacuum Testing Metallized Ceramics Seals*. ASTM, West Conshohocken, PA, USA.

ASTM Standards (2013) *C1161-13: Standard Test Method for Flexural Strength of Advanced Ceramics at Ambient Temperature*. ASTM, West Conshohocken, PA, USA.

Auerkari, P. (ed.) (1996) *Mechanical and Physical Properties of Engineering Alumina Ceramics*. Espoo: Valtion teknillinen tutkimuskeskus (VTT).

AWS Standards (2008) *C32.M/C3.2:2008: Standard Method for Evaluating the Strength of Brazed Joints - Four-Point Bend Testing*. AWS, Miami, FL, USA.

Bahrani, A.S. (1992) The joining of ceramics, *International Journal for the Joining of Materials*, 4(1), 13–19.

Bang, K. and Liu, S. (1994) Interfacial reaction between alumina and Cu-Ti filler metal during reactive metal brazing, *Welding Research Supplement*, March, 54–60.

Belenky, A. and Rittel, D. (2012) Static and dynamic flexural strength of 99.5% alumina: Relation to surface roughness, *Mechanics of Materials*, 54, 91–99.

Boch, P., and Niepce, J.C. (2007) *Ceramic Materials: Processes, Properties, and Applications*. Wiley-ISTE, London, UK and Newport Beach, CA, USA.

Bull, S.J. (2003) On the origins and mechanisms of the indentation size effect, *Zeitschrift für Metallkunde*, 94(7), 787–792.

Byun, W. and Kim, H. (1994) Variations of phases and microstructure of reaction products in the interface of Al_2O_3/Ag-Cu-Ti joint system with heat treatment, *Scripta Metallurgica et Materialia*, 31(11), 1543–1547.

Carim, A.H. (1991) Convergent-beam electron diffraction "Fingerprinting" of M6X phases at brazed ceramic joints, *Scripta Metallurgica et Materialia*, 25(1), 51–54.

Carim, A.H. and Mohr, C.H. (1997) Brazing of alumina with Ti_4Cu_2O and Ti_3Cu_3O interlayers, *Materials Letters*, 33, 195–199.

Carter, E. (2004) *The Effect of Heat Treatment Time on the Structure and Mechanical Properties of Ag-Cu-Ti Based Brazed Alumina Joints*. Sheffield Hallam University, Sheffield, UK.

Chakraborty, R., Mukhopadhyay, A.K., Joshi, K.D., Rav, A., Mandal, A., Bysakh, S., Biswas, S.K. and Gupta, S.C. (2012) Comparative study of indentation size effects in as-sintered alumina and alumina shock deformed at 6.5 and 12 GPa, *International Scholarly Research Network Ceramics*, 1–11.

Chang, Y.A., Goldberg, D. and Neuman, J.P. (1977) Properties of ternary copper-silver systems, *Journal of Physical and Chemical Reference Data*, 6(3), 621–674.

Chen, S., Duan, Y.-H., Huang, B. and Hu, W.-C. (2015) Structural properties, phase stability, elastic properties and electronic structures of Cu-Ti intermetallics, *Philosophical Magazine*, 95(32), 3535–3553.

Cho, H.C. and Yu, J. (1992) Effects of brazing temperature on the fracture toughness of Al_2O_3/Ag-Cu-0.5Ti joints, *Scripta Metallurgica et Materialia*, 26, 797–802.

Coble, R.L. (1961) Sintering crystalline solids: I intermediate and final stage diffusion models; II. Experimental test of diffusion models in powder compacts, *Journal of Applied Physics*, 32, 787–799.

Conquest, D.B. (2003) *Brazing of Alumina for High Temperature Applications*. University of Cambridge, Cambridge, UK.

Cverna, F. (ed.) (2002) *Thermal Properties of Materials*. Materials Park, OH: ASM International.

Dey, A., Bhattacharya, M. and Mukhopadhyay, A.K. (2015) Grain boundary nanohardness of coarse grain alumina, *International Journal of Applied Ceramic Technology*, 12(6), 1199–1209.

Dezellus, O., Arroyave, R. and Fries, S.G. (2011) Thermodynamic modelling of the Ag-Cu-Ti ternary system, *International Journal of Materials Research*, 102(3), 286–297.

do Nascimento, R.M., Martinelli, A.E. and Buschinelli, A.J.A. (2003) Review article: Recent advances in metal-ceramic brazing, *Ceramica*, 49, 178–198.

do Nascimento, R.M., Martinelli, A.E., Buschinelli, A.J.D.A. and Klein, A.N. (1999) Brazing Al2O3 to sintered Fe-Ni-Co alloys, *Journal of Materials Science*, 34, 5839–5845.

Dobedoe, R.S. (1997) *Glass-Ceramics for Ceramic/Ceramic and Ceramic/Metal Joining Applications*. The University of Warwick, Warwick, UK.

Eremenko, V.N., Buyanov, Y.I. and Panchenko, N.M. (1969) Layering region in liquid Ti-Cu-Ag alloys, *Russian Metallurgy*, 5, 129–131.

Eustathopoulos, N. (2015) Wetting by liquid metals—Application in materials processing: The contribution of the Grenoble group, *Metals*, 5, 350–370.

Eustathopoulos, N. (2005) Progress in understanding and modeling reactive wetting of metals on ceramics, *Current Opinion in Solid State and Materials Science*, 9, 152–160.

Eustathopoulos, N. (1998) Dynamics of wetting in reactive metal/ceramic systems, *Acta Materialia*, 46(7), 2319–2327.

Fernie, J.A., Drew, R.L. and Knowles, K.M. (2009) Joining of engineering ceramics, *International Materials Reviews*, 54(5), 283–331.

Finger, L.W. and Hazen, R.M. (1977) 'Crystal structure and compressibility of ruby to 80 kbar' *Carnegie Institution Washington Yearbook*, 76, 525.

Fischer-Cripps, A.C. (2011) *Nanoindentation*, Third Edition, Springer, New York.

Foley, A.G. and Anders, D.J. (1994) Active metal brazing for joining ceramics to metals, *GEC Alsthom Technical Review*, 13, 49–64.

Foresight Materials Panel Advanced Ceramics Task Force (2001) *Engineering our Future Through Ceramics*. DTI.

Galusek, D., Baca, L. and Sajgalik, P. (2002) The role of glass in alumina-based engineering ceramics, *Proceedings of the 6th ESG Conference*. 2–6 June. Montpellier, France, 1.

Gauthier, M.M. (1995) Joining, in *Engineered Materials Handbook Desk Edition*. ASM International, CRC Press, Boca Raton, pp. 846–864.

Ghosh, S., Chakraborty, R., Dandapat, N., Pal, K.S., Datta, S. and Basu, D. (2012) Characterization of alumina-alumina/graphite/monel superalloy brazed joints, *Ceramics International*, 38, 663–670.

Hahn, S., Kim, M. and Kang, S. (1998) A study of the reliability of brazed Al_2O_3 joint systems, *IEEE Transactions on Components, Packaging, and Manufacturing Technology - Part C*, 21(3), 211–216.

Hansen, M. and Anderko, M. (1958) *Constitution of Binary Alloys*. New York: McGraw-Hill.

Hao, H., Jin, Z. and Wang, X. (1994) The influence of brazing conditions on joint strength in Al_2O_3/Al_2O_3 bonding, *Journal of Materials Science*, 29, 5041–5046.

Hao, H., Wang, Y., Jin, Z. and Wang, X. (1997) Interfacial morphologies between alumina and silver-copper-titanium alloy, *Journal of Materials Science*, 32, 5011–5015.

Heikinheimo, L.S.K., Siren, M.J., Kauppinen, K.P. and Auerkari, P.M.S. (1997) Performance of alumina-alumina and alumina-metal joints, in *Mis-Matching of Interfaces and Welds*. Geesthacht: GKSS Research Center Publications, pp. 451–462.

Heiman, R.B. (2010) *Classic and Advanced Ceramics: From Fundamentals to Applications*. Weinheim: Wiley-VCH.

Hirnyj, S. and Indacochea, J.E. (2008) Phase transformations in Ag70.5Cu26.5Ti3 filler alloy during brazing processes, *Chemistry of Metals and Alloys*, 1, 323–332.

Ho, C.Y. and Taylor, R.E. (eds.) (1998) *Thermal Expansion of Solids*. Materials Park, OH: ASM International.

Hosking, F.M., Cadden, C.H., Yang, N.Y.C., Glass, S.J., Stephens, J.J., Vianco, P.T. and Walker, C.A. (2000) Microstructural and mechanical characterization of actively brazed alumina tensile specimens, *Welding Journal*, 8, 222–230.

Ichimori, T., Iwamoto, C. and Tanaka, S. (1999) Nanoscopic analysis of a Ag-Cu-Ti/sapphire brazed interface, *Materials Science Forum*, 294–296, 337–340.

References

Jacobson, D.M. and Humpston, G. (2005) *Principles of Brazing*. Materials Park, OH: ASM International.

Janickovic, D., Sebo, P., Duhaj, P. and Svec, P. (2001) The rapidly quenched Ag-Cu-Ti ribbons for active joining of ceramics, *Materials Science and Engineering*, A304–306, 569–573.

Jorgensen, P.J. and Westbrook, J.H. (1964) Role of solute segregation at grain boundaries during final–stage sintering of alumina authors, *Journal of the American Ceramic Society*, 47(7), 332–338.

Kang, S. and Selverian, J.H. (1992) Interactions between Ti and alumina-based ceramics, *Journal of Materials Science*, 27, 4536–4544.

Kar, A., Mandal, S., Rathod, S. and Ray, A.-K. (2006) Effect of Ti diffusivity on the formation of phases in the interface of alumina-alumina brazed with 97(Ag40Cu03Ti) filler alloy, *Proceedings of 3rd International Brazing and Soldering Conference*. 24th April 2006. San Antonio, TX: ASM International, p. 219.

Kar, A., Mandal, S., Venkateswarlu, K. and Ray, A.-.K (2007) Characterization of interface of Al_2O_3–304 stainless steel braze joint, *Materials Characterization*, 58, 555–562.

Karlsson, N. (1951) Metallic oxides with the structure of high-speed steel carbide, *Nature*, 168, 558–558.

Kelkar, G.P. and Carim, A.H. (1995) Al solubility in M6X compounds in the Cu-Ti-O system, *Materials Letters*, 23, 231–235.

Kelkar, G.P. and Carim, A.H. (1993) Synthesis, properties, and ternary phase stability of M6X compounds in the Ti-Cu-O system, *Journal of the American Ceramic Society*, 76(7), 1815–1820.

Kelkar, G.P., Spear, K.E. and Carim, A.H. (1994) Thermodynamic evaluation of reaction products and layering in brazed alumina joints, *Journal of Materials Research*, 9, 2244–2250.

Kim, J.Y., Hardy, J.S. and Weil, K.S. (2006) High-temperature tolerance of the silver-copper oxide braze in reducing and oxidizing atmospheres, *Journal of Materials Research*, 21(6), 1434–1442.

Kirchner, H.P., Gruver, R.M. and Walker, R.E. (1970) Strength effects resulting from simple surface treatments, *The Science of Ceramic Machining and Surface Finishing*. 2–4 November. p. 353.

Kozlova, O., Braccini, M., Voytovych, R., Eustathopoulos, N., Martinetti, P. and Devismes, M.-F. (2010) Brazing copper to alumina using reactive CuAgTi alloys, *Acta Materialia*, 58, 1252–1260.

Kozlova, O., Voytovych, R. and Eustathopoulos, N. (2011) Initial stages of wetting of alumina by reactive CuAgTi alloys, *Scripta Materialia*, 65, 13–16.

Kristalis, P., Coudurier, L. and Eustathopoulos, N. (1991) Contribution to the study of reactive wetting in the $CuTi/Al_2O_3$ system, *Journal of Materials Science*, 26, 3400–3408.

Kubiak, K.J., Wilson, M.C.T., Mathia, T.G. and Carval, P. (2011) Wettability versus roughness of engineering surfaces, *Wear*, 271, 523–528.

Kyocera Corporation Introduction to Fine Ceramics; What are Fine Ceramics? Available at: http://global.kyocera.com/fcworld/first/about.html (Accessed: 01/21/2015).

Lange, F.F. (1970) Healing of surface cracks in ceramics, *The Science of Ceramic Machining and Surface Finishing*. 2–4 November. pp. 233–236.

Lee, W. and Kwon, O.Y. (1995) Microstructural characterization of interfacial reaction products between alumina and braze alloy, *Journal of Materials Science*, 30, 1679–1688.

Lee, W.E. and Rainforth, M. (eds.) (1994) *Ceramic Microstructures: Property Control by Processing*. Springer Netherlands.

Lee, W.H., Cheon, Y.W., Jo, Y.H., Seong, J.G., Jo, Y.J., Kim, Y.H., Noh, M.S., Jeong, H.G., Van Tyne, C.J. and Chang, S.Y. (2015) Self-consolidation mechanism of nanostructured Ti_5Si_3 compact induced by electrical discharge, *The Scientific World Journal*, 7, 1–8.

Liao, T.W., Li, K. and Breder, K. (1997) Flexural strength of ceramics ground under widely different conditions, *Journal of Materials Processing Technology*, 70, 198–206.

Lin, C., Chen, R. and Shiue, R. (2001) A wettability study of Cu/Sn/Ti active braze alloys on alumina, *Journal of Materials Science*, 36, 2145–2150.

Lin, K., Singh, M. and Asthana, R. (2014) Interfacial chacterization of alumina-to-alumina joints fabricated using silver-copper-titanium interlayers, *Materials Characterization*, 90, 40–51.

Liu, Y., Huang, Z. and Liu, X. (2010) Reaction layer microstructure of SiC/SiC joints brazed by Ag-Cu-Ti filler metal, *Key Engineering Materials*, 434–435, 202–204.

Locatelli, M.R., Dalgleish, B.J., Nakashima, K., Tomsia, A.P. and Glaeser, A.M. (1997) New approaches to joining ceramics for high-temperature applications, *Ceramics International*, 23, 313–322.

Loehman, R.E., Tomsia, A.P., Pask, J.A., Johnson, S.M. (1990) Bonding mechanisms in silicon nitride brazing, *Journal of the American Ceramic Society*, 73, 552–558.

Loehman, R.E. and Tomsia, A.P. (1992) Reactions of Ti and Zr with AlN and Al_2O_3, *Acta Metallurgica et Materialia*, 40, S75–S83.

Lugscheider, E. and Tillmann, W. (1993) Methods for brazing ceramic and metal-ceramic joints, *Materials and Manufacturing Processes*, 8(2), 219–238.

Mandal, S., Ray, A. and Ray, A. (2004a) Correlation between the mechanical properties and microstructural behaviour of Al_2O_3-(Ag-Cu-Ti) brazed joints, *Materials Science and Engineering*, A 383, 235–244.

Mandal, S., Ray, V. and Ray, A. (2004b) Characterization of brazed joint interface between Al_2O_3 and (Ag-Cu-Ti), *Journal of Materials Science*, 39, 5587–5590.

Markets and Markets (2015) Advanced Ceramics Market by Material and by Region - Global Trends & Forecasts to 2020. Markets and Markets.

Massalski, T.B. (1990) *Binary Alloy Phase Diagrams*. ASM International.

Mizuhara, H. and Huebel, E. (1986) Joining ceramic to metal with a ductile active filler metal, *Welding Journal*, October, 43–51.

Mizuhara, H., Huebel, E. and Oyama, T. (1989) High-reliability joining of ceramic to metal, *Ceramic Bulletin*, 68(9), 1591–1599.

Mizuhara, H. and Mally, K. (1985) Ceramic-to-metal joining with active brazing filler metal, *Welding Journal*, 64, 27–32.

Mohammed Jasim, K., Hashim, F.A., Yousif, R.H., Rawlings, R.D. and Boccaccini, A.R. (2010) Actively brazed alumina to alumina joints using CuTi, CuZr and eutectic AgCuTi filler alloys, *Ceramics International*, 36, 2287–2295.

Mondal, S., Pathak, L.C., Venkateswarlu, K. and Das, S.K. (2002) Development and characterization of Ag-Cu-Ti alloys for ceramic brazing, *Indo-Malaysian Joint Workshop on Advanced Materials*, March 2002. Chennai, India: Allied Publishers Pvt. Ltd., 121.

Moorhead, A.J. (1987) Direct brazing of alumina ceramics, *Advanced Ceramic Materials*, 2(2), 159–166.

Moorhead, A.J., Henson, H.M. and Henson, T.J. (1987) The role of interfacial reactions on the mechanical properties of ceramic brazements, in Pask, J.A. and Evans, A.G. (eds.) *Ceramic Microstructures '86: Role of interfaces (Materials Science Research, Vol. 21)*. New York: Plenum Press, pp. 949–958.

Moorhead, A.J. and Keating, H. (1986) Direct brazing of ceramics for advanced heavy-duty diesels, *Welding Journal*, 65(10), 17–31.

Moorhead, A.J. and Kim, H. (1991) Oxidation behaviour of titanium-containing brazing filler metals, *Journal of Materials Science*, 26, 4067–4075.

Moorhead, A.J. and Santella, M.L. (1987) The effect of interfacial reactions on the mechanical properties of oxide ceramic brazements, *Proceedings of BABS 5th International Conference on High Technology Joining*, November 1987. Brighton, England: The British Association for Brazing and Soldering, p. 22.

Moorhead, A.J. and Simpson Jr, W.A. (1993) Effect of surface preparation on strength of ceramic joints brazed with active filler metal, *Ceramic Transactions*, 35, 207–218.

Morgan Advanced Materials plc Properties of TICUSIL®. Hayward, CA: Morgan Advanced Materials plc.

Morrell, R. (1987) *Handbook of Properties of Technical and Engineering Ceramics, Part 2: Data Reviews*. London, England: HMSO.

Morrell, R. (2015) National Physical Laboratory (NPL) Private Communication.

Murray, J.L. (1990) The Cu-Ti (Copper-Titanium) system, in *Binary Alloy Phase Diagrams*. 2nd edn. ASM International, pp. 1494–1495.

Murray, J.L. (1984) The Ag-Cu (Silver-Copper) system, *Metallurgical Transactions A*, 15, 261–268.

Murray, J.L. and Bhansali, K. (1990) The Ag-Ti (Silver-Titanium) system, in *Binary Alloy Phase Diagrams*. 1st edn. ASM International, pp. 102–106.

Murray, J.L. and Wriedt, H.A. (1987) The O-Ti (Oxygen-Titanium) system, *Bulletin of Alloy Phase Diagrams*, 8(2), 148–165.

Naka, M., Tanaka, T. and Okamoto, I. (1985) Joining of silicon nitride to metals or alloys using amorphous Cu-Ti filler metal, *Transactions of the Joining and Welding Research Institute*, 14, 285–291.

Nakamura, S., Tanaka, S., Kato, Z. and Uematsu, K. (2009) Strength-processing defects relationship based on micrographicanalysis and fracture mechanics in alumina ceramics, *Journal of the American Ceramics Society*, 92(3), 688–693.

Nicholas, M.G. (1998) *Joining Processes*. Springer Science & Business Media.

Nicholas, M.G. (1997) The Brazing of Ceramics: Materials Science Aspect of Reactive Braze Alloys, *Interfacial Science in Ceramic Joining*. Springer, 97.

Nicholas, M.G. (ed.) (1990) *Joining of Ceramics (Advanced Ceramic Reviews)*. London, England: Chapman & Hall.

Nicholas, M.G. and Mortimer, D.A. (1985) Ceramic/metal joining for structural applications, *Materials Science and Technology*, 1, 657–665.

Nicholas, M.G. and Peteves, S.D. (1994) Reactive joining: Chemical effects on the formation and properties of brazed and diffusion bonded interfaces, *Scripta Materialia*, 31(8), 1091–1096.

References

Ning, H., Geng, Z., Ma, J., Huang, F., Qian, Z. and Han, Z. (2003) Joining of sapphire and hot pressed Al_2O_3 using Ag70.5Cu27.5Ti2 brazing filler metal, *Ceramics International*, 29, 689–694.

Oku, T., Suganuma, K., Wallenberg, L.R., Tomsia, A.P., Gomez-Vega, J.M. and Saiz, E. (2001) Structural characterization of the metal/glass interface in bioactive glass coatings on Ti-6Al-4V, *Journal of Materials Science: Materials in Medicine*, 12, 413–417.

Oyama, T. and Stribe, K. (1998) Active brazing of Alumina to Copper - Effect of Titanium Concentration on Joint Strength, *Proceedings of 5th International Symposium on Brazing, High Temperature Brazing and Diffusion Welding*. June. Düsseldorf, Germany: DVS-Berichte, 94.

Paiva, O.C. and Barbosa, M.A. (2000) Brazing parameters determing the degradation and mechanical behaviour of alumina/titanium brazed joints, *Journal of Materials Science*, 35, 1165–1175.

Pak, J.J., Santella, M.L. and Fruehan, R.J. (1990) Thermodynamics of Ti in Ag-Cu alloys, *Metallurgical Transactions B*, 21 B, 349–355.

Passerone, A., Valenza, F. and Muolo, M.L. (2013) Wetting at high temperature, in Ferrari, M., Liggieri, L. and Miller, R. (eds.) *Drops and Bubbles in Contact with Solid Surfaces. Progress in Colloid and Interface Science*. CRC Press, Boca Raton, pp. 299–334.

Paulasto, M. and Kivilahti, J. (1998) Metallurgical reactions controlling the brazing of Al_2O_3 with Ag-Cu-Ti filler alloys, *Journal of Materials Research*, 13(2), 343–352.

Paulasto, M., van Loo, F.J.J. and Kivilahti, J.K. (1995) Thermodynamic and experimental study of Ti-Ag-Cu alloys, *Journal of Alloys and Compounds*, 220, 136–141.

Peytour, C., Barbier, F. and Revcolevschi, A. (1990) Characterization of ceramic/TA6V titanium alloy brazed joints, *Journal of Materials Research*, 5(1), 127–135.

Pfeifer, H.U., Bahn, S. and Schubert, K. (1968) Zum aufbau des systems Ti-Ni-Cu und einiger quasihomologer legierungen, *Journal of Less Common Metals*, 14, 291.

Phillips, G.C. (1991) *A Concise Introduction to Ceramics*. New York: Van Nostrand Reinhold.

Profound (2014) Advanced Ceramics – Executive Summary. Profound Report.

Rana, A. (2006) *Comparative Techniques of Joining Alumina*. Sheffield Hallam Univeristy.

Richerson, D.W. (1992) *Modern Ceramic Engineering: Properties, Processing and use in Design*. 2nd edn. New York: Marcel Dekker, Inc.

Rijinders, M.R. and Peteves, S.D. (1999) Joining of alumina using a V-active filler metal, *Scripta Materialia*, 41(10), 1137–1146.

Rodrigues, G., Nunes, C.A., Suzuki, P.A. and Coelho, G.C. (2006) Thermal expansion of the Ti_5Si_3 and Ti_6Si_2B phases investigated by high-temperature X-ray diffraction, *Intermetallics*, 14, 236–240.

Rohde, M., Sudmeyer, I., Urbanked, A. and Torge, M. (2009) Joining of alumina and steel by a laser supported brazing process, *Ceramics International*, 35, 333–337.

Santella, M.L., Horton, J.A. and Pak, J.J. (1990) Microstructure of alumina brazed with a silver-copper-titanium alloy, *Journal of the American Ceramics Society*, 73(6), 1785–1787.

Serizawa, H., Lewinsohn, C.A. and Murakawa, H. (2001) FEM analysis of experimental measurement technique for mechanical strength of ceramic joints, *25th Annual Conference on Composites, Advanced Ceramics, Materials, and Structures: B*. 21–27 January 2001. The American Ceramic Society, p. 635.

Shiue, R.K., Wu, S.K. and Wang, J.Y. (2000) Microstructural evolution at the bonding interface during the early-stage infrared active brazing of alumina, *Metallurgical and Materials Transactions A*, 31 A, 2527–2536.

Singh, M., Martínez-Fernandéz, J., Asthana, R. and Ramirez Rico, J. (2012) Interfacial characterization of silicon nitride/silicon nitride joints brazed using Cu-base active metal interlayers, *Ceramics International*, 38, 2793–2802.

Singh, M., Asthana, R., Varela, F.M. and Martínez-Fernandéz, J. (2011) Microstructural and mechanical evolution of a Cu-based active braze alloy to join silicon nitride ceramics, *Journal of the European Ceramic Society*, 31, 1309–1316.

Statista, T.S.P. Countries with the Largest Bauxite Reserves 2013. Available at: www.statista.com/statistics/271671/countries-with-largest-bauxite-reserves/ (Accessed: 02/04/2015).

Stephens, J.J., Hosking, F.M., Headley, T.J., Hlava, P.F. and Yost, F.G. (2003) Reaction layers and mechanisms for a Ti-activated braze on sapphire, *Metallurgical and Materials Transactions A*, 34 A, 2963–2972.

Stollberg, D.W., Carter, W.B. and Hampikian, J.M. (2005) Nanohardness and fracture toughness of combustion chemical vapor deposition deposited yttria stabilized zirconia–alumina films, *Thin Solid Films*, 211–217.

Suenaga, S., Koyama, M., Arai, S. and Nakahashi, M. (1993) Solid-state reactions of the Ag-Cu-Ti thin film-Al_2O_3 substrate system, *Journal of Materials Research*, 8(8), 1805–1811.

Suenaga, S., Nakahashi, M., Maruyama, M. and Fukasawa, T. (1997) Interfacial reactions between sapphire and silver-copper-titanium thin film filler metal, *Journal of the American Ceramics Society*, 80(2), 439–444.

Suganuma, K., Okamoto, T., Koizumi, M. and Shimada, M. (1985) Effect of thickness on direct bonding of silicon nitride to steel, *Journal of the American Ceramic Society*, 68(12), 334–335.

Suganuma, K., Miyamoto, Y. and Koizumi, M. (1988) Joining of ceramics and metals, *Annual Review of Materials Research*, 18, 47–73.

Tamai, T. and Naka, M. (1996) Ti effect on microstructure and strength of Si_3N_4/Si_3N_4 and SiC/SiC joints brazed with Cu-Ag-Ti filler metals, *Journal of Materials Science Letters*, 15, 1203–1204.

Thompson, D.P. (2003) Forty-sixth Mellor memorial lecture: Euphoria and heartbreak: Two sides of the coin in the development of structural ceramics, *British Ceramic Transactions*, 102(5), 185–192.

Tillman, W., Lugscheider, X.R. and Indacochea, J.E. (1996) Kinetic and microstructural aspects of the reaction layer at ceramic/metal braze joints, *Journal of Materials Science*, 31, 445–452.

Urai, S. and Masaaki, N. (1999) Effect of Sn addition to Cu-Ti filler metals on microstructure and strength of SiC/SiC joint, *Transactions of the Japan Welding Research Institute*, 28(1), 35–40.

Valeeva, A.A., Rempel, A.A. and Gusev, A.I. (2001) Electrical conductivity and magnetic susceptibility of titanium monoxide, *Letters to Jounal of Experimental and Theoretical Physics*, 73, 621–625.

Valette, C., Devismes, M.-F., Voytovych, R. and Eustathopoulos, N. (2005) Interfacial reactions in alumina/CuAgTi braze/CuNi system, *Scripta Materialia*, 52, 1–6.

Verband der Keramischen Industrie (2004) *Breviary Technical Ceramics*. Fahner.

Vianco, P.T., Stephens, J.J., Hlava, P.F. and Walker, C.A. (2003) Titanium scavenging in Ag-Cu-Ti active braze joints, *Welding Journal*, 82(10), 268–277.

Voytovych, R., Ljungberg, L.Y. and Eustathopoulos, N. (2004) The role of adsorption and reaction in wetting in the CuAg-Ti/alumina system, *Scripta Materialia*, 51, 431–435.

Voytovych, R., Robaut, F. and Eustathopoulos, N. (2006) The relation between wetting and interfacial chemistry in the CuAgTi/alumina system, *Acta Materialia*, 54, 2005–2214.

Wenzel, R.N. (1936) Resistance of solid surfaces to wetting by water, *Industrial and Engineering Chemistry*, 28(8), 988–994.

Wiese, J.L. (2001) *Strength of Metal-to-Ceramic Brazed Joints*. Massachusetts Institute of Technology.

Xian, A.P. and Si, Z.Y. (1990) Joining of Si_3N_4 using Ag57Cu38Ti5 brazing filler metal, *Journal of Materials Science*, 25, 4483–4487.

Xiong, H., Chuangeng, W. and Zhenfeng, Z. (1998) Joining of Si_3N_4 to Si_3N_4 using rapidly-solidified CuNiTiB brazing filler foils, *Journal of Materials Processing Technology*, 75, 137–142.

Yadav, D.P., Kaul, R., Ganesh, P., Shiroman, R., Sridhar, R. and Kukreja, L.M. (2014) Study on vacuum brazing of high purity alumina for application in proton synchrotron, *Materials and Design*, 64, 415–422.

Yang, P., Turman, B.N., Glass, S.J., Halbleib, J.A., Voth, T.E., Gerstle, F.P., McKenzie, B. and Clifford, J.R. (2000) Braze microstructure evolution and mechanical properties of electron beam joined ceramics, *Materials Chemistry and Physics*, 64, 137–146.

Yang, J., Fang, H. and Xin, W. (2005) Al_2O_3/Al_2O_3 joint brazed with Al_2O_3-particulate-contained composite Ag-Cu-Ti filler material, *Journal of Materials Science Technology*, 21(5), 782–784.

Yang, J., Fang, H. and Xin, W. (2002) Effects of Al_2O_3-particulate-contained composite filler materials on the shear strength of alumina joints, *Journal of Materials Science Technology*, 18(4), 289–230.

Yang, M., Lin, T. and He, P. (2012) Microstructure evolution of Al_2O_3/Al_2O_3 joint brazed with Ag-Cu-Ti + B + TiH2 composite filler, *Ceramics International*, 38, 289–294.

Yu, H., Tieu, A.K., Cheng, L., Xiong, L., Godbole, A., Li, H. and Kong, C., Qinghua, Q. (2014) A deformation mechanism of hard metal surrounded by soft metal during roll forming, *Scientific Reports*, 4, Article number 5017, 1–5.

Zhu, M. and Chung, D.D.L. (1997) Improving the strength of brazed joints to alumina by adding carbon fibres, *Journal of Materials Science*, 32, 5321–5333.

Zhu, M. and Chung, D.D.L. (1994) Active brazing alloy containing carbon fibers for metal-ceramic joining, *Journal of the American Ceramics Society*, 77(10), 2712–2720.

Index

A

72Ag-28Cu wt.%, 8, 18, 111, 125–127
ABA, 2–3, 7–9, 12–13, 16–17, 33–40, 42–52, 159–160
abrasive, 4, 52, 56–59, 61, 65, 77, 99, 107
acetone, 71
acicular, 89, 94
active braze alloy, 2, 8
activity coefficient, 18
Aerodag Ceramshield, 71, 73
Auger electron spectroscopy, 23, 26
Ag concentration, 17–18, 65, 127
Ag-Cu, 11, 32, 132
Ag-Cu binary phase diagram, 113
Ag-Cu concentration, 16
Ag-Cu-Ti ternary phase diagram, 111, 113, 115, 117, 127, 132
Ag-rich globules, 120–123, 128, 133
Ag-Ti, 111
Akselsen, O.M., 14
Al-O, 15
Al-Ti, 43
Ali, M., 14, 20, 35–37, 40–44, 46, 47, 50–51, 142, 160–161
Alicona InfiniteFocusSL, 69, 85–88, 99, 101
alumina purity, 3, 47–50, 52, 60–61, 82–83, 99, 122, 144, 166–167, 189, 212
Alumina-to-Kovar®, 60, 62
Alumina-to-stainless-steel, 13, 179, 180
Andrieux, J., 125–126, 131–132
anisotropic, 185, 192, 214
anneal, 104, 108
applications, 1, 4, 8, 48, 61, 63, 194
aqueous, 77, 89, 93
argon, 23, 46–47
as-ground, 32, 53, 55–59, 61–67, 82–85, 168–175, 215
as-sintered, 32, 52–54, 58, 61–62
Asthana, R., 2, 7, 12–13, 17, 20, 49–56, 62, 156
asperities, 86, 99
assembly, 45, 70–72
American Welding Society, 62, 76
Auerkari, P., 47–48

B

Bahrani, A.S., 1, 4
Bang, K., 10, 20, 50, 53,
Bayer Method, 4
Belenky, A., 52, 55, 58, 97
Boch, P., 4
Boron, 71
Borosilicate, 32
braze alloy composition, 7–8, 16, 20, 39–40, 46, 83
braze droplet, 7, 9–10, 40, 56
braze foil thickness, 19, 113
braze infiltration, 143–148, 163–165, 169–174, 207–209
braze interlayer thickness, 10, 13, 65, 184, 187–188, 215–216
braze outflow, 128–129, 133–137, 172–177, 180–188, 214–216
brazing atmosphere, 7–8, 28, 32, 46–47, 61, 156

brazing conditions, 28, 31, 46, 122–128, 165, 168, 173
brazing fixture, 2, 45, 59, 66, 71–75, 188
brazing procedure, 7, 66, 71, 75
Bull, S.J., 192
butt-joint assembly, 59, 71, 73, 75
Byun, W., 16, 20, 35–36, 41–44, 50, 53

C

Calcia, 47–48, 85, 153
calibration, 78, 81, 192
Carbolite, 69–71, 77
Carim, A.H., 9, 13, 15–20, 28, 50, 53, 176
Carter, E., 19–20, 50, 53
charge-coupled device, 33, 81
ceramic-to-ceramic, 1, 3–4, 61, 180–182
ceramic-to-metal, 1, 3–4, 11, 13, 32, 180
ceramography, 107
CeramTec UK Ltd, 5
Chakraborty, R., 192
chamfer, 52, 67–68
Chang, Y.A., 111
chemical activity, 8–9, 17, 151
chemical etching, 77, 89, 93–94
Chen, S., 203
Cho, H.C., 20, 23, 25–26, 41–44, 50, 53, 60–62
cladding, 8, 110, 172
coaxial feedthrough, 5
Coble, R.L., 48
coefficient of thermal expansion, 1, 11–12
  values of typical phases, 12
commercially available, 5, 8–9, 12, 16, 67, 71, 76
composite braze alloys, 132, 179–180, 188
compressive stress, 63, 69, 96, 104, 108, 215
concentration gradient, 19
conductivity, 69, 72
consumption in reaction, 15, 17, 28, 128, 134, 137, 174, 179
continuity of reaction layers, 25, 40, 42, 44–46, 58, 65, 146, 149
conventional methods, 5, 7, 40, 47, 55, 63, 188, 191
Coorstek Ltd, 67, 69, 192
corrosion, 4, 8
Chromium, 7
coefficient of thermal expansion mismatch, 1, 11–13, 48–49, 179
Cu-Ti phase formation, 122–128, 132, 134, 173, 215
Cu-Ti phases, 17, 120–137, 172–174, 184–188, 205–216
Cu concentration, 16–19, 65, 127–128
Cverna, F., 11

D

D-96 alumina, 69, 82, 85–108
D-100 alumina, 69. 82, 85–108
defects, 52, 55, 58–63, 103–107, 165, 178, 184
dendrites, 125, 185
densification, 47–48, 59
design of experiments, 82

Dey, A., 192
Dezellus, O., 127, 149
Diamond, 28, 53–54, 57, 76–77, 81, 139
dielectric, 4
diffraction patterns, 137, 139–142, 145–161
diffusion coefficient, 19, 32
discontinuous grain growth, 48, 103
dislocation density, 145, 199
dissolution, 16, 33, 122, 126, 128, 132, 134, 173
do Nascimento, R.M., 4, 14, 17, 46, 205
Dobedoe, R.S., 180
ductility, 12–13, 17, 180, 184, 187, 216
dye penetrant inspection, 58, 104, 165
Dynallox, 67, 69

E

elastic modulus, 3, 68–69, 72, 81, 191
electron probe microanalysis, 78, 85, 93, 153, 169–170
elemental profiles, 153, 169–170
elevated temperature, 8, 47, 59–60
entrapped porosity, 52, 92, 94
equations, 7, 11, 14–16, 32, 36, 39, 55, 156
Eremenko, V.N., 111
etching, 77, 89, 91–94, 107
Eustathopoulos, N., 1, 9–10, 20
eutectic temperature, 8, 48, 111
evaporation, 51, 89, 106, 170, 174
excess Ti, 115, 122–136, 172–174

F

Fernie, J.A., 1, 4, 7
Finger, L.W., 137
Fischer-Cripps, A.C., 191
fissured surface, 106, 108, 164–165, 169, 174, 208–209, 215
focussed ion beam milling, 78–79, 154–155
Foley. A.G., 1, 7, 12
fractography, 58, 75, 104–105, 175, 178, 181, 186, 207–209
fracture toughness, 3, 68–69
free energy, 8, 14–15, 38, 77
furnace, 46–47, 69–72, 74, 77

G

Galusek, D., 48–49
Gauthier, M.M., 1, 46
Geometries, 2, 47–48, 52–53, 55, 61, 67, 191
Ghosh, S., 2, 4, 11–14, 17, 20, 50, 53, 62, 205
Gibbs free energy, 14–15
globular, 126, 128, 134
gradation in coefficient of thermal expansion, 11, 216
grain pull-out, 56, 58, 76, 91
grinding damage, 52, 56–60, 65, 97, 99, 103, 108, 169, 207
ground-and-heat-treated, 32, 62, 69, 82–83, 99, 101, 168, 206, 215
ground-and-polished, 32, 61–62

H

high-angle annular dark-field, 51, 141, 143, 148–150, 159–170
Hafnium, 1, 7
Hahn, S., 4, 20, 24, 27, 29, 50, 52–53
Hansen, M., 17, 127

Hao, H., 19–20, 37, 41, 43–45, 50, 53, 62–63, 176
hardness, 1, 69, 72, 81, 179, 188, 191–199, 204
Heikinheimo, L.S.K., 20, 50, 53, 62
Heiman, R.B., 4, 7
Helium, 45, 57
hermeticity, 1, 8, 45–46, 57
high Ti concentration, 21–22, 27–28, 30–31, 44, 71
Hirnyj, S., 111, 116–117
Ho, C.Y., 11
Hosking, F.M., 20, 50, 53, 61
hydrofluoric acid, 77, 89, 93
Hysitron Inc., 81–82

I

Ichimori, T., 20, 24, 27, 29, 36, 50, 53, 56–57, 144
immiscibility, 111, 115
industrial, 1, 4, 8, 211
inert, 1, 7–8, 46
inhomogeneous, 36–37, 111–112
interdependent variables, 2–3, 40
interfacial chemistry, 7, 16, 52, 64–65, 154
intergranular, 60, 94, 108, 148, 165, 167, 169, 172–174
intermediate Ti concentration, 20–25, 27, 29, 31, 42, 44, 117
intermetallic, 111, 115, 124, 127
isothermal, 32–33, 72, 126–127, 132

J

Jacobson, D.M., 8, 21, 112
Janickovic, D., 20, 35–36, 50, 53
Jeol, 81
joining mechanism, 3, 12, 14–16, 28, 36, 117
joint edges, 10, 78, 121–136, 163–167, 177–188, 202–216
joint performance, 13, 63–65, 175–216
Jorgensen, P.J., 48

K

Kang, S., 51, 156
Kar, A., 11–12, 19–20, 28, 50, 53
Karlsson, N., 28
Kelkar, G.P., 9, 14–17, 27–28, 31, 151
Kim, J.Y., 16, 35–36, 41–44, 53, 59
Kirchner, H.P., 59–60, 104, 165
Kovar, 32, 60
Kozlova, O., 3, 16–17, 24, 29, 31–34, 39–42, 50, 53, 125–127
Kristalis, P., 8, 36
Kubiak, K.J., 55

L

Lange, F.F., 165
lapping, 46, 52–53, 58–59, 61
Lee, W., 3, 16, 19–20, 24, 27–29, 48, 50, 54, 156
Liao, T.W., 52, 58
Lin, C., 3, 11–12, 31, 39, 40,
Lin, K., 12, 15, 16, 23, 28, 41, 42, 46, 50, 53, 116, 139
line scans, 78, 153–154, 160, 169
liquidus, 12, 21, 71–72, 111, 113, 115–116, 122, 125, 172
Liu, Y., 10, 53, 156, 159
Locatelli, M.R., 1, 8, 20, 50, 54

Index

Loehman, R.E., 14, 23, 25–26, 31, 46, 49–50, 156
low Ti concentration, 20, 22, 25–27, 40, 44
Lugscheider, E., 46

M

macro images, 75, 128
Magnesia, 23–24, 41, 47–50, 85, 153
Mandal, S., 11–12, 18–20, 28, 50, 54, 128
Massalski, T.B., 161
matrix, 132, 179, 186, 203, 205–206, 214, 216
metal, 3, 7, 11, 52, 61, 64, 78, 179, 186
metallic, 1, 9, 12, 21–22, 33, 110–111
metastable, 151
Mg, 49, 51, 78, 89, 93–94, 107–108, 154, 157–159, 171–174
microcracks, 58
microhardness, 62, 179
microstructural evolution, 109–174, 184, 188, 206, 211, 215–216
migration of secondary phase, 106, 162, 164, 169, 172–174, 208, 210
Mizuhara, H., 1, 3, 20, 24, 32, 50, 52, 54, 58, 60–62, 165
Mohammed Jasim, K., 3, 20, 24, 28, 30, 32, 50, 53, 62
molar, 14, 17–18, 127
molybdenum, 80
Mondal, S., 50, 54
Monocrystalline alumina, 10–11, 20–28, 33–34, 39, 43, 45–50, 54, 56–57, 61
monolithic, 47, 55, 59, 63, 65–66, 178, 211, 214
Moorhead, A.J., 1, 12, 14, 20–24, 31, 41, 46, 50, 54, 58–59, 62, 165
Morgan Advanced Materials plc, 11–12, 71–72
morphology, 43, 89, 94, 126
Morrell, R., 47–48, 77, 104
multi-layered Cu-Ti structure, 129, 131–137, 165–167, 172–174, 184, 205, 214
Murray, J.L., 113–114, 151–152

N

NaCl-type structure, 142, 146
Naka, M., 156–157, 159
Nakamura, S., 58, 104
Nanoindentation
    Berkovich Indenter, 13, 81, 190–191
    depth Profile, 81, 192–195, 199–201
    displacement-controlled nanoindentation, 81–82, 194
    nanohardness distribution plot, 201–205
    nanoindenter, 82
    targeted indents, 188, 192, 193–195, 199–201, 204
    Triboindenter, 81–82
Niobium, 7
Nickel, 7–8, 32
Nicholas, M.G., 8, 20, 111, 126, 160
Ning, H., 20, 41, 45, 50, 54, 56–57, 176
Nitrogen, 7

O

O-Ti binary phase diagram, 151–152
Oku, T., 185
OriginLab, 82, 201–205
Oxygen, 7, 14–15, 25, 46, 72, 115
Oyama, T., 60–61

P

Paiva, O.C., 4
Pak, J.J., 7, 17–18, 128
Passerone, A., 55
Paulasto, M., 2, 20, 50, 54, 111
Peytour, C., 3, 20, 24, 28, 30, 50, 54, 56
Pfeifer, H.U., 149
Phase diagram, 111, 113–117, 126–127, 132, 151–152, 160–161
Phillips, G.C., 3
physicochemical, 7
plastic deformation, 145, 176, 191–192, 199
Poisson's ratio, 72
polycrystalline, 47, 137, 139, 142, 145–146, 170–171
polycrystalline alumina, 1, 50, 65, 67
porosity, 47–48, 52, 56, 91–92, 94, 108, 148, 205
precipitation of phases, 9, 48, 59, 113, 115, 125–126, 132
profilometry, 68–69, 85–86
Platinum, 79–80

R

Rana, A., 20, 50, 52, 54
rapid cycle proton synchrotron machines, 4
reaction kinetics, 17–18, 39, 65, 179
reaction layer formation, 3, 9–12, 16, 18–19, 31, 40, 52, 65–66, 68
reaction product formation, 7, 14, 17, 33, 40
refractory, 4, 71
reinforcement of braze interlayer, 188, 204
relative Ti concentration, 20, 110, 113, 122, 128, 132, 136–137, 149, 172
resistivity, 9, 69, 72
Richerson, D.W., 47
Rijinders, M.R., 20, 50, 54
Rittel, 52, 55, 58, 97
Rodrigues, G., 185
Rohde, M., 4

S

Santella, M.L., 16, 20, 24–25, 29, 50, 54, 62
scanning electron microscopy, 76, 78, 85
schematics
    bend test fixture, 75
    brazing temperature, 44
    brazing time, 34
    brazing fixture, 73
    CTE and secondary phase interaction, 154
    multi-layered Cu-Ti structure, 126, 133
    summary of joint strengths, 190, 213
    summary of microstructural evolution, 138
    wetting behaviour, 31
secondary phase, 49, 51, 92–93
secondary phase interaction, 65–66, 153–160, 162, 173, 184–186
Serizawa, H., 181–182, 184, 187
Shiue, R., 3, 20, 31, 33, 39–41
short test bars, 67–71, 73, 89, 99, 164
Silicon Carbide, 53–54, 57, 156–157, 159, 186
Silica, 47–48, 50, 77, 85, 93, 107, 153, 156, 211
Silicon Nitride, 1, 7, 45, 71, 159, 186
Singh M., 2, 7, 13, 17, 20, 49, 51, 53, 56, 62, 156

sintering, 47–48, 52, 56–57, 60, 66, 103
sintering temperature, 48, 103, 164, 207
solidus, 12, 40, 72
solubility, 8, 15–17, 28, 127, 172–173
spatial distribution of Ti_5Si_3, 161, 185–186, 216
scanning probe microscopy, 192, 194–197, 199, 205
stainless steel, 71, 73, 159, 179–180
standard test bars, 67, 72, 75–76, 85, 89, 99, 104, 207, 215
Stephens, J.J., 14–16, 23, 25, 27, 29, 41–42, 46, 50, 54, 60–61, 139, 142, 144
stoichiometric, 15, 17, 25, 78
Stollberg, D.W., 192
stress concentrators, 52, 208–209
stress contours, 180–181
Struers Ltd
    Durocit, 76
    MD-Allegro, MD-Largo, MD-Piano, 76–77
substrate, 9–10, 40, 125, 131
Suenaga, S., 15, 23, 25–26, 28
Suganuma, K., 156, 180–181
surface asperities, 56, 86–88, 99, 101
surface condition, 7, 32, 46–47, 52–58, 60–61, 65–66, 82, 97, 107, 173, 189, 212
surface defects, 48, 52, 58, 104
surface roughness
    Ra values, 50, 53–58, 68, 85–86, 89–90, 96–100, 108
    Rsh_{faying}, 89, 99
    Rsh_{outer}, 89, 99
    Rst_{outer}, 89, 99

T

Taguchi method, 82
Tamai, T., 156–157
test bar geometries, 70
testing methods
    ASTM C1161-13, 66–67, 72, 75, 85–87, 107, 185
    ASTM F-19, 58, 62–64
    double-bonded shear, 62
    double cantilever beam, 62
    four-point bend testing, 13, 52, 59, 62–64, 66, 68, 72, 180
    peel testing, 32
    shear testing, 13, 32, 45, 62–63
    single-edge notched beam, 62
    tensile testing, 45, 58, 62–64
    three-point bend testing, 46, 55, 58, 62
thermal etching, 77, 89, 92, 94, 103, 106
thermally induced residual stress, 1, 8, 11–12, 63, 65, 145, 176–177, 179, 184, 186, 199, 208, 214–215
Ti_2O, 15–16, 22
Ti concentration factor, 110, 122, 128, 132, 149, 172
Ti getters, 72, 74
Ti-Cu-O layer, 9–10, 25, 27, 40, 42, 44, 65
Ti-O layer, 10, 15, 25, 27, 40, 42, 44, 65
Titanium Carbide, 157, 159
Titanium Nitride, 157

Titanium Silicide, 156–161, 172, 174, 185–186
Tillman, W., 46, 156–157
Titanium activity diagram, 152
transmission electron microscopy, 76, 79, 81, 137
triple pocket grain boundary, 49, 60, 79, 89, 154–165, 170–174, 185–186, 207–216
triple point, 55
Tungsten, 80
typical microstructure, 10

U

Urai, S., 159, 186

V

vacuum furnace, 46–47, 51, 72, 74
Valeeva, C., 151
Valette, C., 3, 16–17, 20, 23, 25, 29, 41–42, 50, 54
Verband der Keramischen Industrie, 3
Vianco, P.T., 3, 20, 49–50, 54, 60–61
Voytovych, R., 10–11, 16, 20–36, 44–50, 54–56, 125

W

Weibull statistics, 94, 96–97, 101, 103–104
Wenzel, R.N., 55
Wesgo Metals, 8, 71–72, 109
wetting behaviour
    Dynamic contact angle, 9, 33–34
    Equilibrium contact angle, 7, 9–10, 19–22, 31, 33–34, 39–40, 46, 48–50, 55–56
    Reactive wetting, 3, 7, 9, 15, 21–22, 55–56
    wettability, 7–10, 21–22, 31, 34, 39, 47, 55–56, 112
Wiese, J.L., 14

X

Xian, A.P., 45, 156, 159, 186
Xiong, H., 45, 59
X-ray diffraction, 22–30, 34–36, 41–43, 159

Y

Yadav, D.P., 5, 20, 50, 54, 63–64
Yang. P., 152
Yang, J., 20, 50, 54, 179, 188, 203
Yang, M., 4, 20, 50, 54
Young-Dupré equation, 7, 55
Yu, H., 25, 42, 44, 60, 62, 109

Z

Zeiss Surfcom, 68–69, 78
Zhu, M., 13, 110, 132, 179–180, 188, 203
Zirconium, 1, 7